Modeling and Optimization of Signals Using Machine Learning Techniques

Scrivener Publishing
100 Cummings Center, Suite 541J
Beverly, MA 01915-6106

Publishers at Scrivener
Martin Scrivener (martin@scrivenerpublishing.com)
Phillip Carmical (pcarmical@scrivenerpublishing.com)

Modeling and Optimization of Signals Using Machine Learning Techniques

Edited by

Chandra Singh

Rathishchandra R. Gatti

K.V.S.S.S.S. Sairam

Manjunatha Badiger

Naveen Kumar S.

and

Varun Saxena

Scrivener Publishing

WILEY

This edition first published 2024 by John Wiley & Sons, Inc., 111 River Street, Hoboken, NJ 07030, USA and Scrivener Publishing LLC, 100 Cummings Center, Suite 541J, Beverly, MA 01915, USA

For more information about Scrivener publications please visit www.scrivenerpublishing.com.

Wiley Global Headquarters
111 River Street, Hoboken, NJ 07030, USA

For details of our global editorial offices, customer services, and more information about Wiley products visit us at www.wiley.com.

Library of Congress Cataloging-in-Publication Data

ISBN 978-1-119-84768-7

Front cover images supplied by Pixabay.com
Cover design by Russell Richardson

Set in size of 11pt and Minion Pro by Manila Typesetting Company, Makati, Philippines

Printed in the USA

10 9 8 7 6 5 4 3 2 1

Contents

Preface

Organization of the Book

The chapters are organized as concepts of modeling and optimization of signal processing in machine learning.

Signal processing captures, interprets, describes, and manipulates physical phenomena. Mathematics, statistics, probability, and stochastic processes are among the signal processing languages we use to interpret real-world phenomena, model them, and extract useful information. This book comprises chapters on key problems in machine learning (ML) and signal processing arenas. This book also presents the different kinds of signals that humans and machines use to communicate, and their treatments and applications. It discusses various mathematical methods involved in signal processing and ML, thereby enabling students to design their own models and optimize them efficiently. The book focuses on mathematical principles, and contains coding-based assignments for implementation. Prior exposure to ML is not required. The book also focuses on applications in signal processing and communication. This book is intended for advanced undergraduate and postgraduate students, researchers, and practitioners engaged with signal processing and its applications.

Chapter 1 Land Use and Land Cover Mapping of Remotely Sensed Data Using Fuzzy Set Theory Related Algorithm

This chapter focuses on remote sensing theory, differences between practical and ideal remote sensing systems, Landsat 8 data characteristics, and concepts of electromagnetic radiation. Remote sensing can be broadly defined as the science of obtaining useful information about physical objects and/or environment, through recording and measuring energy patterns from sensor systems.

Chapter 2 Role of AI in Mortality Prediction in Intensive Care Unit Patients

This chapter focuses on intensive care unit mortality rate and risk prediction, both of which are major concerns in any hospital. Prediction is a term highly dependent on the correct analysis of available data.

Chapter 3 A Survey on Malware Detection Using Machine Learning

This chapter presents ML structures that have been tried, investigated, and created to portray and order malware into their malware families, utilizing attributes separated and obtained from static and live examination of noxious programming.

Chapter 4 EEG Data Analysis for IQ Test Using Machine Learning Approaches: A Survey

This chapter presents an analysis of various approaches utilized for IQ test as a study. From this, the intention is to understand the various methodologies used for intelligence level estimation and use them to distinguish the downsides of existing framework.

Chapter 5 Machine Learning Methods in Radio Frequency and Microwave Domain

This chapter presents a comprehensive review on recent research advancements and ML techniques used for the optimization of RF circuits.

Chapter 6 A Survey: Emotion Detection Using Facial Reorganization Using Convolutional Neural Network (CNN) and Viola-Jones Algorithm

This chapter includes mental health and related problems. Keeping this in mind, we came up with an idea where we create an app that detects user's emotions and suggests a binaural beats playlist for them to help calm their emotions.

Chapter 7 Power Quality Events Classification Using Digital Signal Processing and Machine Learning Techniques

This chapter discusses feature extraction based on wavelet transformation in detail. Feature selection and optimization play a vital role in optimizing datasets to the classifier after feature extraction. We use a genetic algorithm (GA) for this purpose. Then, machine learning models like probabilistic neural network (PNN), back propagation neural (BPN) network, and support vector machine (SVM) are used as classifiers to classify various PQ events.

Chapter 8 Hybridization of Artificial Neural Network with Spotted Hyena Optimization (SHO) Algorithm for Heart Disease Detection

In this chapter, we focus on the proposed method's (SHO algorithm) effectiveness verified by different patients considering 13 constraints as the dataset. These constraints are evaluated for training and testing each data in the dataset. The efficiency of the proposed approach is shown in comparison with other methods, SLO, PSO, GDA, and GA.

Chapter 9 The Role of Artificial Intelligence, Machine Learning, and Deep Learning to Combat the Socio-Economic Impact of the Global COVID-19 Pandemic

This chapter discusses the severe impacts of COVID-19 and the role of technology particularly artificial intelligence, machine learning, and deep learning in combating the virus. Various statistical data for this survey are also presented.

Chapter 10 A Review on Smart Bin Management Systems

This chapter provides a review of various approaches used in smart bin management systems.

Chapter 11 Unlocking Machine Learning: 10 Innovative Avenues to Grasp Complex Concepts

This chapter provides a mathematical statement that expresses facts in the context of a problem, a model often known as a machine-learning algorithm. We look at 10 alternative approaches for demystifying ML and offering a knowledge route used by individuals unfamiliar with fundamental ideas, each with clear descriptions, images, and examples.

Chapter 12 Recognition Attendance System Ensuring COVID-19 Security

This chapter focuses on a face detecting, temperature sensing, report generating, and all-in-one attendance system that is low cost and easily integrated into the current system to re-innovate an existing project methodology and make it available at a low cost.

Chapter 13 Real-Time Industrial Noise Cancellation for the Extraction of Human Voice

This chapter on processing and playing uses the concept of multiprocessing in processors with each core of the processor performing a different function. The user is protected from surrounding noises by an additional passive attenuator.

Chapter 14 Machine Learning-Based Water Monitoring System Using IoT

This chapter aims to provide a machine learning-based water monitoring system using multiple sensors. The system can collect a variety of data from water, including temperature and pH, and then, water clarity is measured.

Chapter 15 Design and Modeling of an Automated Driving Inspector Powered by Arduino and Raspberry Pi

This chapter focuses on the effort made to secure way of inspecting test drive during COVID-19 pandemic using new technologies for the societal benefits. The Secured and Automated Driving Inspector powered by Arduino and Raspberry Pi is a new technology, which can be embedded in the test cars for producing eligible drivers by assisting the inspector during driver's test.

Chapter 16 Kalman Filter-Based Seizure Prediction Using Concatenated Serial-Parallel Block Technique

This chapter focuses on a method that combines two distinct techniques, the Kalman filter and a concatenated serial-parallel block technique, to predict epileptic seizures. The Kalman filter is a mathematical tool often used for estimating and predicting the state of dynamic systems. In the context of seizure prediction, it is employed to analyze and model the dynamics of brain activity data.

Chapter 17 Current Advancements in Steganography: A review

This chapter focuses on the details of the datasets used for steganography and also contains some details of steganography's counterpart called steganalysis.

Chapter 18 Human Emotion Recognition Intelligence System Using Machine Learning

This chapter investigates a speech emotion detection system that outperforms previous systems on the basis of data, feature selection, and methodology.

Chapter 19 Computing in Cognitive Science Using Ensemble Learning

This chapter focuses on human–computer interaction to find a technique to identify human activity and apply ensemble learning from an individual level to produce a more accurate prediction of a whole that fully mirrors social-cognitive science in a very efficient computational approach.

Overall, editing this book was an incredible opportunity. We are indebted to many individuals for their help and support, including our families, friends, and fellow members. We are especially grateful to the chairman, principal, faculty, and fraternity of the Sahyadri College of Engineering and Management, Mangaluru and NMAM Institute of Technology, Nitte.

Chandra Singh
Rathishchandra R. Gatti
K.V.S.S.S.S. Sairam
Manjunatha Badiger
Naveen Kumar S.
Varun Saxena

1

Land Use and Land Cover Mapping of Remotely Sensed Data Using Fuzzy Set Theory-Related Algorithm

Adithya Kumar* and Shivakumar B.R.

NMAM Institute of Technology, NITTE, Udupi, Karnataka, India

Abstract

All bodies, planets, living beings, and inanimate objects emit electromagnetic radiation, and the amount and type of radiation emitted depend largely on their temperatures. Electromagnetic radiation may be emitted by an object or may come from another body and could be reflected by it. There are operational satellite systems that sample virtually every region of the electromagnetic spectrum, with spatial resolutions from 0.5 to 5,000 m. The scientific community's interest in spatiotemporal studies of global change, environmental monitoring, and human impacts involves the use of remote sensing data. Remote sensing systems, particularly those located on satellites, provide a repetitive and synoptic vision of Earth, which is of great interest in monitoring and analyzing human activities and their impacts. The activities include the evaluation and monitoring of the environment like urban growth and hazardous waste; detection and monitoring of global changes, deforestation, global warming, and exploration of non-renewable resources and their land use; civil engineering;, the acquisition of satellite imagery from relevant sources; the preprocessing of satellite imagery as per requirement; and the development and classification of satellite data using fuzzy C-means (FCM), modified FCM, K-means, and fuzzy inference system (FIS) techniques.

***Keywords*:** FCM, modified FCM, K-means, FIS techniques

**Corresponding author*: adithya.kumar53@gmail.com

Chandra Singh, Rathishchandra R. Gatti, K.V.S.S.S.S. Sairam, Manjunatha Badiger, Naveen Kumar S. and Varun Saxena (eds.) Modeling and Optimization of Signals Using Machine Learning Techniques, (1–22) © 2024 Scrivener Publishing LLC

1.1 Introduction

In this chapter, an introduction to remote sensing theory, the differences between practical and ideal remote sensing systems and the Landsat 8 data characteristics and concepts of electromagnetic radiation are presented [1, 2]. Remote sensing can be broadly defined as the science of obtaining useful information about physical objects and/or the environment through recording and measuring energy patterns from the sensor systems. The information about distant objects and the environment is obtained through the electromagnetic radiations emitted or reflected by the Earth's surface features, which are of interest. Multiple data users that make use of the ideal remote sensing system would be able to acquire the required data at a higher speed with less or no expense of any required area of interest [3, 4]. Given this information, multiple data users would be able to make decisions on managing and observing the Earth's surface features. A real remote sensing system encounters a lot of problems in each stage of the remote sensing system. The following section explains the limitations of all ideal remote sensing systems. The spectral distribution of reflected and/or emitted electromagnetic energy is not uniform, and it varies from time and place on different Earth features. Passive remote sensing systems entirely depend on the energy from the sun. Solar energy is non-uniform with respect to time and place. The sources used in the remote sensing systems are also non- uniform with respect to wavelength and time. Therefore, there is always a need to calibrate these sources depending on the mission, time, and location. In a real remote sensing system, there is always a need for atmospheric and radiometric corrections due to atmospheric errors. The atmosphere always modifies the energy emitted and reflected by the Earth's surface features. Eliminating the atmospheric effects is important in those applications involving observations of the same geographical area. Different materials do not reflect and emit energy uniquely. There is spectral overlapping in the recorded data. The spectral response plays an important role in detecting, identifying, and analyzing the Earth's surface materials. A good understanding of energy interactions is required to obtain information about the required features. From the above points, it is clear that super sensors do not exist. The sensors cannot be sensitive to all possible wavelengths and have limits with respect to sensitivity. The sensors record the objects according to their spatial resolution. The ability of the sensors to differentiate the smallest object and separate it from its surroundings is called spatial resolution. The spatial resolution depends on the heterogeneity of the ground area being sensed. The digital images obtained consist of mixed pixels causing problems in identifying the land cover classes. Depending on the application, the choice of sensors varies from airborne vehicles to space stations. Photographic systems produce images of higher spatial resolution, but spectral sensitivity is absent in those sensors. Therefore, to obtain information on all the land cover classes, sensors with higher spectral resolution are required rather than sensors with higher spatial resolution. However, sensors having higher spectral resolution are expensive. Proper image processing techniques are required to visualize if the sensor data are of low resolution. Therefore, there are always tradeoffs between cost and resolution. Remote sensing systems generally are of two types: passive and active. An active remote sensing system possesses an illumination system or source of energy. It directs this energy to the objects to measure the energy. Some examples of active sensor systems are radars and laser scanners. The passive remote sensing system utilizes the sun's energy to record the emitted and/or reflected values of the surface features. Landsat series sensors are an example for passive remote sensing systems. The sensors that generate artificial radiation to the Earth's surface features are called active sensors. Radio detection and ranging or radar is used to detect the Earth's surface features using the pulses of electromagnetic radiation in microwave ranges. These radiations are not affected by weather conditions such as clouds, fog, and wind. The reflected signal intensity is used to extract the information. These types of sensors using self-generated radiation are useful for specific satellite missions and operations [5, 6].

Sensors that use natural light and the thermal radiation of the Earth's features to extract the information are called passive. This information is greatly affected by the seasons, time of capture, and height of the satellite. The Landsat satellite is equipped with spectrometers that measure the signals at a spectrum of bands. Therefore, multispectral channels are obtained, which provide greater information. Satellite imagery consisting of pixel values are the measures of the intensity of the electromagnetic radiation of the Earth's surface features. The values of brightness differ in the image based on time, type of area, and sensors. Certain properties of imagery like resolution should be known. The individual elements of the digital image are called pixels. The size of the pixel and the type of sensor determine the spatial resolution. The length of the edges of the pixels is used to measure resolution. The number of spectral channels is defined as the spectral resolution of the image. The reflection of Earth surface features occurs in different wavelengths. The human eyes can recognize only the visible spectrum of radiations. The land area, which has high reflections in particular wavelengths, is sensed by the satellite sensors. Most of the Earth observation satellites have between three and eight bands. The linear imaging self-scanning sensor III (LISS-III) has four spectral bands. The higher spectral resolution allows larger features to be extracted.

The time interval the satellite takes to reach the same area is called temporal resolution. The temporal resolution is determined by the orbit of the satellite, the altitude, and the sensor's characteristics. Generally, Earth observation satellites have a repetition rate of 14–16 days. Landsat 7 has 16 days. The highest temporal resolution is for meteorological satellites, which is 15 minutes [7, 8].

The surface reflectance characteristics of three main Earth surface features are discussed here:

- **Vegetation:** The chlorophyll pigment in leaves absorbs radiation in red and blue wavelengths but reflects in the green wavelength. The reflectance characteristics vary with wavelength [9, 10, 16, 18].
- **Water:** The water absorbs most of the radiation in the longer wavelengths. The water reflects in shorter wavelengths and hence blue in appearance. In the near-infrared wavelengths, the water is darker in color. The reflectance varies due to the depth of the water, the materials in the water, and the type of water [12].
- **Soil:** The nature of the soil determines the reflection and emission. The presence of moisture decreases the reflectance. The reflectance of soil is present in a large range of wavelengths depending upon the soil [13].

1.1.1 Overview on Landsat 8

Landsat 8 is an American Earth observation satellite launched in 2013. It is the seventh satellite to reach orbit successfully. The operational land imager (OLI) sensor consists of nine capturing devices, and thermal infrared sensor (TIS) has two capturing devices. The thermal bands measure the surface temperature and lighter pixels represent hot temperatures. Thermal bands provide a lot of information regarding heat units present in urban regions. The Landsat 8 orbit height is 705 km above the Earth's surface. It collects 400 scenes of 185-km wide regions in one day, and an archive is made available on the United States Geographical Survey (USGS) website within 24 hours of acquisition. Generally, the first three band combinations of the initial nine bands are used as the inputs to the algorithms. Other band combinations can also be used as per the analyst's choice.

These band combinations are useful to extract certain land cover classes more precisely [14].

This conveys information on the various stages of remote sensing systems and the techniques used to capture remotely sensed (RS) data. The choice of the sensor was discussed, which is an important decision in obtaining proper RS data with good-quality images. In the next section, an introduction to the image classification process will be discussed.

1.2 Image Classification

Image classification, in remote sensing perspective, is said to be a collection of three processes:

- Image preprocessing,
- Classification, and
- Image post-processing.

In the upcoming sections, a detailed explanation of each of the abovementioned processes is presented.

Preprocessing steps are applied to compensate for systematic errors and enhance the image for a higher level of interpretability, and are not necessarily implemented in all classification procedures. The most basic types of corrections used for preprocessing are radiometric and geometric corrections. Similarly, the most basic enhancement techniques are categorized into linear and non-linear enhancement techniques. Image corrections may be employed to remove the errors caused by the atmosphere, seasonal variations, and imperfections in scanning instruments. In the following sections, an explanation of the two types of corrections is presented. The sun's azimuth, elevation, and atmospheric conditions cause radiometric errors. The digital numbers are error-prone and must be corrected. Radiometric correction is important when data are obtained at different time intervals. The pixel value obtained not only includes the emitted and/or reflected radiation of the Earth's surface, but also the radiation scattered by the atmosphere. Because the interest is in the actual surface values, the corrections are required. The sensor records the magnitude of the electromagnetic radiations as a digital number (DN). These numbers can be transformed into radiance and reflectance. Metadata information is required to perform this conversion. This process improves the quality and interpretability of RS data. The geometric distortions are caused by the variations in terrain elevation and the Earth's spinning motion. Since the Earth's rotation on its axis is from west to east and satellite orbits occur from the Earth's pole to pole, the errors in elevation occur. The correction of geometric errors can be done through a process called rubber sheeting. It involves the warping and stretching of the image to georegister control points. Image enhancement deals with converting raw satellite data into better, more interpretable images. The commonly used image enhancement techniques are discussed here. Image enhancement techniques are broadly classified into two ways: global enhancement and spatial enhancement. Examples for global enhancement are linear contrast stretch, histogram equalization, and piece-wise contrast stretch. Spatial enhancement techniques are concerned with smoothing and sharpening images. The contrast enhancement is done to distinguish the finer differences of the features. The higher DN values and additive effects cause a reduction in contrast to the blue and the infrared range of the electromagnetic spectrum. This effect is called a haze, and contrast enhancements are used to minimize the effects. The back scattering to space by the molecules in the atmosphere causes path radiance. This radiance is detected by the sensor, which must be eliminated. In the histogram plot, the low-end DN values are assigned to extreme black, and high-end DN values are assigned to extreme white. The pixels in the medium range are distributed linearly. The drawback of linear stretch rarely occurs. DN values are assigned many levels as it does to frequently occurring DN values. The linear contrast in most cases produces a dull image. The single land cover class can be discriminated properly using this method. The detailed differences within the regions of interest are done, and remaining areas are assigned a single gray tone. Histogram equalization allows areas of lower contrast to obtain higher contrast. This method is largely used to enhance the image. The image intensities are better distributed to perform equalization. The frequently repeating intensity values are spread efficiently, and therefore increase global contrast. Image classification is an important part in digital image processing since it is necessary to have an image displaying a magnitude of colors representing different land cover features of the Earth's surface. To obtain this imagery, many classification algorithms are presented in the literature. Image classification can be categorized based on multiple factors. The most basic categorization is based on the requirement of training sites. The results of this categorization are supervised and unsupervised classification techniques. The first step in the RS image classification stage is the acquisition of imagery. The satellite imagery obtained must be visually interpreted for corrections. There will be a requirement for both the removal of noises and the enhancement techniques. These are the preprocessing stages before classification. The preprocessing stages like noise removal and enhancement techniques are done if the input data are found to have errors; otherwise, according to requirement, only necessary preprocessing steps can be followed such as radiometric or geometric corrections. After all these operations, the classification process begins. The classification procedure has two main categories, unsupervised and supervised.

1.3 Unsupervised Classification

The ability of the algorithm to statistically group the data without training data is called unsupervised classification. The unsupervised classifier performs clustering or division of datasets with the help of similarity and dissimilarity features [11]. The classes that result after grouping are called

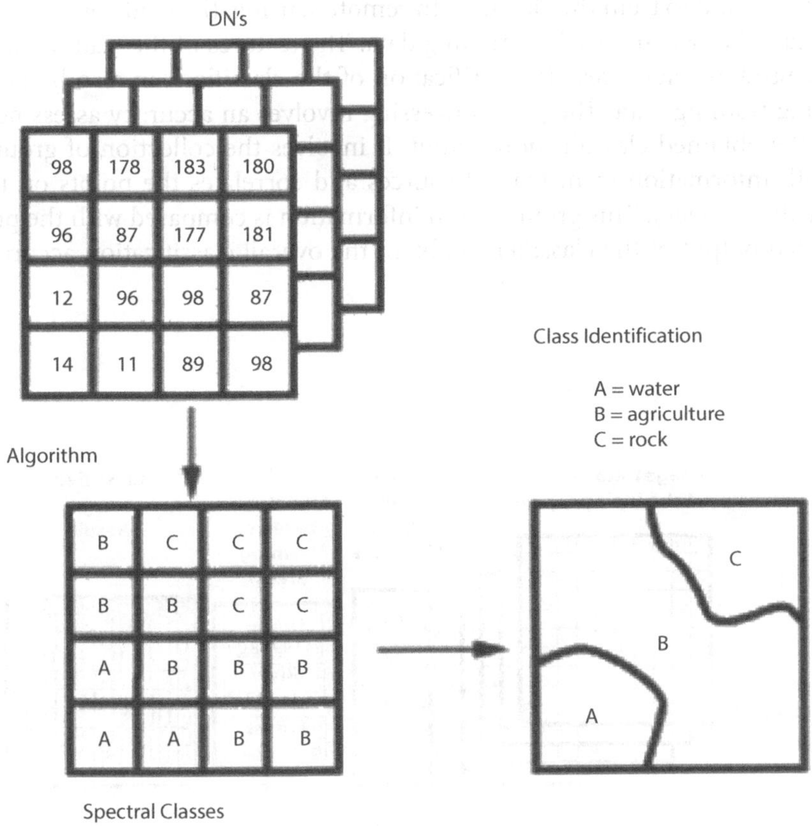

Figure 1.1 Illustration of an unsupervised classification [15].

spectral classes. The pixels closer to each other are more likely to belong to the same class. Therefore, as shown in Figure 1.1, the pixels are grouped by evaluating their features using the algorithm. The main aim of classification in remote sensing is to represent the different features of the Earth's surface in a different magnitude of colors for environmental monitoring, change detection, and urban planning.

1.4 Supervised Classification

Supervised classification algorithms use reference classes as training data. This additional information must be provided by the user. The pixels extracted from the different channels are provided as inputs to the trained classifier as shown in Figure 1.2.

The supervised classification techniques require time to collect the training data and to train the classifier. In remote sensing, the land cover classes already known are used as training data. The features of the training data are used as references. The verification of the classifier can also be done using training data. The post-processing involves an accuracy assessment of the obtained classification output. It involves the collection of ground truth information from trusted sources and correlates the points on the satellite imagery. This ground truth information is compared with the predicted output of the classifier to obtain the overall classification accuracy [17].

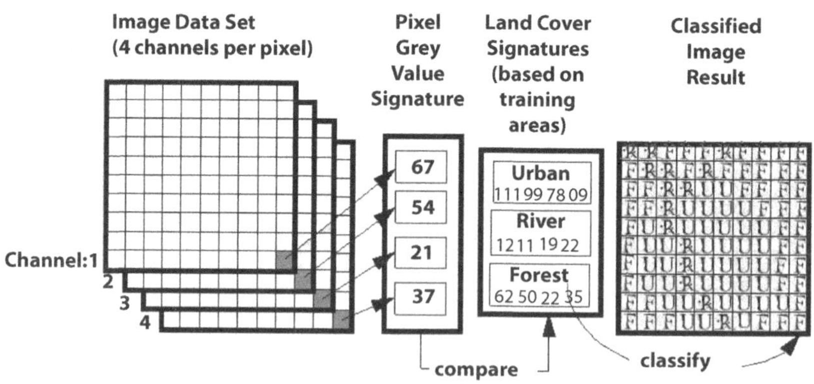

Figure 1.2 Illustration of a supervised classification [19].

1.5 Overview of Fuzzy Sets

In the previous section, the image classification procedure and preprocessing steps were explained. This section explains the concept of fuzzy sets and their applicability in RS data.

Lotfi A. Zadeh initiated the fuzzy logic at the University of California, Berkeley in 1965. Boolean logic or crisp logic is the opposite of fuzzy logic. The element of a set belongs to only one set in Boolean logic, but fuzzy logic evaluates the probability of the element belonging to every set. Notions that are rather small or very slow can be written mathematically and processed by the algorithms in more human-like thinking.

The membership functions are used to give the grades to these input variables. The membership functions are of different types such as Gaussian, triangular, and trapezoidal. They are chosen based on the requirements of the application. Due to the presence of spectrally similar land cover classes, the traditional maximum likelihood classifier does not properly differentiate within the classes. The misclassifications occur between the building and the road classes. The spatial features can be properly utilized by fuzzy pixel-based classifiers. The classifier uses the initial classes by the maximum likelihood classifier to perform classification. The fuzzy rules are then used to perform the decision on which class the pixels must be assigned. The process of converting a crisp quantity to fuzzy is called fuzzification. Many quantities like speed and temperature may be deterministic, but they have uncertainty. The uncertainty in the variable causes unclear information. This can be represented better using fuzzy sets. The membership functions are used to represent this information. Similarly, the pixels are also not crisp or deterministic in satellite imagery. In this study, fuzzy-related algorithms were used to perform classification. The classification process involves the assignment of pixels to a land cover class. The categorization of pixels in the image is performed by classification algorithms. The process of converting a fuzzy quantity to a crisp one is called defuzzification. The output obtained through the operation of fuzzy sets is fuzzy. Therefore, there is a need to convert it to crisp logic. The result of the land cover classification has to be converted back to crisp value to finally assign the pixel to one particular class. After all the pixels are converted to crisp logic, it is displayed as a thematic map. This provides a brief overview of fuzzy theory and its concepts. Properties and operations of fuzzy sets are also presented with relevant mathematical equations. A brief introduction to fuzzy classification is also presented. In the next section, the classification techniques used in the study are discussed in detail.

In the previous section, the fuzzy theory and its properties were discussed. In this section, the implementation of fuzzy C-means (FCM), K-means, and the proposed FCM are presented and discussed. This section also includes the application of fuzzy inference systems (FIS) for land cover classification.

1.5.1 Fuzzy C-Means Clustering

RS data can be partitioned by initializing the number of partitions or clusters required, which is *c*. For example, if three classes need to be classified, then *c* is three. Then three cluster centers are initialized randomly with the number in the range of 0 to 65,535. For example, $c_1 = 20{,}000$, $c_2 = 40{,}000$, and $c_3 = 50{,}000$. These cluster centers are used to calculate the membership grades (u_{ij}), which lie in the range of 0 to 1.

Then, the cluster centers are recalculated through an iterative process that attempts to minimize the following objective function.

$$\mathrm{j} = \sum_{\mathrm{i}=1}^{N} \sum_{\mathrm{j}=1}^{c} u_{ij}^{m} \, | x_i - c_j |^2 \tag{1.1}$$

1.5.2 Algorithm of Fuzzy C-Means

The fuzzy C-means algorithm consists of the following steps:

Step 1: Convert the input image into a one-dimensional array.
Step 2: Select the number of clusters, *c*, and initialize randomly.
Step 3: Compute the membership matrix for pixel *i* and class *j*.

$$u_{ij} = \frac{1}{\sum_{k=1}^{c} \frac{|| x_{\mathrm{i}} - c_{\mathrm{j}} ||}{|| x_{\mathrm{i}} - c_k ||}} \tag{1.2}$$

Step 4: Recalculate the *c* cluster centers.

$$cj = \frac{\sum_{\mathrm{i}=1}^{N} u_{\mathrm{ij}}^{m} . x_i}{\sum_{\mathrm{i}=1}^{N} u_{\mathrm{ij}}^{m}} \tag{1.3}$$

Step 5: Calculate the objective function as in Section 1.5.1.
Step 6: Repeat steps 3 to 5 until the objective function is minimized. Otherwise, exit.

1.6 Methodology

Landsat 8 satellite imagery of a region in Kumta Taluk was obtained from the USGS Earth Explorer website. The preprocessing of the satellite imagery was performed by employing histogram equalization and radiometric corrections. The implementation of FCM, modified FCM, K-means, and FIS was done in MATLAB® R2015a. Performing land use land cover (LULC) mapping of the considered study area by using the above techniques was done to classify the required satellite image by feeding the inputs required. It is learned that the manual feature extraction of a 1 km^2 area using high-resolution data like IKONOS is very expensive. Particularly, urban areas are highly dynamic, which are strongly involved in inter-urban land use pattern changes.

1.6.1 Modified Fuzzy C-Means Technique

The advantage of the proposed algorithm over conventional FCM is that it uses all the input bands of a multispectral imagery. This eliminates the effects of spectral correlation on the dataset. Figure 1.4 shows the mean values of each LULC class plotted against a band number. It is observed from Figure 1.4 that using the multispectral imagery at the input of FCM suffers from the effects of spectral overlapping, band correlation, and mixed pixel issues. Hence, it is advantageous to classify each multispectral band separately using FCM algorithm for an efficient classification process. The flowchart of modified FCM is depicted in Figure 1.3. When a three-band combination image is used, the pixel values are very close to each other, which is inaccurate. The plot of pixel means against bands as in Figure 1.4 shows that the Evergreen Forest, Water Body, and Vegetation have good separability in band 7 compared to other bands. It also shows that band 1 and band 2 together are of no use because of the high correlation in their pixel values.

1.6.2 Construction of a Fuzzy Inference System

The Gaussian function utilizes the mean (x_c) and the standard deviation (σ) to calculate the membership grade (μ_x) as shown in Figure 1.5. The

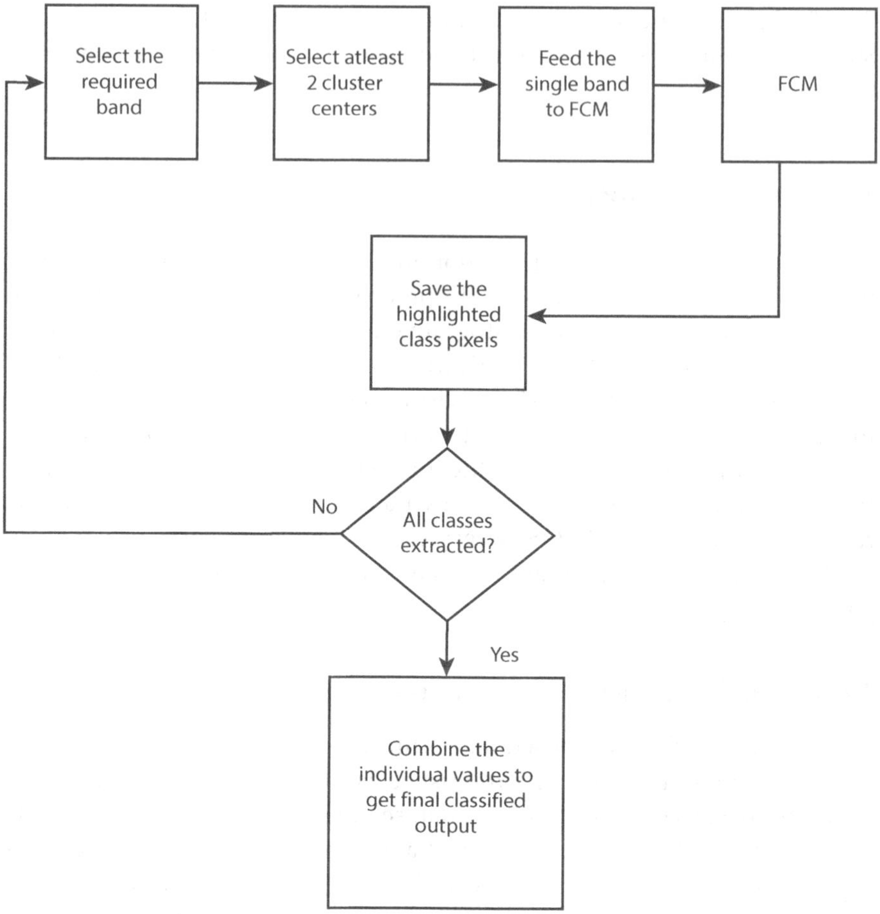

Figure 1.3 Flowchart of the modified FCM.

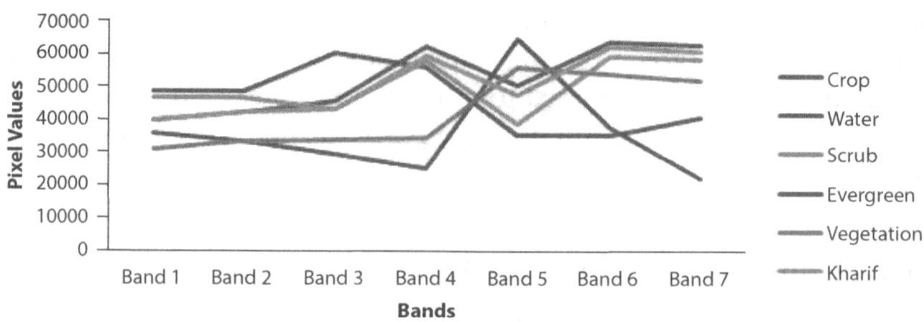

Figure 1.4 Plot of the pixel mean values v/s band numbers.

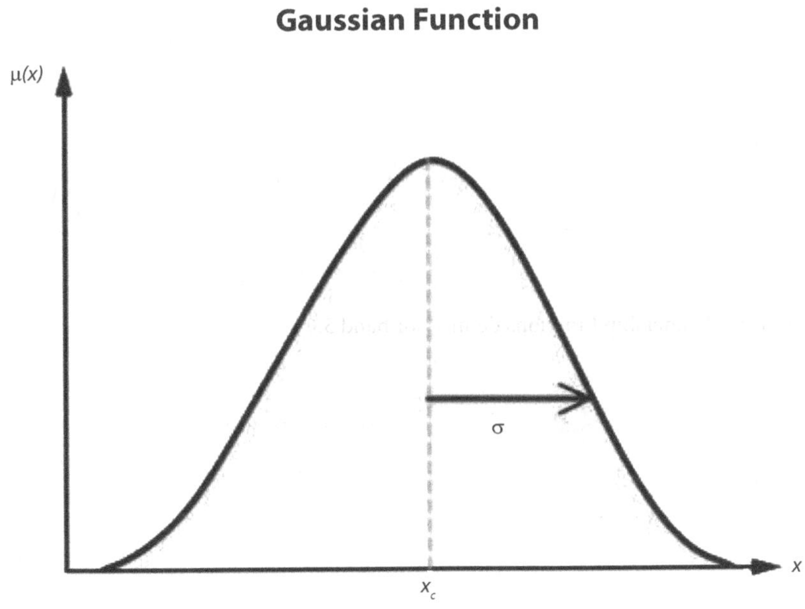

Figure 1.5 Gaussian membership function.

training areas are used to obtain the mean and the standard deviation from which the membership function of the land cover class is built. For example, the membership function of water is obtained from the training areas by calculating the mean and the standard deviation. Similarly, all the membership functions of the required land cover classes are obtained as shown in Figure 1.6 for band 6. By calculating the mean and the standard deviation from the required bands, the database of the membership functions is built as shown in Figures 1.6, 1.7 and 1.8. The rule base contains a number of fuzzy if–then rules.

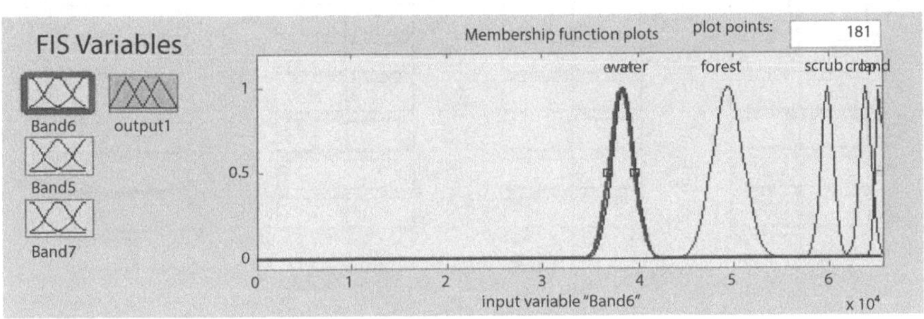

Figure 1.6 Membership functions defined for band 6.

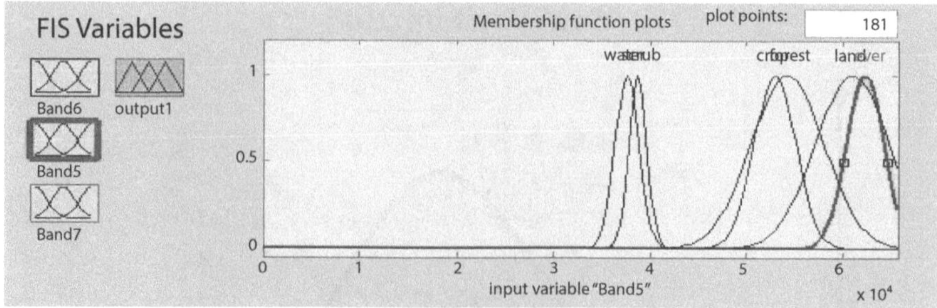

Figure 1.7 Membership functions defined for band 5.

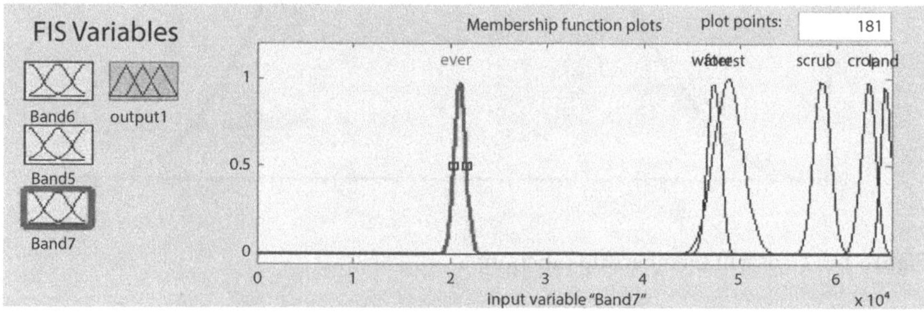

Figure 1.8 Membership functions defined for band 7.

After the membership functions are prepared by using the Gaussian membership function, the rule base is prepared by using IF–THEN rules. These rules connect the membership functions using AND and OR operators as shown in Figure 1.9. In the rule viewer, the AND stands for

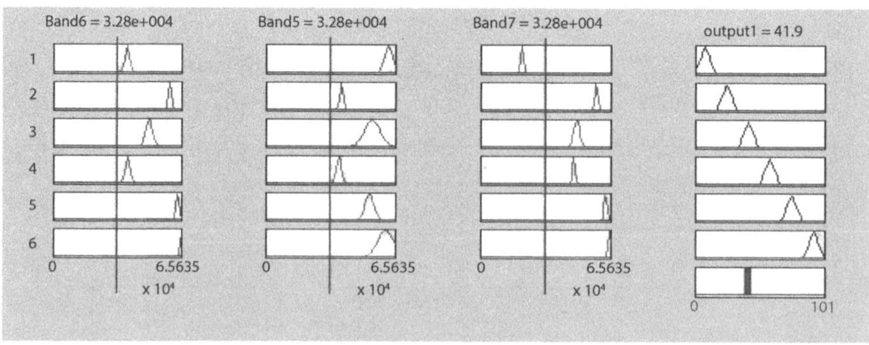

Figure 1.9 Rule viewer showing the connection between input and output.

minimum fuzzy set operation and the OR stands for maximum fuzzy set operation.

The inputs provided to the rule base are fuzzified using the Gaussian membership function, and by applying the minimum and maximum rules, the fuzzy output is obtained. This fuzzy output is converted to a crisp output by applying the defuzzification operation.

1.6.3 K-Means Algorithm

The K-means algorithm is an unsupervised algorithm used for grouping pixels in a simple and easy way. The procedure is to assign centroids for each cluster and calculate the distance of the pixels with all the cluster centers. New cluster centers are calculated at the end of each iteration. Suppose that there are n samples: $x_1, x_2, x_3, \ldots, x_n$. The objective is to develop k clusters and k should be less than n. The procedure for developing k clusters is as follows:

Step 1: Select k random points in the signal space; these points represent the initial cluster centers.
Step 2: Assign each pixel to the cluster to which it has the least distance.
Step 3: When all the pixels are assigned, recalculate the cluster centers to produce new centers.
Step 4: Repeat steps 2 and 3 until the centers become constant or until the difference in old and new cluster centers is negligible. After the centers are optimized, the algorithm stops and exits.

The objective function used to terminate the algorithm from its iterations is given by:

$$J = \sum_{j=1}^{k} \sum_{i=1}^{n} || x_i^{(j)} - c_j || \tag{1.4}$$

where $|| x_i^{(j)} - c_j ||$ is the term that calculates the distances of the pixels with the cluster center c_j. The optimized cluster centers are then used to assign once again all the pixels to specific classes and display a classified image. Classification techniques used in the study are qualitatively and quantitatively presented. Three unsupervised clustering algorithms are discussed: FCM, K-means, and modified FCM. The supervised technique presented in the study is the FIS. In the next section, the results obtained by the classification techniques are discussed.

1.7 Results and Discussion

A detailed discussion of the classification techniques considered for the study has been provided. In this section, the results of the classification of Landsat 8 RS data using FCM, modified FCM, FIS, and K-means techniques are presented. The FCM technique has been implemented with two different approaches. In the first approach, a three-band multispectral image was provided as the input. This approach was not able to classify the land covers when diverse classes were present because of highly correlated pixel values. The second approach involved the use of all input data bands at the FCM input and showed promising classification results. Results for FIS are also presented in the chapter. The Landsat 8 data used in the study were acquired from the USGS website (www.earthexplorer.usgs.gov).

1.7.1 FCM Technique Results

Figures 1.10 and 1.11 show the classified output map of the study area. For example, 13 pixels belonging to Crop Land and Kharif have been misclassified to Thin Vegetation. Similarly, 418 pixels belonging to Evergreen Forest have been misclassified to Crop Land and Kharif. The overall accuracy of this classification method is 54.01%. The overall accuracy was obtained using Equation 1.5:

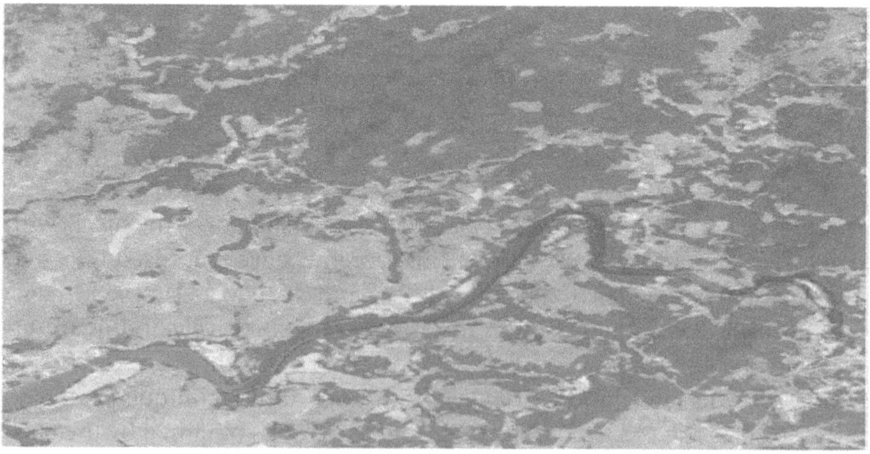

Figure 1.10 Study area used for the research.

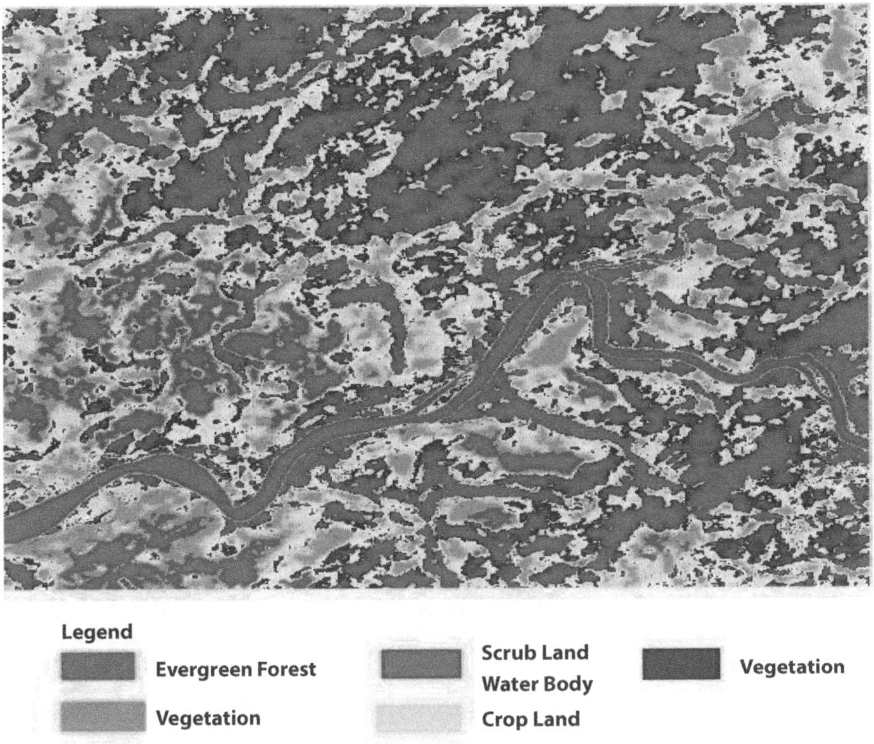

Figure 1.11 FCM classified map of the study area for band combination 574.

$$\text{Overall accuracy} = \frac{\text{Total number of correctly classified pixels}}{\text{Total number of pixels}} \times 100$$

$$\text{Overall accuracy} = \frac{5498}{10168} \times 100 = 54.01\% \qquad (1.5)$$

1.7.2 Modified FCM Technique Results

The second dataset uses individual bands at the input of FCM successively. Table 1.1 shows the confusion matrix obtained for FCM using this method. The output classified map is shown in Figure 1.12. The overall classification accuracy of this classification is 97.83%. The land cover classes such as the Evergreen Forest, Scrub Land, Thin Vegetation, Kharif, Crop Land, and Water Body have been classified. A single-band classification properly

Table 1.1 Confusion matrix for conventional FCM.

A \ P	Water body and scrub land	Crop land and Kharif	Evergreen forest	Thin vegetation
Water Body and Scrub Land	86	0	0	0
Crop Land and Kharif	0	28	0	13
Evergreen Forest	0	418	5355	4227
Thin Vegetation	0	12	0	29
Overall Classification Accuracy = 54.01%				

A: Actual, P: Predicted

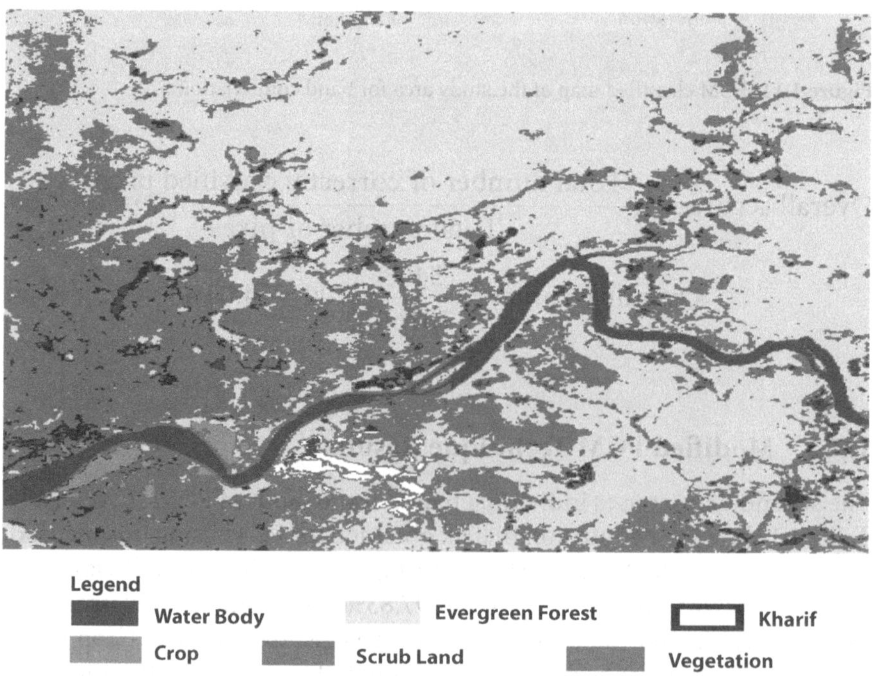

Figure 1.12 Modified FCM classified map for the study area.

classifies the land covers because each band is used based on its reflectance of the land features. The single band is classified to obtain a certain land cover class. Similarly, all the land cover classes are classified. This method avoids the pixel mixing caused by the three-band combination. The FIS is implemented by building a classifier. The process of building a classifier is by obtaining training areas of the required land cover classes for the specific channel. The membership functions of the FIS are built using the mean and the standard deviation of the land cover classes. The values obtained from the training areas build the Gaussian membership functions. These are obtained from the specific land cover classes. In FIS rules, the AND operator corresponds to the minimum value and the OR operator corresponds to the maximum value. All rules are connected using the OR operator, which takes the maximum value from the fuzzified inputs. The classified output of the FIS method is shown in Figure 1.13. The land cover classes have been well separated by the FIS system. The overall classification accuracy is found to be 91.19%. It is observed that the accuracy of the FIS classifier depends on the training values provided.

Figure 1.13 Classification output of the FIS method.

The advantage of the FIS is the ability to implement the classification using simple IF-THEN rules. This system allows adding any number of input bands. It helps to make use of all the Landsat 8 bands, which contain more information than the conventional three-band combination input. The classification output of the FIS method is depicted in Figure 1.13.

1.7.3 K-Means Technique Results

The K-means algorithm identified the following classes from the study area: Evergreen Forest, Water Body, Scrub Land, and Thin Vegetation. The classification of the K-means method is depicted in Figure 1.14. The Water Body and Scrub Land have been categorized as a single class because the pixel values are very close to each other. The overall classification accuracy obtained was 56.40%. This chapter discusses the results obtained for FCM, modified FCM, FIS, and K-means techniques. The classification accuracy obtained for modified FCM was better than the other techniques used. In the next section, conclusions drawn from the study are discussed.

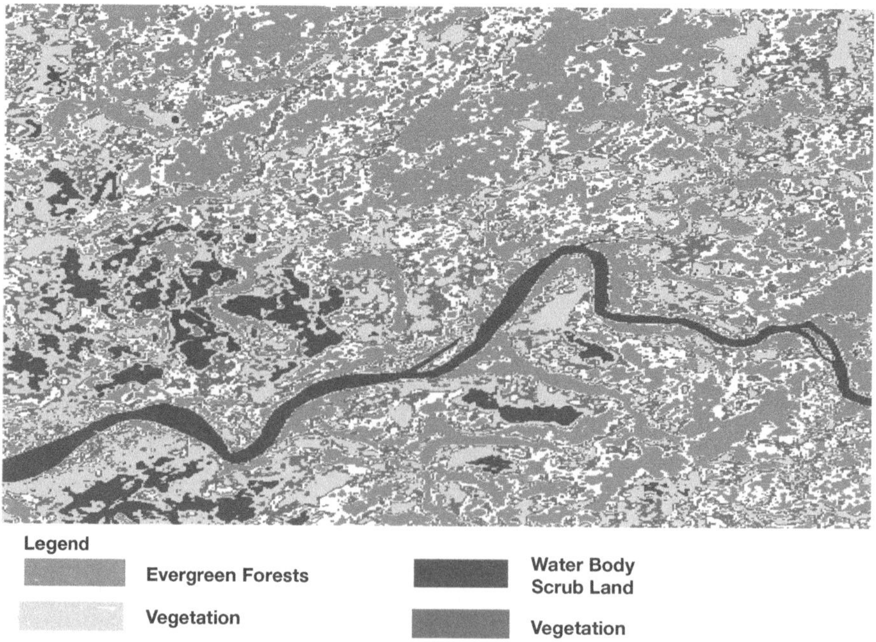

Figure 1.14 Classification of the K-means method.

1.8 Conclusion

The study illustrated the classification capability of the abovementioned classifiers in the presence of mixed pixels and spectrally overlapping classes. Conventional FCM and K-means classifiers failed to provide satisfactory results in the presence of spectrally overlapping classes. The proposed modified FCM technique tackled the presence of spectral overlapping in the study area excellently and produced significantly better results as compared to the conventional FCM and K-means techniques. It is believed that the proposed modified FCM produced better results because it used all the spectral bands at the input. The study also presented classifications of the study area using supervised FIS. FIS required training data to classify the RS data. It was observed during the study that the accuracy of the FIS technique depends on the training provided. Comparing the results of all the classification techniques considered for the study, the proposed technique of the modified FCM produced the best classification accuracy for the study area considered. Hence, it can be concluded that the proposed modified FCM technique is a valid contender for classifying RS data with high efficiency.

References

1. Bezdek, J.C., "*Pattern Recognition with Fuzzy Objective Function*", Plenum Press, New York, 1981, pp. 43-93.
2. L. A. Zadeh, "Probability Measures of Fuzzy Events", *J. Math. Anal. Appl.*, 1968, pp. 421-427.
3. R. Jeansoulin, Y. Fonyaine and W. Frei, "Multitemporal segmentation by means of Fuzzy sets," in *Proc. 1981 Machine Process Of Remotely Sensed Data Symp. (West Lafayette, IN)*, 1981, pp. 336-339.
4. R. L. Cannon, J. V. Dave, J. C. Bezdek and M. M. Trivedi, "Segmentation of a Thematic Mapper Image using the Fuzzy c-means clustering algorithm," *IEEE Trans. Geosci. Remote Sens.*, vol. GE-24, May 1986, pp. 400-408.
5. S. D. Zenzo, R. Bernstein, S. D. Degloria and H. G. Kolsky, "Gaussian maximum likelihood and contextual classification algorithms for multicrop classification," *IEEE Trans. Geosci. Remote Sens.*, vol. GE-25, Nov. 1987, pp. 805- 814.
6. J. T. Kent and K. V. Mardia, "Spatial classification using Fuzzy membership models," *IEEE Trans. Pattern Anal. Mach. Intell.*, vol. 10, no. 5, 1988, pp. 659-671.
7. Ramstein, G. and M. Raffy, "Analysis of the structure of radiometric remotely sensed images", *Int. J. Remote Sens.*, Vol. 10, 1989, pp. 1049- 1073.

8. Raimundo and M. Juraez, "Classification of Remotely Sensed Data with Ongoing Learning Capability", *Int. J. Remote Sens.*, Vol. 50, No. 22, 2002, pp. 4965-4970.
9. F. Melgani and L. Bruzzone, "Classification of Hyperspectral Remote Sensing Images with Support Vector Machines," *IEEE Trans. Geosci. Remote Sens.*, Vol. 42, No. 8, 2004, pp.1778-1790.
10. Crippen, R. E, "The regression intersection method of adjusting image data for band ratioing", *Int. J. Remote Sens.*, 1987, pp.137-155.
11. Ganesan P, K. Palanivel, B. S. Sathish, V. Kalist and Khamar Basha.Shaik, "Performance of Fuzzy based Clustering Algorithms for the Segmentation of Satellite Images", *IEEE Seventh National Conference on Computing, Communication and Information Systems*, 2015, pp. 23-27.
12. A. W. Abbas, n. Minallh, N. Ahmad, S.A.R. Abid, M.A.A. Khan, "K-Means and ISODATA Clustering Algorithms for Landcover Classification Using Remote Sensing", *Sindh Univ. Res. J.*, 2016, pp. 315-318.
13. Wang and Gong, P. "Hyperspectral Remote Sensing and Its Applications", Higher Education Press, Beijing, 2000, pp. 40-44.
14. C.J. Van der Sande, S.M. de Jong, A.P.J. de Roo, "A segmentation and classification approach of IKONOS-2 imagery for land cover mapping to assist flood risk and flood damage assessment", *Int. J. Appl. Earth Obs. Geoinf.*, 2003, pp. 217-229.
15. Jiawei Lu and Micheline Kamber, "*Data Mining Concepts and Techniques*", 2nd ed., Elsevier, 2006, pp. 318-322
16. Patel N. K., Medhavy T. T., Patnaik C. and Bandyopadhyay., "Multi-temporal ERS-1 SAR data for identification of rice crop", *J. Indian. Soc. Remote Sens.*, 2005, pp. 33–39.
17. Jianchao Fan, Min Han, Jun Wang, "Single point iterative weighted Fuzzy C- Means clustering algorithm for remote sensing segmentation", *J. Pattern Recognit.*, 2009, pp. 2527-2540.
18. Chengfan Li, Jingyuan Yin, Junjuan Zhao. "Extraction of urban vegetation from high resolution remote sensing image", *2010 International Conference on Computer Design and Applications*, vol.1, Qinhuangdao, Hebei, China, June25-27, 2010, pp. 4403-4406.
19. Shetty, Shruthi H., *et al.* "Supervised Machine Learning: Algorithms and Applications." *Fundamentals and Methods of Machine and Deep Learning: Algorithms, Tools and Applications* (2022): 1-16.

2

Role of AI in Mortality Prediction in Intensive Care Unit Patients

Prabhudutta Ray[1], Sachin Sharma[2*], Raj Rawal[3] and Dharmesh Shah[2]

[1]*Institute of Advanced Research, Gandhinagar, Gujarat, India*
[2]*Indrashil University, Mehsana, Gujarat, India*
[3]*Gujarat Pulmonary and Critical Care Medicine, Ahmedabad, Gujarat, India*

Abstract
Mortality rate and risk prediction in the intensive care unit (ICU) are two major concerns in hospitals. Prediction is highly dependent on the correct analysis of available data. Many data analysis techniques can provide sufficient information to correctly predict the survival of ICU patients during their admission time in the hospital. Despite methodological advances in machine learning algorithms, clinicians often use black box model. This flaw creates a difficulty in translating complex clinical parameters to corresponding model features. Regarding mortality prediction, it is found that different types of machine learning algorithms have different features that can be properly selected for the respective problem. This chapter focuses on the uses of the types of machine learning algorithms relevant to correct prediction related to survival of the patients.

***Keywords*:** Intensive care unit (ICU), emergency departments (EDs), electronic medical records (EMR), machine learning (ML), sepsis-related organ failure assessment (SOFA), artificial intelligence (AI), early mortality prediction for intensive care unit (EMPICU), data mining (DM), directional coronary atherectomy (DCA), simplified acute physiology score (SAPS), acute physiology and chronic health evaluation (APACHE), length of stay (LOS), intravascular coronary atherectomy (ICA), mortality probability model (MPM)

**Corresponding author*: sachin.sharma@iar.ac.in

Chandra Singh, Rathishchandra R. Gatti, K.V.S.S.S.S. Sairam, Manjunatha Badiger, Naveen Kumar S. and Varun Saxena (eds.) Modeling and Optimization of Signals Using Machine Learning Techniques, (23–40) © 2024 Scrivener Publishing LLC

2.1 Introduction

In any hospital, emergency departments (EDs) often play an important role in developing the front line of healthcare systems, which can ensure an efficient and good quality service for the acute conditions patient. In the case of first evaluations, the patient's condition and acuity level are defined, which help identify an emergency method for severely ill patients who need immediate care [1]. Nowadays, hospitals are running with multiple constraints, which include shortage of funds, unavailability of experienced doctors and radiologists, and costly resources used in healthcare like the intensive care unit (ICU), where critically ill patients are treated by medical practitioners. Using electronic equipment, ICU patients are monitored regularly to measure not only physiological data but also several clinical data, which provide a good opportunity for measuring the severity of diseases and a valuable data analysis to measure the survival rate of the ICU patients during the admission time [2]. Patients in ICU are in critical condition, so quick and accurate decisions should be taken by medical practitioners or intensivists regarding mortality prediction for the advantage of patients and for the uses of medical care resources. Septic and severe sepsis conditions are still major causes related to death for ICU patients. Regarding treatment outcomes, a timely diagnosis is crucial. ICUs continuously generate huge amounts of data that require proper analysis [3]. Several deep learning methods are used in the field of artificial intelligence (AI) to analyze large quantities of electronically stored medical records (EMR) that offer good discoveries in the field of medicine. Due to poor storage structure and different kinds of heterogeneity in raw EMR data, a major difficulty arises when introducing AI techniques with ICU data [4]. Labor-intensive work at different levels include manual data entry, such as sorting personal medical records and interpreting laboratory results that suppress the progress of AI techniques [5].

2.2 Background

In hospitals, an important problem to address is the correct mortality prediction for patients. Once patients are transferred to the ICU, challenges come up, not only related to cost and time, but also about predictions related to early death rate for patients in the ICU [6]. In the past few decades, scoring systems related to disease severity and organ dysfunction were developed like acute physiology and chronic health evaluation II (APACHE II), simplified acute physiology score II (SAPS II), sepsis-related organ failure assessment (SOFA), logistic organ dysfunction score (LODS), an multiple organ dysfunction score (MODS), that could be used with a machine learning (ML) approach for mortality prediction and outcome regarding hospital mortality. Most research mainly focused on the development of data mining (DM) models for the severity of illness scoring systems, which were especially designed for risk estimation for 24 or 48 hours after ICU admission [7]. However, the abovementioned scoring systems have provided information related to the survival group, death group, and statistical measurements like the mean and covariance, leading to the important parameters that are gender, age, and health complications such as coronary heart disease, heart failure, hypertension, arrhythmia, anemia, and a bedridden status, which are very helpful for clinical diagnosis and mortality prediction. Clinical manifestations greatly influence the day-to-day assessment of critically ill patients who are suffering from long duration in the ICU, and all these patients are mostly assessed based on the subjective knowledge and experience of doctors specializing in intensive care. However, some kind of early assessment regarding the improvement of the patients is required, which can help mitigate the cost and expenditure.

2.3 Objectives

This chapter focuses on the respective data challenges related to the early mortality prediction for admitted ICU patients and the proposals on the recent ML algorithm-based skeleton for ICU patients related to early mortality prediction for intensive care unit (EMPICU) [8, 9]. Most of the models are totally focused on the data, and they are purely statistically derived. Currently, machine computing power has been enormously increased to introduce several new ML methods that are used in binary classification regarding the analysis of survival issues of the patients [10]. Further, it is necessary to include biostatistics:

- ✓ In the field of AI, it is found that ML methods must be employed to minimize each patient's follow-up time.
- ✓ Exploring the reliability and validity of data collection methods.

Logistic regression algorithm was also employed several times to predict survival and to carry out the ranking of features as a function for follow-up time. This chapter does not go into the details of feature-ranking techniques [11].

2.4 Machine Learning and Mortality Prediction

A traditional computer programming approach is different from an AI-based approach. An AI algorithm is based on a principle, which is similar to how children absorb knowledge in their day-to-day lives by learning from different exposures and examples [12, 13]. AI is basically putting intelligence into the machines so that they work, operate, and react like human beings. Further, an artificially intelligent machine can read data in real-time, observe the situation, and act accordingly. Seven major disciplines where AI can be used are the following:

- ✓ Image processing and computer vision,
- ✓ Local languages improvement,
- ✓ Processing of natural language,
- ✓ Algorithms related to ML and deep learning,
- ✓ Robotics,
- ✓ Knowledge representation, and
- ✓ Automated reasoning.

The major difference between ML and AI is that ML is considered as a part in which machines find useful information from generated sets of data, but inactive on available information [14]. Several statistical techniques are available in the field of ML, which help to develop useful algorithms that are capable of "statistical learning." Basically, ML algorithms are supervised and unsupervised. Supervised learning requires an effort when training and testing the network from a large dataset. Further, supervised learning is divided into two categories: regression and classification [15]. Generally, regression techniques are used to test hypotheses that involve causal relationships between variables, with the actual choice of models that are being based on their own significance and goodness of fit. Classification algorithms, on the other hand, are trained with data where lines have inputs and provide an output based on the data [16]. In the continuations of unsupervised ML, on the other hand, users do not need to supervise the model with the known input. It allows the model to work independently to find patterns and information that were not previously detected. It mainly deals with unlabeled data [17]. Due to the absence of designated information, the task of an unsupervised algorithm is to find the utility of the results and

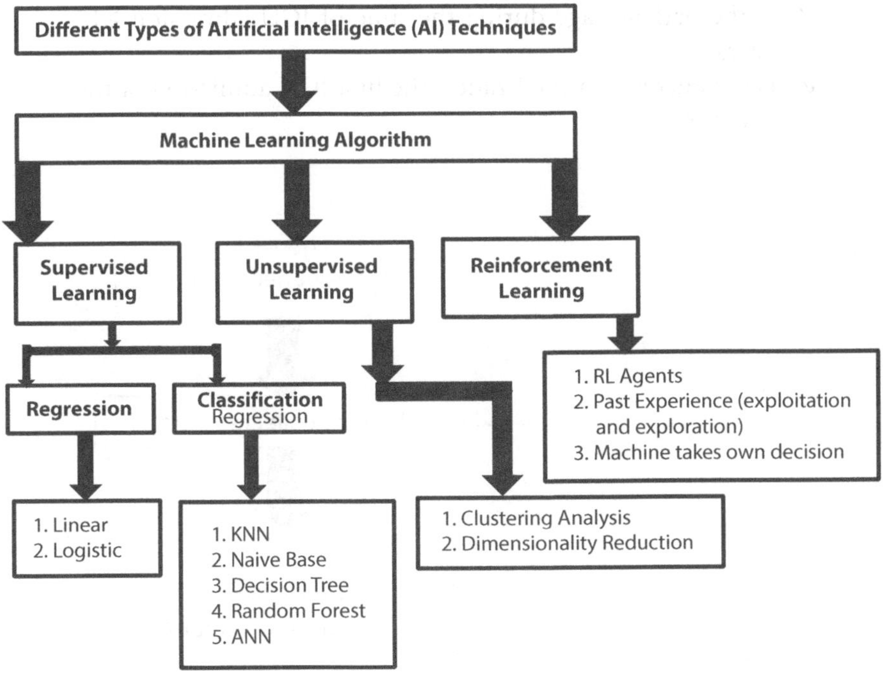

Figure 2.1 Different ML algorithms.

to determine whether the right output has been achieved. Unsupervised algorithms are useful in many cases as research attempts are made to understand collections of huge data. The processes are used in dimensionality reduction, clustering, and anomaly detection. Figure 2.1 summarizes the different ML algorithms [18].

2.4.1 Model Selection

Several kinds of models are available for mortality prediction besides founded models, which try to advance the mortality prediction for ICU patients; they are the mortality probability model (MPM) and SAPS that are purely statistically derived as the names suggest. AI-based models are also used, but the correct prediction of mortality and customization of models to use remain very difficult to achieve due to the complexity of the ICU environment where patients may suffer from more than one critical condition [19]. The types of model selection depend on various parameters, such as:

- ✓ Is the patient's age during the time of ICU admission ≥18 years;
- ✓ Is the current hospitalization the first ICU admission of the patient?

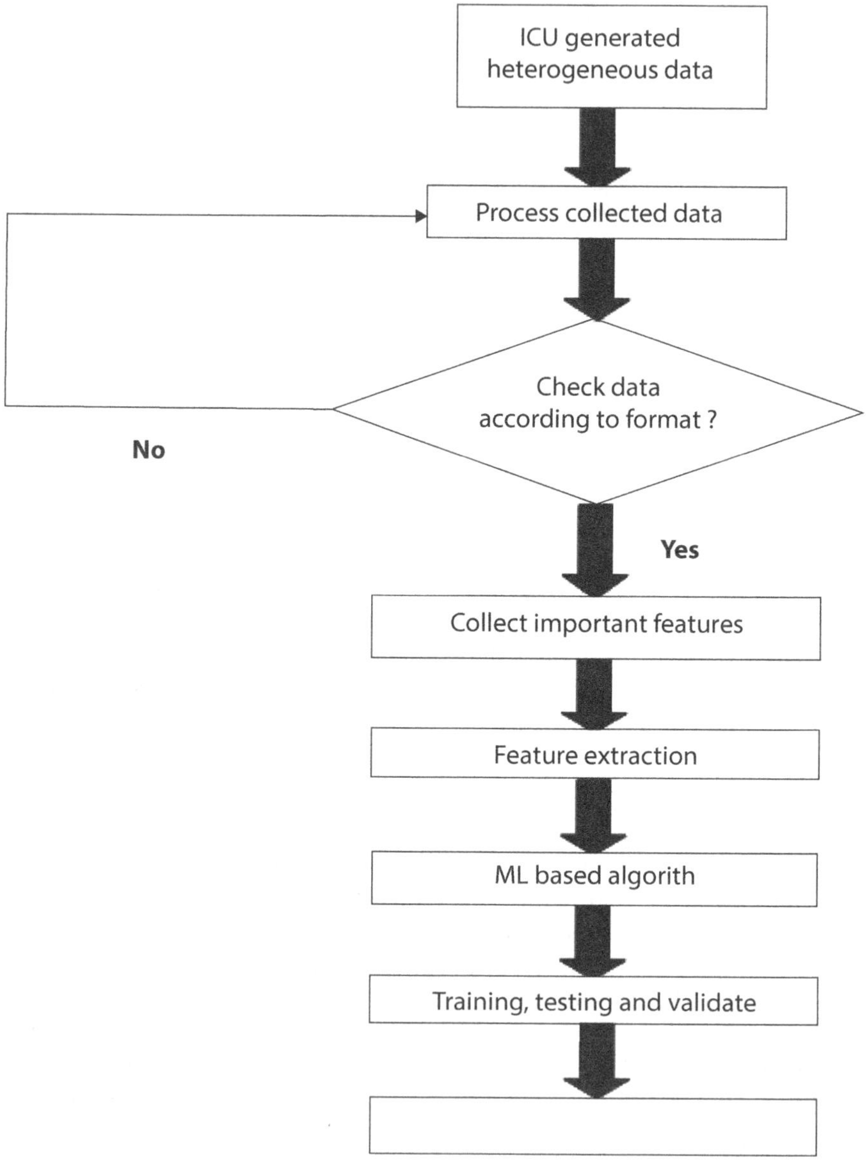

Figure 2.2 Model selection techniques.

✓ Has the patient been cared for more than 4 hours in the ICU?
✓ Does the patient have trauma burns or had undergone a coronary bypass graft surgery?.

Other criteria also need to be included, such as balloon angioplasties, directional coronary atherectomy (DCA), intravascular coronary atherectomy (ICA), and stent placement (BareMetal or Drug Eluting). The above-mentioned parameters gave the basic criteria for model selection, but collecting important features from the collected data is an important criterion also. Figure 2.2 gives a brief representation of the model selection.

2.4.2 Mortality Prediction for ICU Patients

There are several factors involved in the mortality prediction for ICU patients, such as:

i) The influence or impact of the ICU environment; and
ii) The improved scoring systems that help with mortality prediction.

This can be further categorized into two parts, namely:

✓ The conventional scoring systems related to mortality prediction and
✓ The early scoring systems that deal with mortality prediction.

iii) The pattern of continuous data collection.

All these activities deal with the generation and collection of heterogeneous data for the patients. These heterogeneous collections of data require a selection of necessary attributes, which is very difficult to decide regarding the type of dataset that should be considered for constructing the respective model for analysis. There are different types of challenges to be faced because of the characteristics of the available heterogeneous ICU data [20, 21]. Any developed ML-based model needs to address these three important criteria:

1. The missing or unrecorded values in the data,
2. The selection of proper attributes, and
3. The class imbalance problem.

Other than cardiovascular disease patients, those with cancer, respiratory diseases, and diabetes are also admitted to the ICU, and the above-mentioned criteria are very important during the time of data collection and the recording of day-to-day important observations for predicting mortality from developed ML-based models.

Further, the exclusion and inclusion of attributes are also important criteria for getting an efficient output from the execution of the ML model. Choosing all the features for the model is very inefficient and time-consuming for emergency cases [22].

2.4.3 Datasets Generation and Preprocessing

After collecting the data, the major task is to manage the data for further use, which requires a detailed examination of several data variables that are also needed with major focus on important parameters, variety of choices, and assumptions. At this stage, the patients' ages must be considered. Their ages should be 18 years or older with single ICU stays for cardiovascular, surgical, and medical ICU. Different types of lab-test variables are then collected in the first 6 hours of a patient's admission regarding the mortality prediction in the ICU unit [23, 24]. Further, a considerable variation in mortality has also been observed in many cases among different kinds of ICU patients, but after the admission, the characteristics present at the time of admission [25]. In many cases, data regarding ICU patients were collected from different hospitals and were required as inputs provided in the models that help to predict the mortality rate of the ICU patients [26]. From the discussion, it is clear that fixing some kind of inclusion and exclusion criteria is required for the ICU patients before providing inputs to the model to find the outcome of the respective models, which are described in Table 2.1 and Table 2.2.

2.4.3.1 A > Inclusion Criteria

Parameters such as serum creatinine, ejection fraction, serum sodium, high blood pressure, and anemia do not give enough clinical variation for risk assessments compared to patients aged 18 years or older.

Further, staying in the ICU for less than 6 hours does not give sufficient improvement and enough variation of clinical data to get a clear picture of the improvement of the patients during the initial treatment.

Table 2.1 Inclusion criteria.

SR no.	Criteria	Inclusion
1	Aged 18 years or older	Yes
2	Stayed in the ICU for at least 6 hours	Yes

Table 2.2 Exclusion criteria.

SR no.	Criteria	Exclusion
1	Burn patients	Yes
2	Trauma patients	Yes
3	Patients having coronary artery bypass grafting (CABG) surgery	Yes
4	Patients suffering from the myocardial infarction (MI)	Yes

2.4.3.2 B > Exclusion Criteria

The following are the exclusion criteria that also need to be considered:

The exclusion criteria are important because there are limited chances in getting documentation that explicitly indicates the principal operative procedure performed on these patients before being transferred to the ICU [27]. Further, other explicit details regarding the principal operative procedure occurred in the immediate context of any of the following:

- ✓ Bites,
- ✓ Blunt trauma,
- ✓ Burns (thermal, chemical, or electrical),
- ✓ Falls, and
- ✓ Fights.

Besides the inclusion and exclusion criteria, other criteria also need to be taken into consideration like readmission and transfers from another acute care hospital [28].

2.4.4 Structure of Datasets

One of the major considerations is the type of datasets being used for the necessary prediction, as the collected datasets containing the medical

Table 2.3 Most important features.

SR no.	Feature	Explanation	Measurements	Range
1	Age	Patient's age	In years	[18–90]
2	Anemia	Shortage of red blood cells or hemoglobin in the body	Boolean value	0, 1
3	High blood pressure	If a patient has hypertension	Boolean	0, 1
4	Creatinine phosphokinase (CPK)	Level of the CPK enzyme in the blood	mcg/L	[23, 7861]
5	Fasting blood sugar (FBS)	Level of fasting blood sugar	sugar > 110 mg/dL	1 = true. 0 = false
6	Details of ejection fraction	Percentage of blood leaving the heart at each contraction	%	[12, 82]
7	Angina (chest pain)	Chest pain has four types	0: Typical, 1: atypical, 2: non-anginal 3: asymptomatic	0 to 3
8	Trestbps	Resting blood pressure	In mm Hg	120/80
9	Cholesterol	Serum cholesterol	mg/dL	less than 200 mg/dL
10	Restecg	Resting electro cardiographic results have three levels	0: normal	10
11	Thalach	Maximum heart rate achieved	1: flat; 2: up sloping: 0: down sloping	0 to 2

(Continued)

Table 2.3 Most important features. (*Continued*)

SR no.	Feature	Explanation	Measurements	Range
12	Exercise-induced angina	Exercise-induced angina	0= no, 1= yes	0 to 1
13	Slope	Peak exercise ST segment	0 = up, sloping	13
14	Thallium stress level test	Thallium stress test	3 = normal; 6 = fixed defect; 7 = reversible defect	3, 6, 7

records of different kinds of patients are based on several features. The datasets contain 14 most important features, which report clinical, body, and lifestyle information, which are described in Table 2.3 as follows. Vital measurements and laboratory test values fluctuate as time progresses. For example, glucose levels may increase or decrease after 1 hour, but these are temporal trends. Changes in blood pressure, creatinine phosphokinase (CPK), ejection fraction (EF), or chest pain level are very crucial for the diagnosis in the case of critically ill patients [29].

Some features are represented in the binary term like anemia, high blood pressure, diabetes, gender, and smoking. Doctors consider a patient to have anemia if the patient's hemoglobin count levels were lower than 36%. Regarding the important features, CPK describes the margin of the CPK enzyme in the blood [30, 31]. Basically, the CPK starts to flow into the blood when muscle cells become damaged. So, increasing the levels of CPK in the patient's blood might point out heart failure or severe damage in an internal body part. The throwing fraction states the percentage of the quantity of blood pumped out by the left ventricle with each contraction. Creatinine is a waste product that comes from the digestion of protein in your food and the normal breakdown of muscle tissue [32, 33]. Majority of doctors specifically investigates serum creatinine levels to check the functionality of kidneys. Patients with an increasing level of serum creatinine in the blood may indicate renal dysfunction. Another important factor is the sodium level in the blood, which impacts the workings of muscle cells and nerves. Sodium tests must be performed routinely to find the natural levels of sodium in the blood. In any case, sodium found at an abnormally low level in the blood might cause heart failure [34–36].

The features related to death are also included as relevant targets related to binary classification that describes if the patient survived or died before the end of the follow-up period, which was approximately 145 days on average. Here, the survived patient's death event is '0', while the dead patient's death event is '1'. Information regarding datasets have been given here though other sets of data also need to be considered for the assessment of the patient's condition in the ICU [37, 38].

As mentioned previously, the 14 most important features are useful for the prediction of mortality using ML algorithm, but sometimes more features need to be considered by doctors to get direction and improve treatment. Clinical observations and the variability in clinical diagnosis outputs result in vast and diverse ICU data. However, attempting to encompass all parameters is impractical, as it would inefficiently utilize machine computing power for emergency predictions [39].

2.5 Discussions

The MPM and SAPS models are dependent on the collection of accurate data from the patients. During the time of admission in the ICU, the recorded data gave the basis for further clinical diagnosis. As ICU environments generate a huge number of heterogeneous kinds of data, collecting important features from those data is also an important stage before inputting these data to the corresponding model for critical predictions. Further, during the collection of data, it is also important to collect these data accurately so that it will also provide the model for correct predictions [40].

2.6 Conclusion

Today's ICUs are information-intensive, and a major challenge is to deal with a continuous information flow; few are useful, the majority are not. However, objectives must be considered in all the relevant information required to correctly predict the mortality of patients. Further, a more difficult task is to improve intelligent monitors, which are based on ML and proficient regular measures of human reactions to serious illnesses with a high degree of certainty.

2.7 Future Work

There are many areas where correct uses of expert technology can easily guide us to deal with overloaded information. These ML algorithms, on the other hand, have been helping us to experiment with the data stored in electronic medical records for the prediction of mortality and duration of stay in the ICU. In this chapter, a brief overview of several factors is pointed out. For future work, the development of knowledgeable ML monitors, with a high degree of accuracy and certainty for responses to critical illnesses of the ICU patients, is worth pursuing.

2.8 Acknowledgments

This study was performed through the cooperation of different hospital ICUs and several diagnostic centers in India.

2.9 Funding

This work is supported under the SHODH Scheme of the Education Department, Gujarat, India, bearing grant no. 202001440003.

2.10 Competing Interest

The authors declare that they have no competing interests.

References

1. Cong Li, Zhuo Zhang, Yazhou Rena, Hu NieYuqing Lei aHang, QiuabZenglinXu da XiaorongPu Machine learning based early mortality prediction in the emergency department, *Int. J. Med. Inf.*, Volume 155, 5th September 2021, 104570, https://doi.org/10.1016/j.ijmedinf.2021.104570.
2. Beatriz NistalNuño, Machine learning applied to a Cardiac Surgery Recovery Unit and to a Coronary Care Unit for mortality prediction, *J. Clin. Monit. Comput.*, volume 214, 15th April 2021 https://doi.org/10.1007/s10877-021-00703-2
3. Ximing Niel, Yuan Cai, Jingyi Liu. Mortality Prediction in Cerebral Hemorrhage Patients Using Machine Learning Algorithms in Intensive

Care Units, *J. Front. Neurol.*, 20 January 2021, https://doi.org/10.3389/fneur.2020.610531

4. Thanakron Na Pattalung, SitthichokChaichulee , Comparison of machine learning algorithms for mortality prediction in intensive care patients on multi-center critical care databases, *International Conference on Engineering and Technology May 21, (ICET 2021, IOP Conf. Series: Mater. Sci. Eng.*, 1163(2021) 012027, doi:10.1088/1757-899X/1163/1/012027
5. Kyongsik Yun, Jihoon Oh, Tae Ho Hong, Eun Young Kim. Prediction of Mortality in Surgical Intensive Care Unit Patients Using Machine Learning Algorithms, *Front. Med.*, 8:621861., Volume 8, Article 621861, 31 March 2021 doi: 10.3389/fmed.2021.621861
6. Martina Barchitta, Andrea Maugeri, Giuliana Favara, Paolo Marco Riela, Giovanni Gallo, Ida Mura, Antonella Agodi. Early Prediction of Seven-Day Mortality in Intensive Care Unit Using a Machine Learning Model: SPIN-UTI Project. *J. Clin. Med.*, March 2021, 10, 992, https://doi.org/10.3390/jcm10050992.
7. ChonghuiGuo, Mucan Liu, Menglin Lu A Dynamic Ensemble Learning Algorithm based on K-means for ICU mortality prediction , 5th February 2021, Elsevier, *Appl. Soft Comput. J.*, 103, 1568-4946/2021, https://doi.Org/10.1016/j.asoc.2021.07166
8. Michael Moor, Bastian Rieck, Max Horn, Catherine R. Jutzeler ,and KarstenBorgwardt Early Prediction of Sepsis in the ICU Using Machine Learning: A Systematic Review, 28 May 2021 *Front. Med.*, 8:607952. doi: 10.3389 /fmed.2021.607952
9. Zhengyu Jiang, Lulong Boa, ZhenhuaXu, Yubing Song, Jiafeng Wang, Pingshan Wen, Xiaojian Wan An explainable machine learning algorithm for risk factor analysis of in-hospital mortality in sepsis survivors with ICU readmission, Elsevier *J. Comput. Methods Programs Biomed.*, March 2021, 106040 01692607, https://doi.org/10.1016/j.cmpb.2021.106040.
10. Khalid Alghatani, NarimanAmmar Predicting Intensive Care Unit Length of Stay and Mortality Using Patient Vital Signs: Machine Learning Model Development and Validation, *JMIR Med. Inform.*, Published on 5.5.2021 in Vol 9, No 5 (2021): May 2021 doi:10.2196/21347.
11. Fuhai Li, Hui Xin, Jidong Zhang, Prediction model of in-hospital mortality in intensive care unit patients with heart failure: machine learning-based, retrospective analysis of the MIMIC-III database *BMJ Open*, june 2021;11:e044779. doi:10.1136/bmjopen-2020-044779.
12. Fawad Ahmad, HumaAyub, RehanLiaqat Mortality Prediction in ICU Patients Using Machine Learning Models, 2021 *International Bhurban Conference on Applied Sciences and Technologies (IBCAST)*, 12-16 Jan. 2021, INSPEC Accession Number: 20633348, DOI : 10.1109/ IBCAST51254.2021.9393012.
13. Jong Ho Kim, Young Suk Kwon, Moon SeongBaek Machine Learning Models Predict 30-Day Mortality in Mechanically Ventilated Patients, May 2021, *Clin. Med.*, 2021, 10, 2172. https://doi.org/10.3390/jcm10102172.

14. NaghmehKhajehali, ZohrehKhajehali The prediction of mortality influential variables in an intensive care unit: a case study, *Pers. Ubiquitous Comput.*, (2021), 26 February 2021, https://doi.org/10.1007/s00779-021-01540-
15. Hodgson CL, Udy AA, Bailey M, Barrett J, Bellomo R, Bucknall T, Gabbe BJ, Higgins AM, Iwashyna TJ, Hunt-Smith J, *et al.* The impact of disability in survivors of critical illness. *Intensive Care Med.*, 2017;43(7):992–1001.
16. Dinglas VD, Aronson Friedman L, Colantuoni E, Mendez-Tellez PA, Shanholtz CB, Ciesla ND, Pronovost PJ, Needham DM. Muscle weakness and 5-year survival in acute respiratory distress syndrome survivors. *Crit. Care Med.*, 2017; 45(3):446–53.
17. Hill AD, Fowler RA, Pinto R, Herridge MS, Cuthbertson BH, Scales DC. Long-term outcomes and healthcare utilization following critical illness—a population-based study. *Crit. Care*, 2016; 20:76.
18. Herridge MS, Chu LM, Matte A, Tomlinson G, Chan L, Thomas C, Friedrich JO, Mehta S, Lamontagne F, Levasseur M, *et al.* The RECOVER Program: disability risk groups and 1-year outcome after 7 or more days of mechanical ventilation. *Am. J. Respir. Crit. Care Med.*, 2016;194(7):831–44.
19. Mebazaa A, Casadio MC, Azoulay E, Guidet B, Jaber S, Levy B, Payen D, Vicaut E, Resche-Rigon M, Gayat E. Post-ICU discharge and outcome: rationale and methods of the The French and euRopean Outcome reGistry in Intensive Care Units (FROG-ICU) observational study. *BMC Anesthesiol.*, 2015; 15:143.
20. Garland A, Olafson K, Ramsey CD, Yogendran M, Fransoo R. Distinct determinants of long-term and short-term survival in critical illness. *Intensive Care Med.*, 2014; 40(8):1097–105.
21. Yende S, Linde-Zwirble W, Mayr F, Weissfeld LA, Reis S, Angus DC. Risk of cardiovascular events in survivors of severe sepsis. *Am. J. Respir. Crit. Care Med.*, 2014; 189(9):1065–74.
22. Marino R, Struck J, Maisel AS, Magrini L, Bergmann A, DiSomma S. Plasma adrenomedullin is associated with short-term mortality and vasopressor requirement in patients admitted with sepsis. *Crit. Care*, 2014; 18 (1):R34.
23. Yancy CW, Jessup M, Bozkurt B, Butler J, Casey Jr. DE, Drazner MH, Fonarow GC, Geraci SA, Horwich T, Januzzi JL, *et al.* 2013 ACCF/AHA guideline for the management of heart failure: a report of the American College of Cardiology Foundation / American Heart Association Task Force on Practice Guidelines. *J. Am. Coll. Cardiol.*, 2013;62(16):e147–239.
24. Lassus J, Gayat E, Mueller C, Peacock WF, Spinar J, Harjola VP, van Kimmenade R, Pathak A, Mueller T, Disomma S, *et al.* Incremental value of biomarkers to clinical variables for mortality prediction in acutely decompensated heart failure: the Multinational Observational Cohort on Acute Heart Failure (MOCA) study. *Int. J. Cardiol.*, 2013;168(3):2186–94.
25. Needham DM, Davidson J, Cohen H, Hopkins RO, Weinert C, Wunsch H, Zawistowski C, Bemis-Dougherty A, Berney SC, Bienvenu OJ, *et al.*

Improving long-term outcomes after discharge from intensive care unit: report from a stakeholders' conference. *Crit. Care Med.*, 2012; 40(2):502–9.
26. Desai SV, Law TJ, Needham DM. Long-term complications of critical care. *Crit. Care Med.*, 2011; 39(2):371–9.
27. Winters BD, Eberlein M, Leung J, Needham DM, Pronovost PJ, Sevransky JE. Long-term mortality and quality of life in sepsis: a systematic review. *Crit. Care Med.*, 2010; 38(5):1276–83.
28. Braun A, Chang D, Mahadevappa K, Gibbons FK, Liu Y, Giovannucci E, Christopher KB. Association of low serum 25-hydroxyvitamin D levels and mortality in the critically ill. *Crit. Care Med.*, 2011;39(4):671–7.
29. Orwelius L, Nordlund A, Nordlund P, Simonsson E, Backman C, Samuelsson A, Sjoberg F. Pre-existing disease: the most important factor for health related quality of life long-term after critical illness: a prospective, longitudinal, multicentre trial. *Crit. Care*, 2010;14(2):R67.
30. Wunsch H, Guerra C, Barnato AE, Angus DC, Li G, Linde-Zwirble WT. Three-year outcomes for Medicare beneficiaries who survive intensive care. *JAMA*, 2010;303(9):849–56.
31. Hlatky MA, Greenland P, Arnett DK, Ballantyne CM, Criqui MH, Elkind MS, Go AS, Harrell Jr. FE, Hong Y, Howard BV, *et al.* Criteria for evaluation of novel markers of cardiovascular risk: a scientific statement from the American Heart Association. *Circulation*, 2009;119(17):2408–16.
32. Cuthbertson BH, Rattray J, Campbell MK, Gager M, Roughton S, Smith A, Hull A, Breeman S, Norrie J, Jenkinson D, *et al.* The PRaCTICaL study of nurse led, intensive care follow-up programmes for improving long term outcomes from critical illness: a pragmatic randomised controlled trial. *BMJ*, 2009; 339:b3723.
33. Williams TA, Dobb GJ, Finn JC, Knuiman MW, Geelhoed E, Lee KY, Webb SA. Determinants of long-term survival after intensive care. *Crit. Care Med.*, 2008; 36(5):1523–30.
34. Hofhuis JG, Spronk PE, van Stel HF, Schrijvers GJ, Rommes JH, Bakker J. The impact of critical illness on perceived health-related quality of life during ICU treatment, hospital stay, and after hospital discharge: a long-term follow-up study. *Chest*, 2008;133(2):377–85.
35. Rey G, Jougla E, Fouillet A, Hemon D. Ecological association between a deprivation index and mortality in France over the period 1997–2001: variations with spatial scale, degree of urbanicity, age, gender and cause of death. *BMC Public Health*, 2009;9:33.
36. Pencina MJ, D'AgostinoSr RB, D'AgostinoJr RB, Vasan RS. Evaluating the added predictive ability of a new marker: from area under the ROC curve to reclassification and beyond. *Stat. Med.*, 2008;27(2):157–72.
37. Yende S, D'Angelo G, Kellum JA, Weissfeld L, Fine J, Welch RD, Kong L, Carter M, Angus DC, Gen IMSI. Inflammatory markers at hospital discharge predict subsequent mortality after pneumonia and sepsis. *Am. J. Respir. Crit. Care Med.*, 2008; 177(11):1242–7.

38. Cook NR. Use and misuse of the receiver operating characteristic curve in risk prediction. *Circulation*, 2007;115(7):928–35.
39. Bagshaw SM, Mortis G, Doig CJ, Godinez-Luna T, Fick GH, Laupland KB. One-year mortality in critically ill patients by severity of kidney dysfunction: a population-based assessment. *Am. J. Kidney Dis.*, 2006;48(3):402–9.
40. Wright JC, Plenderleith L, Ridley SA. Long-term survival following intensive care: subgroup analysis and comparison with the general population. *Anaesthesia,* 2003;58(7):637–42.

3

A Survey on Malware Detection Using Machine Learning

Devika S. P.*, Pooja M. R. and Arpitha M. S.

Vidyavardhaka College of Engineering, Mysore, India

Abstract

With an emphasis on the Android platform and related technologies, the exponential rise in cybercrimes has resulted in a notable increase in cyberattack occurrences, many of which have catastrophic outcomes. An important component of Android framework security is malware identification. To overcome this risk, the machine learning algorithm which has a high degree of accuracy in malware detection was used. An important component of the Android application's security is its ability to distinguish between benign and malicious apps. Malware has become a big computer threat in recent times due to the web's rapid expansion and advancement. Ultimately, friendly programs falling under a particular class will typically have a similar set of features. Despite what is generally expected, pernicious applications will, in general, have unusual highlights, which are extraordinary for the classification that they have a place with. Machine learning structures have been tried, investigated, and created to portray and order malware into their malware families, utilizing attributes separated and obtained from static and live examination of the noxious programming.

Keywords: Cybercriminals, machine learning, malware detection

3.1 Background

Genuinely, Android is the primary focus on stage by malware creators trying to assume the responsibility for more than one billion portable devices on the planet. Android's conspicuousness was a result of being

**Corresponding author*: devikasp5@gmail.com

Chandra Singh, Rathishchandra R. Gatti, K.V.S.S.S.S. Sairam, Manjunatha Badiger, Naveen Kumar S. and Varun Saxena (eds.) Modeling and Optimization of Signals Using Machine Learning Techniques, (41–54) © 2024 Scrivener Publishing LLC

open-source. Because of the open climate, malware creators can create malicious applications with a piece of malevolent code. Cell phones are rapidly turning into the essential method for correspondence and data access for billions of individuals. Cell phone security is more significant. One of the primary reasons is the truth that applications do not need to come straightforwardly from the Play Store. They can be "side-stacked" from elective business sectors or then again straightforwardly introduced from APK (Android Application Package) records. By far, most of these gadgets are not state-of-the-art. Updates are made to fix security openings that are found in the working framework that can be abused by assailants. With the fame of Android applications, it can likewise welcome digital crooks to create vindictive applications that can take data from cell phones [1, 2].

3.2 Introduction

The absence of safety review on Android applications escalates the spreading of noxious applications (malware) on the Android stage. Lately, different exploration endeavors have been made to address the security and protection dangers forced by malware in Android stage, among which static investigation and dynamic examination are the two significant exploration headings. Program decompilation and the following information stream are the two significant strategies utilized in static investigation. Then again, a dynamic investigation by and large plans to screen the practices of versatile applications in terms of getting to private information or utilizing confined API calls. ML works by ordering a class in a characterized dataset. Characterization is utilized to recognize two classes to be specific considerate and malware. A significant advancement in all ML-based approaches is the highlighted choice. Acquiring an ideal list of capabilities would not just assist in improving experimentation results; however, it will likewise help in lessening the scourge of dimensionality related with most ML-based algorithms. Malware is the briefest term utilized for pernicious programming which is an unsafe malevolent piece of code. Malwares are executable codes; sometimes they go about as downloader for malware as well—for example, PDF and PHP connect, which deals with the framework and download more malignant programming to execute on a framework. In the event that the product coordinates with a mark of known malware, that product can be securely isolated or erased from the framework—ideally before any harm is done. The act of signature-based malware discovery is adequate for ensuring a framework against known malwares. Antivirus organizations investigate newfound malware and delivery marks to their clients to secure them against new dangers, yet until a mark is delivered for a recently recognized malware there will be no safeguard against the disease. Progressed constant threat entertainers (APT) have regularly abused long-standing security weaknesses for various years without recognition. Plainly, we can see that protection including as it were realized dangers are lacking to guarantee the classification, accessibility, and honesty of processing frameworks, applications. Normally, it misuses the weaknesses of each target framework. For example, a few programmers can use a bug in real programming (e.g., a program or web application module) to unlawfully capture an organization worker for taking significant information or cash [3, 4].

Types of Malwares

Malware can be classified into a couple of classes depending upon its motivation.

- Ransomware: It is a kind of malware that encodes a mishap's files. The attacker by then requests an installment from the mishap to reestablish authorization to the information upon segment.
- Adware: The term adware is habitually used to portray a type of malware (malicious software) that presents undesirable ads to a PC client. The promotions delivered by adware are in some cases as a spring up or once in a while in an "unclosable window".
- Spyware: Spyware is appalling programming that assaults your conduct device, attractive your web use data in addition to sensitive information. Spyware has assigned a sort of malware—malevolent encoding planned to contact or mischief your computer much of the time without your understanding.

  Worms: A PC worm is a kind of malware that spreads copies of itself starting with one PC then onto the next. A worm can mirror itself with no human participation, and it does not need to interface itself to an item program to cause hurt.
- Virus: "Malware" was short for malicious software, a single term suggesting contamination, spyware, worm, etc. Malware is expected to create problems in an independent

PC or a network of PCs. Thus, in any place that the malware term was used, it infers a program that is intended to harm your PC. It very well might be an infection, worm, or Trojan.

- Trojan: A Trojan pony or Trojan is a kind of malware that is often concealed as authentic programming. Trojans can be used by computerized cheats and software engineers endeavoring to get to the customers' structures. Customers are routinely hoodwinked by some sort of friendly plan into stacking and executing Trojans on their structures.

3.3 Literature Survey

The main concept behind this malware identification scheme is to combine the acquisition of permissions, the number of unsafe APIs, and their weights to create a feature vector and measure the similarity score between Android applications. As of late, there have been some examination works that put attention on identifying malevolent applications for Android stages. They gave an outline of Android malware with respect to its hazardous practices and weaknesses. ML methods can give astounding identification precision rates relying upon the nature of the highlights which are utilized for preparing the setup. Android malware identification has three principal forms, which are static, dynamic, and a hybrid. The static investigation method analyzes the source code of the application without execution to identify vindictive codes and examples. Then again, dynamic investigation inspects the application during runtime in a controlled virtual climate also, thus screening the application's dynamic direct and the system's responses. For the two methodologies, the information is gathered to prepare ML classifiers to assemble a partition demonstrating the favorable and malignant qualities of the applications [5]. Typically, a static investigation reports a high exactness rate in recognizing malware, and it is moderately modest in contrast to dynamic examination as far as exertion, time, and computational assets are concerned. Though the exact paces of the classifiers increased with expanding the nature of the highlights, we relate between the applications' highlights and the highlights that are expected to convey its classification's usefulness to identify vindictive examples. All in all, a malware recognition classifier is prepared for every classification independently. Unexpectedly, vindictive applications will, in general, demand strange highlights, less or more than what is normal for the classification that they have a place with.

The wild development of Android malware makes it more complex to keep away from discovery and harder to order utilizing regularly utilized ML calculations. Human exertion of naming the malware as malignant or amiable cannot stay aware of the speed of the voluminous age of Android malware, bringing about a lopsidedness of substantially more unlabeled information than marked information [6]. Indeed even Virus Total could set aside a long effort to conclude the expectation of a huge measure of unlabeled information. Furthermore, this can cause managed to learn calculations to debase in execution. Moreover, because of this lopsidedness of unlabeled and named information, the circulation seen from the marked information cannot be the same as the real information dispersion. This is seen in malware location, as the study proposes that malware families show polymorphic practices. Converted into ML, it infers that, at the testing stage, the test information can be correspondingly disseminated as the preparation information, yet it is not indistinguishably appropriated as the preparation information. Android frameworks have an authorization-based instrument to implement security limitations on applications. The establishment and startup of an Android application require various authorizations like read contacts, web access, empower or debilitate application segments, and so on. The installer bundle shows the arrangements of authorizations where a client can permit or deny these consents. At the point when a client sets the consent, then it can, without much of a stretch, access the asset, and it is beyond the realm of imagination to expect to deny these authorizations until the application is uninstalled. Essentially, a learning classifier is prepared on a marked arrangement of tests, utilizing component vectors that address syntactic or semantic properties of the applications. The design of Android (applications) gives important syntactic data, for example, strategy and bundle names, consents, and set up records. Thus, statically determined highlights permit the identification and order of diverse Android applications. Comprehensively speaking, Android malware investigation is of two kinds: static analysis and dynamic analysis. Static investigation essentially includes dissecting the code structure without executing it, while the dynamic examination will be an assessment of the runtime conduct of Android apps in a compelled climate. Static examination-based characterization goes through two approaches: one was consent-based, while the other included a portrayal of the source code as a sack of words. Another methodology was dependent on distinguishing the most critical consents and applying AI to it for assessment. A significant advance in all ML-based methodologies included determination. Getting an ideal list of capabilities would not just assist in improving experimentation results, yet it will likewise help in diminishing the scourge of dimensionality related to most AI-based calculations. Static examination method program deterioration and the following information stream are two significant procedures utilized in the static investigation. The static investigation method analyzes the source code of the application without execution to recognize vindictive codes and examples. Static examination reports a high exactness rate in recognizing malware. Static discovery depended on explicit strings from the dismantled code without executing the twofold document. A static examination can rapidly catch the punctuation, yet it was effortlessly upset by code obscurity and encryption innovation [7, 8]. A dynamic investigation, by and large, plans to screen the conduct of portable applications as far as getting to private information or utilizing limited API calls. Dynamic examination analyzes the application during the runtime in a controlled virtual climate. The double surface investigation gives equivalent and comparable outcomes to dynamic examination methodology with respect to precision. Dynamic location investigation the malware conduct, for example, network exercises, framework calls, and recording tasks by executing the malware. This procedure can recognize recently made malware; be that as it may, it requires more execution time. There are various sorts of malware. In some cases, genuine applications can have malware infused into them that will run behind the scenes while the application is playing out its planned work. The client may never realize that the malware is really running and performing unapproved activities. Malware can additionally be downloaded at runtime. This is known as an "update assault". On different occasions, the client can be tempted to download malware publicized for another reason. This is known as a "drive-by download". Machine learning algorithms and techniques have been used to detect malware on mobile devices in a variety of ways. These techniques range from support vector machines (SVM), neural network techniques, Android behavior-based malware detection, model-based semi-supervised, naïve Bayes techniques, decision trees, random forest, K-nearest neighbor, genetic algorithm, supervised algorithm, and semi-supervised. These days, the quantity of brilliant gadgets has been expanding fundamentally because of the quick advancement of the Internet of Things (IoT) innovation. There have been various sorts of utilizations identified with IoT gadgets, including savvy transportation, brilliant industrial facilities, shrewd homes, and keen labs. When all is said in done, the expression "web of things" addresses an organization of numerous detecting gadgets with restricted assets that can do wired/remote correspondences with cloud administrations or others. Given billions of IoT gadgets presently utilized in numerous parts of day-by-day lives, IoT has been bit by bit getting probably the most fragile piece of the PC organization. Programmers increasingly devise more objective IoT gadgets as they can rapidly assault them and make more gigantic harms than customary PCs. It would seem that malware discovery has been assuming a crucial part in numerous frameworks, particularly in portable stages that have an enormous volume of new applications submitted to the applications stores [9].

Malware Detection Techniques

Support vector machine (SVM) was a representation of the preparation data as centers in the space that combination dependent on their classification in kind of social events that are confined by an unmistakable particular hole called a hyperplane. In the preparation stage, SVM develops a model of examples from the preparation information which is utilized as a space for the arrangement stage. In the grouping stage, the new info focuses are planned into the prepared space and ordered dependent on which side of the hole they fall on. SVM structures a hyperplane or set of hyperplane for information arrangement and relapse as shown in Figure 3.1. For more certainty and less speculation mistake, the hyperplane should be chosen by a utilitarian edge that makes the distance between the closest preparing

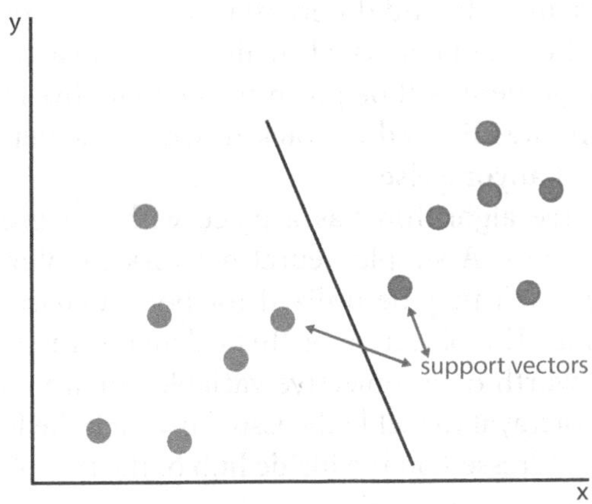

Figure 3.1 A simple example of SVM.

information focuses in any class however much bigger as could reasonably be expected. The SVM was chosen as a result of its protection from over-fitting even in the exceptionally high dimensional factors space like our datasets.

An artificial neural association learning computation or neural association or essentially neural net is a computational learning structure that employs an organization of abilities to grasp and decipher a data commitment of one design into the principle yield, as a rule in another structure. The idea of the fake neural organization was roused by human science and the path neurons of the human cerebrum cooperate to grasp contributions as of human detects. Neural organizations is only one of numerous instruments and approaches utilized in ML calculations. The neural organization itself might be utilized as a piece in various ML calculations to deal with complex information contributions to a space that PCs can comprehend. Supervised learning, as the name indicates, has the presence of a boss as an educator. Fundamentally directed learning was the point at which we instruct or, on the other hand, train the machine using data that was all-around marked. This implies that some data is as of now named with the correct answer. From here on, the machine was set with another plan of examples (data) so the directed erudition calculation investigates the readiness data and creates the right result from marked in order. Unsupervised learning was the collection of a machine using information that is not one or the other characterized nor name and permitting the calculation to circle back to that in order lacking heading. At this time, the task of the machine was to bunch unsorted records as per resemblances, models, and differentiations with no previous arrangement of data. Rather than oversaw learning, no instructor was given that suggests no arrangement will be given to the machine [10]. Along these lines, the machine was limited to track down the secret design in unlabeled data without anyone else.

The decision tree algorithm has a place with the group of administered ML calculations. A simple neural network topology is depicted in Figure 3.2. It very well may be utilized for both a characterization issue and a relapse issue. The objective of this calculation is to make a model that predicts the worth of an objective variable, for which the choice tree utilizes the tree portrayal to tackle the issue in which the leaf hub relates to a class name and addressed on the inside hub of the tree. A simple example of a decision tree is depicted in Figure 3.3.

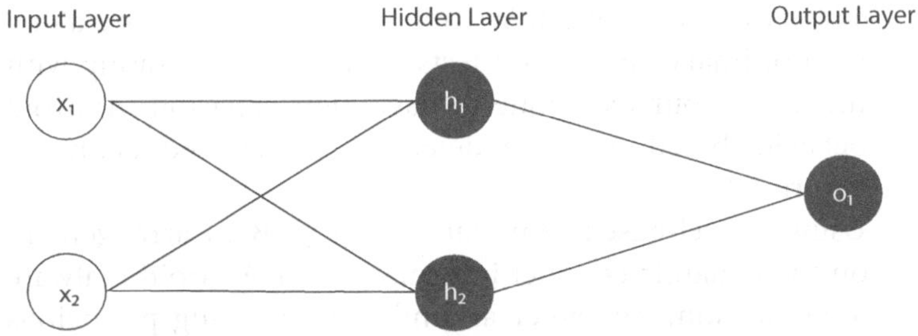

Figure 3.2 A simple example of neural network technology.

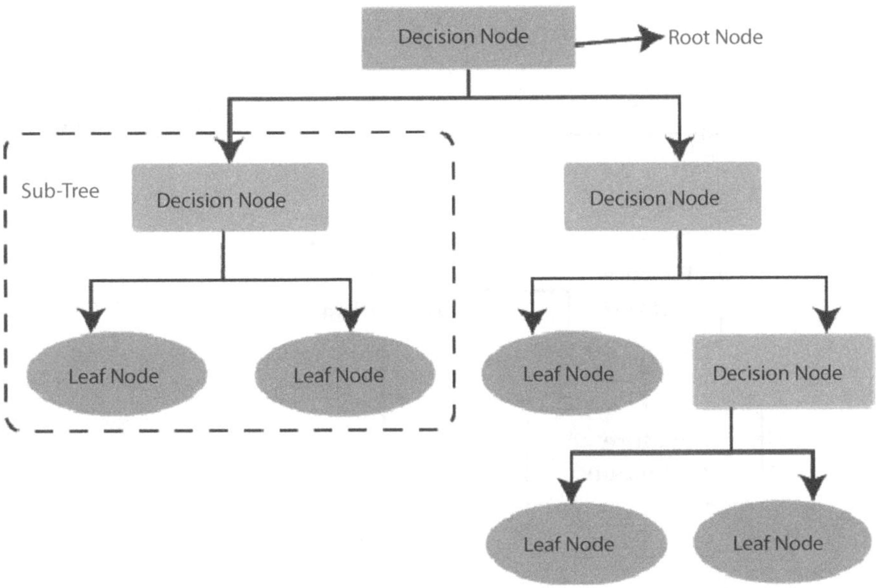

Figure 3.3 A simple example of a decision tree.

- An Android behavior-based malware detection method utilizing machine learning: They progress an Android application sandbox, Droidbox, by embedding a VIEW-ID programmed elicit program that can click versatile applications in a significant request. Exploiting Droidbox result, we gather the conduct like organization exercises, the record read/compose, and consent as the element information and utilize diverse ML calculations to arrange malware and also assess the presentation. We utilize countless malware along with ordinary application tests toward demonstrating with the aim of our technique to have high precision. Android behavior-based malware detection is depicted below in Figure 3.4.

- Using model-based semi-supervised (MBSS) arrangement on the dynamic conduct information for Android malware identification, we center around distinguishing pernicious conduct at runtime by utilizing dynamic conduct information. The primary benefit of semi-regulated grouping is the

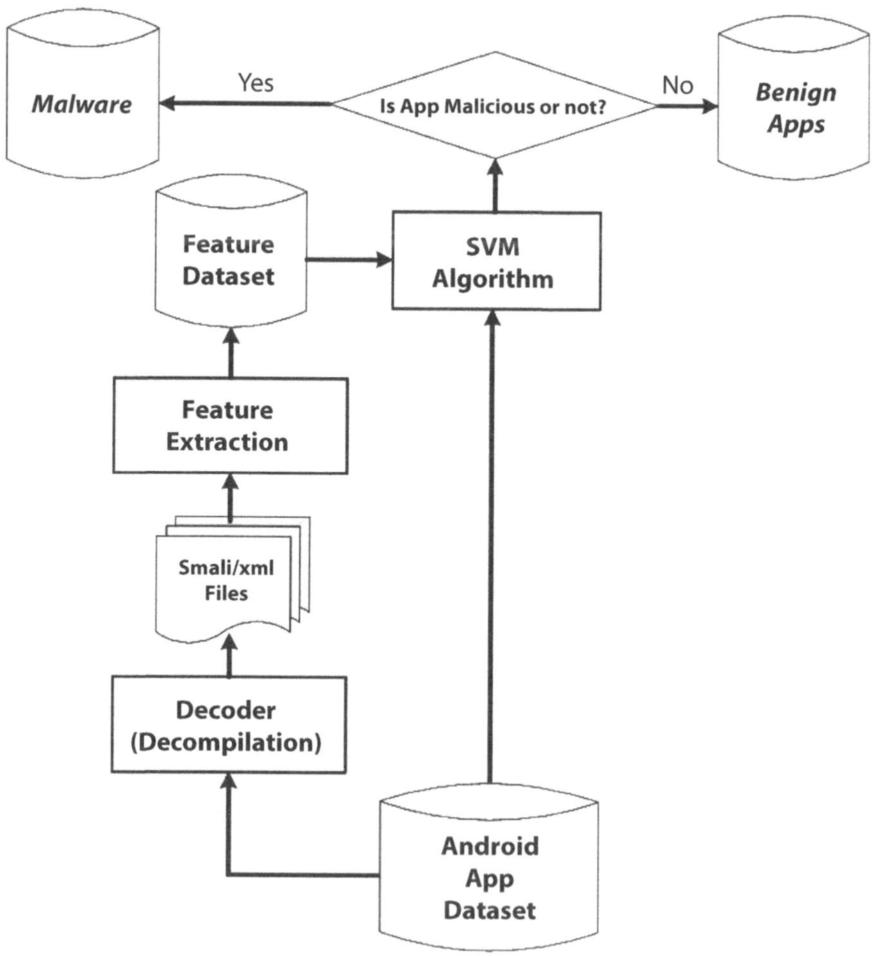

Figure 3.4 A simple example of Android behavior-based malware detection.

> solid strength in execution for out-of-test testing. The model-based semi-managed arrangement utilizes both marked and unlabeled information to assess the boundaries since unlabeled information typically conveys significant data on the model fitting methodology. In particular, we use blend displaying to accomplish the arrangement task. Blend demonstrating is a sort of ML strategy, which accepts that each segment of the blend addresses a given arrangement of perceptions in the whole assortment of information. Gaussian blend demonstrating is the most prevalently utilized applied and considered procedure [11].

Naive Bayes estimation was an administered learning computation, which relies upon the Bayes speculation and was used for dealing with gathering issues. Blameless Bayes classifier was one of the direct and best classification computations which helps in building the speedy ML models that can make quick assumptions. Random forest was a standard ML estimation that has a spot with the oversaw learning system. It might be used for both classification and regression issues in ML. It depends on acknowledging a method of integrating different classifiers to address a startling problem and to improve the projection of the model. As the name proposes, "random forest is a classifier that contains different decision trees on various subsets of the given dataset and takes the typical trend to work on the judicious exactness of that dataset." Naive Bayes is shown in Figure 3.5. Rather than relying upon one decision tree, the subjective forest anticipates each tree's assumption and relies upon the prevailing part votes of estimates, and it predicts the last yield.

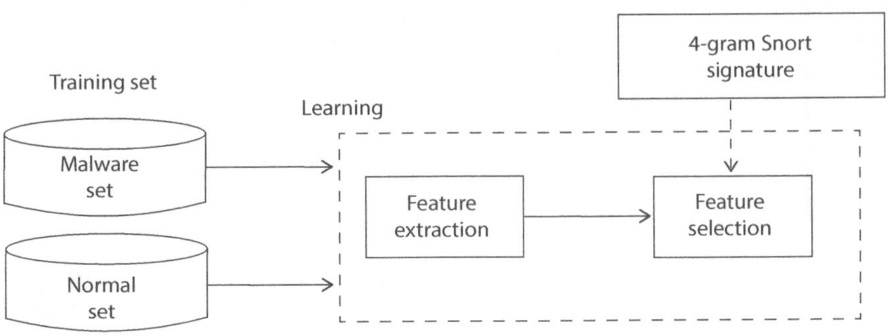

Figure 3.5 A simple example of naive Bayes.

- K-nearest neighbor was one of the most un-complex machine learning estimations subject to the supervised learning approach. K-NN computation acknowledges the resemblance between the original case/data and available cases and places the original case into the class that is, by and large, similar to the open orders. K-nearest neighbor is depicted in Figure 3.6.
- A genetic algorithm (GA) is an inquiry put together as enhancement procedure-based concerning the standards of genetics and natural selection. It is habitually used to tackle enhancement issues in research and in ML. The evaluation of the tests and starter results expected that this affirmation of thought is extraordinarily productive and fit for perceiving malware.
- In supervised learning, you instruct the machine using data that was well "stamped." Machine learning was an ML tactic, wherever you do not need to coordinate the models, overseeing learning licenses to assemble data or produce data yield from experience. Independent ML assists you in finding a wide scope of dark models in data. For pattern, you will be encouraged to choose the time taken and to reach back based on environmental conditions, times of days, and events—for example, the baby can perceive a variety of

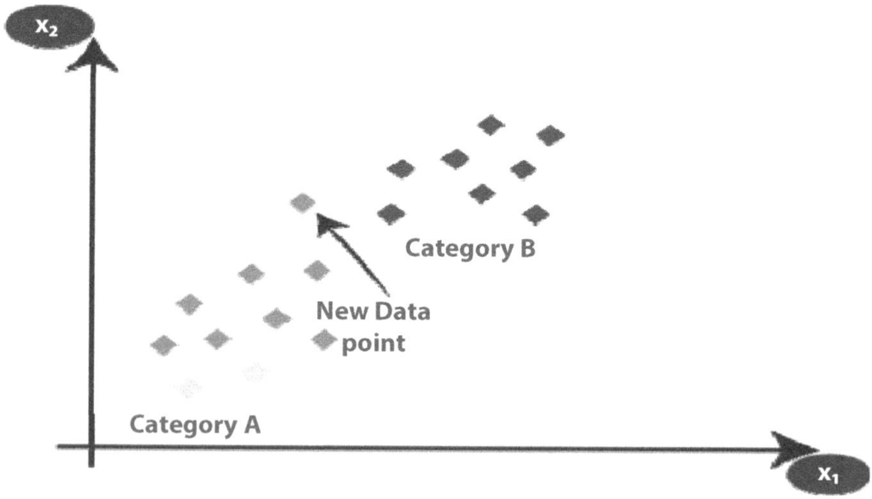

Figure 3.6 A simple example of K-nearest neighbor.

canines subject to long-ago coordinated learning. Backslide and classification are two types of coordinated AI methodology. Batching and association are two types of unsupervised learning. In a coordinated learning model, information and yield parts will be given, while in a performance learning model, only facts will be given [12].

3.4 Discussion

The extent of this paper is to introduce a malware identification approach utilizing ML. In this paper, we will focus on Android applications. Due to the unusual development of these malevolent programs, we need to utilize different robotized ways to deal with finding the contaminated files. There are various territories in this work that are as yet open for additional examination work. As indicated by the investigation and perception, we can see the capability of ML calculations over conventional techniques that are utilized against infection instruments for malware discovery. Furthermore, we have additionally talked about various ML calculations that can be of extraordinary assistance in recognizing malware. Likewise with the speedy turn of events and progression of the web, malware is a significant danger. Introducing just legitimate applications from the play store is significant.

3.5 Conclusion

The outcomes show that ML-based malware identification is compelling. We were likewise ready to depict the distinction between administered and solo ML as far as their identification exactness. Albeit in this work, practically all the exhibition measurements utilized were valuable in deciding the best ML calculation; some were more helpful than others. The preparation time, for example, was not pertinent as the vast majority of the models were worked in under 1 s. Another procedure is not to permit applications to be introduced over USB. It was significant not to follow any questionable connections and routinely checking the gadgets with hostile to infection programming can assist with facilitating a decrease in the danger of introducing malware on an Android gadget. More examination should be done to grow new strategies for identifying malware before it is even installed.

References

1. W. Han, J. Xue, Y. Wang, L. Huang, Z. Kong, MalDAE: Detected as well as explained malware based on correlation and fusion of static and dynamic characteristics, (2019) 208–233, http://dx.doi.org/10.1016/j.cose.2019.02. 007.
2. P. Burnap, R. French, F. Turner, K. Jones, Malware classified using machine activity data, and self-organising feature maps. *Comput. Secur.,* 73 (2017) 399–410, http://dx.doi.org/10.1016/j.cose.2017.11.016:
3. A. Damodaran, F.D. Troia, C.A. Visaggio, T.H. Austin, M. Stamp, A comparison between static, dynamic, and hybrid analysis for malware detection is done, *J. Comput. Virol. Hacking Tech.* (2017) 1–24, http://dx.doi.org/10.1007/s11416-015-0261-z.
4. E.M. Dovom, A. Azmoodeh, A. Dehghantanha, D.E. Newton, R.M. Parizi, H. Karimipour, Fuzzy pattern tree for edge malicious files detection and categorization in Iot, *J. Syst. Archit.,* 97 (March) (2019) 1–7, http://dx.doi.org/10.1016/j.sysarc. 2019.01.017.
5. M. Ficco, F. Palmieri, Leaf : cybersecurity training platform for realistic edge-iot scenarios, *J. Syst. Archit.,* 97(September2018)(2019)107–129, http://dx.doi.org/10.1016/j.sysarc.2019.04.004.
6. K. Khan, A. Mehmood, S. Khan, M.A. Khan, Z. Iqbal, W.K. Mashwani, A survey on intrusion detection and prevention in wireless ad - hoc networks, *J. Syst. Archit.,* (2019) 101701, http://dx.doi.org/10.1016/j.sysarc.2019.101701.
7. AV-TEST, Malware statistics and trends report, AV-TEST, 2020, https://www. avtest.org/en/statistics/malware.
8. A. Bushby, F. Cybersecurity, How deception can change cyber security defences, *Comput. Fraud Secur. Bull.* 2019 (1) (2019) 12–14, http://dx.doi.org/10.1016/ S1361-3723(19)30008-9.
9. E. Gandotra, D. Bansal, S. Sofat, Malware analysis as well as classification: A survey, *J. Inf. Secur.* 05 (02) (2014) 56–64, http://dx.doi.org/10.4236/jis.2014.52006, http://www.scirp.org/journal/PaperDownload.aspx?DOI=10.4236/jis.2014. 52006.
10. S.M. Muzammal, M.A. Shah, S.J. Zhang, H.J. Yang, security risks and authentication techniques for smart devices: *Int. J. Autom. Comput.* 13 (2016) 350–363, http://dx.doi.org/10.1007/s11633-016-1011-5.
11. Shetty, Shruthi H., *et al.* "Supervised Machine Learning: Algorithms and Applications." *Fundamentals and Methods of Machine and Deep Learning: Algorithms, Tools and Applications* (2022): 1-16.
12. Karkera, Thejas, and Chandra Singh. "Autonomous bot using machine learning and computer vision." *SN Comput. Sci.* 2 (2021): 1-9.

4

EEG Data Analysis for IQ Test Using Machine Learning Approaches: A Survey

Bhoomika Patel H. C.*, Ravikumar V. and Pavan Kumar S. P.

Department of Computer Science and Engineering, Vidyavardhaka College of Engineering, Mysore, India

Abstract

The thinking skill of humans were based on their problem-solving skill, basically a cognitive process which includes improvement and thinking. It is important to find a different number of intelligences in students and children according to their ages. Normal strategies for estimating IQ anyway are presented to biasness issues. An elective answer for order IQ levels through an Electroencephalogram (EEG)-based insight classifier model is proposed. Multiple tests are done to classify a ratio of IQ including multiple datasets using machine learning classification via neural network. Support Vector Machine (SVM), Artificial Neural Networks (ANN), and various division calculation and classifiers are utilized to distinguish the degrees of knowledge of people groups of different ages utilizing distinctive tests. In this survey, we investigate various approaches utilized for IQ test as a study. From this, we would be able to understand the various methodologies used for intelligence level estimation and furthermore used to distinguish the downsides of existing framework.

***Keywords*:** Electroencephalogram (EEG), intellectual quotient (IQ), machine learning, nervous system, alpha band, beta band, support vector machine (SVM), artificial neural networks (ANN)

**Corresponding author*: bhoomijanu9@gmail.com

Chandra Singh, Rathishchandra R. Gatti, K.V.S.S.S.S. Sairam, Manjunatha Badiger, Naveen Kumar S. and Varun Saxena (eds.) Modeling and Optimization of Signals Using Machine Learning Techniques, (55–74) © 2024 Scrivener Publishing LLC

Introduction

Intelligence quotient (IQ) levels of peoples are found in all ages, and it is a problem-solving cognitive process. It refers to mental age (MA) expressed as a ratio of chronological age (CA) multiplied by 100. For IQ to remain stable, MA must increase with CA over time. This is true until around 18 years, when intellectual abilities are usually fully developed. The IQ level is based on a total score out of a maximum score in testing memory or designed to focus on human intelligence. From prior measurements, the score of IQ is found on isolating a psychological age, and it is obtained by using human smart test. IQ scores are generally utilized in instruction or in organizations to check for scholarly inability. By considering expertise, people depend on their critical thinking and normal psychological interactions. There are numerous different sorts of IQ test which join diverse subjects; a few subjects are visual and brain testing.

The IQ test should be possible with an electroencephalogram (EEG). An EEG can compute an electrical interaction of the brain. Moreover, signals are put down utilizing the traditional scalp or MATLAB. The fNIRS give guidance of the direction of the mind and specific places that have been initiated. The process of extracting IQ results happens by the electrical pursuit of the brain from the scalp. The sign highlights that Delta has a recurrence of 3 Hz or below, which fundamentally centered around an immense rest of grown-ups. The Theta of recurrence is at 3.5 to 7.5 Hz, which focused on the rest of the youngsters. The Alpha sign has a recurrence somewhere in the range of 7.5 and 10 Hz, which is mindful over the head on two sides. It shows when the eyes are shut and rest them, and it disappears when the eyes are opened and one begins thinking and figuring. What is more, the Beta sign has a recurrence of 14 Hz. It is noteworthy that middle when human open their eyes and begin offering work to mind. To describe a particular assortment being development, behavior and cognition where chiefly based on resting state functional MRI (fMRI). EEG considers that the analyzed youngsters with learning incapacities have more noteworthy theta and less alpha (and beta) waves according to their ages while they are in normal stage. These discoveries are frequently viewed as proof for the alleged evolutional deferral theories. This information has been utilized to manage a particular neurotherapy treatment that helps to improve the learning inabilities.

Electroencephalography gives an unobtrusive strategy which acts on a different state engaged with knowledge and learning. EEG is touchy to an element of mental handling and also, accordingly, the impacts of data

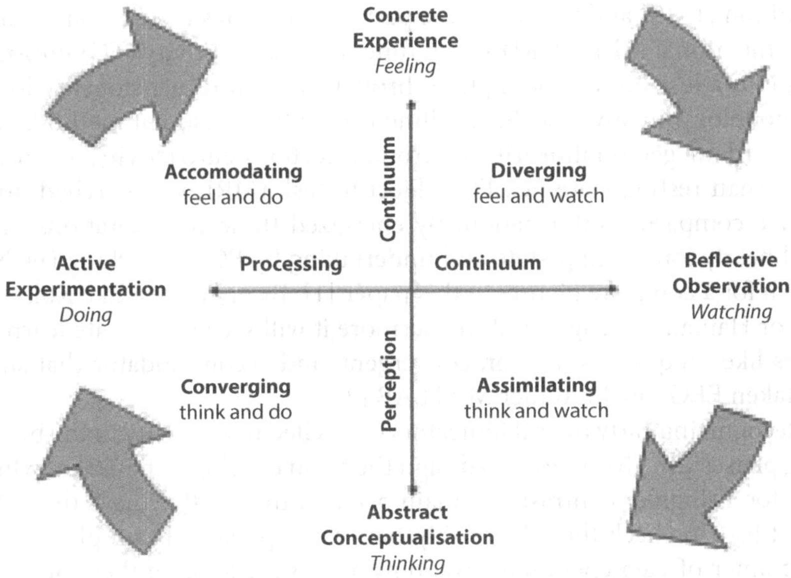

Figure 4.1 Learning style classifications.

encoding on cerebrum frameworks. In an upcoming EEG innovation permit, brain movement will be written down during regular conduct, increasing the chance of constant admittance to the cerebrum rhythms supporting data encoding. Where different datasets taken from various papers in marvel of interest may yield high instructive information compared to other individual datasets, in light of the fact that the numerous datasets can give corresponding data of the noticed wonder from a few unique perspectives. In such connection, conspire is very delicate to the scaling of the information. The learning style classifications are depicted in Figure 4.1.

While in checking a result of IQ the execution of machine learning or deep learning classification to analyze a solitary individual experiencing scholarly individual IQ issue. The EEG catches the data of the individual exposed to different activities of the mind while performing, and all the information are inspected by a machine which previously learned about the EEQ grouping.

4.1 Related Work

The numerous proposals for detecting an IQ level of students using different techniques are A unique individual's character like behavior and cognition [1–17] and to explain the past examinations focused on resting-state functional MRI (fMRI)-based functional connectivity (FC) information into a less-dimensional place through straight dimensionality lower methodology. In any case, linear dimensionality assessment methods that disregard to get nonlinearity of frontal cortex neuroactivity. Likewise, other than resting-state FC, FC subject to task fMRI will be relied upon to give comparing information. By energized these considerations, they nonlinearly join resting-state and undertaking by FC networking (FCN)a search for a complete picture in this paper [1]. By giving test like Kolb and Vak or Hamann strength and furthermore it will see an alternate learning styles like unique, assimilator, convergent, and accommodator that signs are taken EEG tool kit under MATLAB [2].

Recognizing early neural indicators of intellectual capacity in the beginning phases will give an idea to design the brain developmental trace which help for a singular contrast as a main priority. In this, they have three different lists to check the relationship between a putative brain physiological pointer of data encoding (swapping a forward-looking theta during a video) at the start and then after the mental result [3].

The brain physiological underpinnings of learning disease stay undisclosed. By these neural assessments, the written-down EEG to a gigantic illustration of students with learning disabilities (LD) and healthy control kids during rest state in which the eyes may be open or closed. They decided to take six repeat gatherings (delta, theta, lower and upper alpha, and lower and upper beta) to rethink whether kids with LD show a theta power growing high and alpha band reduces dependent on their instances of electroencephalogram influencing [4].

The EEG signs of youngsters were recorded during a mental pile of various sorts as per the proposed plan of analysis [5]. Intelligence level is used to check the composite mental constituents of mental execution resting; EEG is recorded from the left prefrontal cortex [6]. Intellectual developmental disorder issue (IDD) is described by the imperfect mental capacities of individuals likewise, such as intelligence level, physical appearance, behavior, and eager development. It basically occurs in the posterity of non-modern countries where prosperity and feeding necessities are hazardous [7].

Neural processing functioning of the human frontal cortex change as demonstrated by the trouble level of the issue and the understanding level of people. So, they used a two-structure containing functional near-infrared spectroscopy (FNIRS) and electroencephalogram (EEG) for looking at these particular differentiations. A reliable-wave 32-channel fNIRS from OxyMon fNIRS device and a 19-channel EEG from g.tec's association were used to think about the oxygenation strategy similarly as the electrical activity of cerebrum while these issues of Raven's Progressive Matrix (RPM) knowledge assessments. By utilizing these data to gauge the intelligence quotient (IQ) of the person not playing out a total intelligent numerical knowledge test in a long-term period and analyzing an appropriate response of individuals, the corresponding question [8].

The paper targets concocting a novel non-invasive technique for estimating human knowledge. A recently concocted scoring plan is utilized to eventually create a mark for the children. Wavelet parcel change approach to contain extraction is applied to a five-channel EEG information. This methodology utilizes db-8 as the mother wavelet [9].

Cerebral paralysis (CP) is quite possibly the most debilitating engine sorts of CP and has been traditionally connected with injury to the basal ganglia and thalamus. However, mental brokenness is fundamental in CP; there is an absence of appropriated quantitative assessments exploring the association between white matter (WM) microstructure and insight in these CP type [10].

(a) largest front facing theta change inside a video was identified with the highest nonverbal marks of a year, yet not verbal marks. (b) Expulsion of the highest limit front facing theta change marks delivered the relationship with nonverbal abilities minor [r (33) = 0.26, p = 0.1]. (c) Theta power changed by situation, area, and one of them has a half-video.

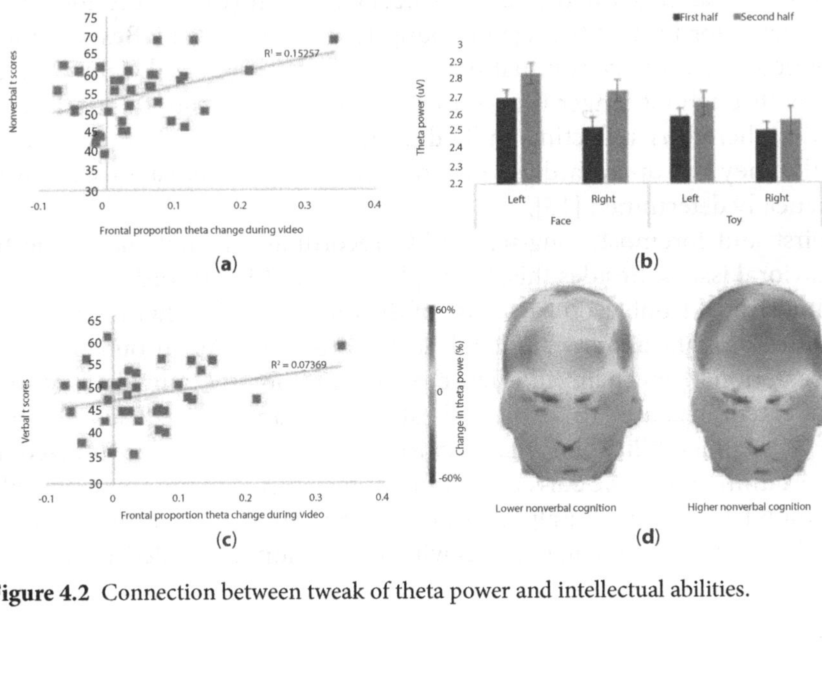

Figure 4.2 Connection between tweak of theta power and intellectual abilities.

The mistake rates are ±1 SE. (d) Outlining a geology of progress in theta control throughout a recording in youngsters with low and high intellectual capacity [3]. A connection between tweak of theta power and intellectual abilities is depicted in Figure 4.2.

The current examination researched the neuroscience work of regular day-to-day existence leader work (EF) shortfalls in grown-ups with ADHD with high IQ [11]. It has been set up that the prefrontal and back parietal cerebrum flaps, which are for the most part identified with insight, have numerous associations with the cerebellum. In any case, there is a restricted examination exploring cerebellum's connection with psychological cycles. In this examination, the organization of cerebellum was dissected altogether to explore its general association in people with lowest and highest liquid memory remainder (IQ). Functional attractive reverberation imaging (fMRI) information and were chosen from 136 people in resting state from the Human Connectome Project (HCP) information base [12]. In perceptual decision-making has given that two different brain segments separate test order, just one followed a proof managing the decision process. Then, the intellectual writing, they are a differentiation between a front-focal evoked potential estimated with EEG starting at 350 ms that appears to show commonality and a late parietal evoked likely which top at 600ms that tracks recollection [13]. The point of this examination was to research the connection between singular electroencephalogram (EEG) attributes in the resting state and the degree of nonverbal knowledge [14].

They researched whether electroencephalogram (EEG) accounts were reasonable for predict the age of youngsters and teenagers. Besides, they explored whether overestimation or underestimation old enough was steady throughout longer time-frames, as steady expectation mistake can be deciphered as reflecting individual maturational level of cerebrum. Lastly, they set up regardless of whether the age-forecast mistake was genetically determined [15].

First and foremost, long-term EEG recordings are not many due to behavioral issues. Besides this, the explanation of EEG recording has been demonstrated troublesome because of the unpredictable EEG signal irregularities brought about by mental health problems. As an outcome, the presentation of mechanized seizure recognition for ID individuals is to a great extent obscure [16]. A brain beat utilizing a neurofeedback training (NFT) is vulnerability on the subject that different test plans are utilized in past examinations. The current examination expected to improve a flexible distant NFT structure for alpha disposition and then to favor the impact of the NFT system on intelligent level with a deception-controlled assembling [17].

Support vector machine (SVM) grouping is completed dependent on the low-dimensional embeddings [1, 16]. For EEG recording, we used electroencephalograph "brain scalp" by Brain Products [4, 5]. A conventional method is used to find the IQ level of students [6, 12]. Machine learning classification and deep learning machine classification [7, 9]. Neuropsychological method and neurofeedback training method [10, 17]. Regression model and linear regression methods [8, 14].

4.1.1 Signal Pre-Processing, Filtering, and Feature Extraction

The signals are first pre-handled utilizing MATLAB 2015a. EEG epochs increasing ±100 μV are viewed as electrooculogram commotions and, in this manner, naturally dismissed. At further examination, the sign is restricted to a span of 2 min and 30 s. The pre-prepared signal is then isolated into theta, alpha, and beta signals via band-pass channels. Power spectral density (PSD) for every recurrence bandies then procured by means of Welch's strategy.

The system of examination techniques associated with this investigation is appeared. The work involves test determination and EEG securing, signal pre-preparing and separating, power proportion extraction and IQ level forecast utilizing EEG-based knowledge characterization model is depicted in Figures 4.3(a, b).

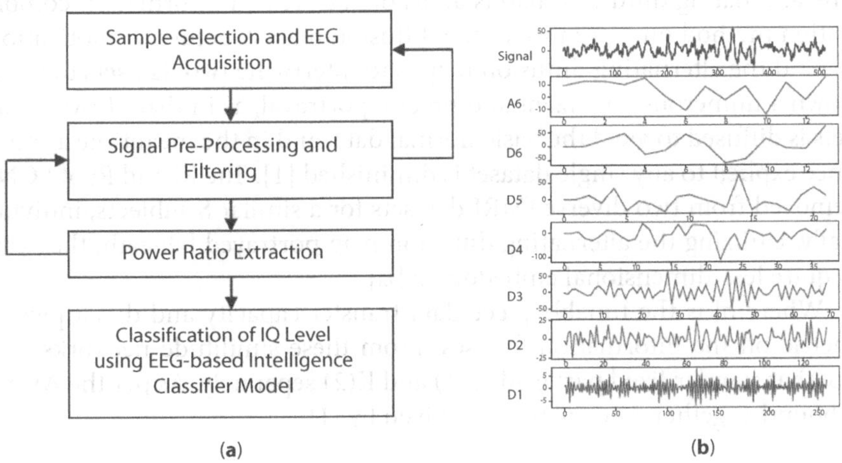

Figure 4.3 (a) Flowchart of IQ level order utilizing the EEG-based insight classifier model [6]. (b) Test EEG signal deterioration utilizing DWT [7].

From the cleanest dataset, we eliminated wavelet coefficients using a discrete wavelet technique (DWT) for each part. DWT will get little nuances, rapid changes, complex moreover, likenesses in EEG signal likenesses in EEG signal [7].

4.2 Equations

Energy spectral density (ESD) is figured as energy conveyance under the power spectral density [PSD] bend. The data between the EEG groups are then standardized utilizing power proportion connections communicated by Equation [6].

$$\text{Theta Ratio} = \frac{\theta}{\theta + \alpha}$$

$$\text{Alpha Ratio} = \frac{\alpha}{\alpha + \beta}$$

$$\text{Beta Ratio} = \frac{\beta}{\alpha + \beta}$$

4.2.1 Alternating a Diffusion Map-Based Combination of Two FCN Datasets

The alternating diffusion map is an as of late created information combination method on a premise of the diffusion map structure. A motivation behind the alternating diffusion map is to intertwine two datasets to track down a numerous cognizant and precise portrayal, as in that of two datasets is diffused to yield the basic normal data, and in the meantime annoyance explicit to any single dataset is diminished [1]. The Ri and Rj of FCNs removed from two diverse fMRI datasets for a similar S subjects, individually. Utilizing the alternating diffusion map portrayed beneath, they can acquire less-dimensional embeddings {zi}.

Where σl is the tunable piece data transfer capacity and d is a picked metric on the information focuses. From these similitude networks, thy got that standardized part grids k(1) and K(2) separately. As per the ADM, a bound-together piece network is given by [1]

$$\widetilde{K} = K^{(1)}\left(K^{(2)}\right)^{T} + K^{(2)}\left(K^{(1)}\right)^{T} ..$$

4.2.2 Information Examination

They examined time-recurrence portrayals of EEG signals acquired by means of a ceaseless wavelet technique (CWT), which has lately become an extremely well-known procedure to examine elements of brain mind movement. CWT is a convolution of EEG signal with fundamental work [5].

$$W_n(f,t) = \sqrt{f} \int_{-\infty}^{+\infty} x_n(t) \psi^*(f,t) dt,$$

where p = 1, 2 … P is the quantity of EEG channel (p = 31); what is more, "*" represents complex formation. As the mother wavelet of CWT then utilized the difficult Morlet wavelet [5]

$$\psi(\eta) = \frac{1}{\sqrt[4]{\pi}} e^{j\omega_0 \eta} e^{-\eta^2/2}$$

The key is to become familiar with the vector of loads that segregates the two kinds of upgrades.

This technique accepts the information dispersed as calculated capacity:

$$p(c = old \mid x) = \frac{1}{1 + e^{-w^T x}} = f(y)$$

The degree of knowledge is decidedly influenced by the EEG abundancy in the privilege lead, decrease of the alpha-beta list slope from the correct fundamental to the fleeting lead, and the abundancy of the EEG auto brain signals in the left focal lead. The declaration of theta beat in the right front facing lead and the force of the great recurrence segment of beta mood in the correct back worldly lead behind the scenes EEG of the subject are contrarily connected with the degree of insight. We checked how well the model can partition the subjects into two gatherings: subjects with normal knowledge or more normal insight [14].

$$IQ = IQ = 100.19\text{-}0.18\ p\text{-}(0.36q$$

$$\text{-}0.25c)\text{-}0.3r + 0.18s + 13.43t,$$

where p is the power of theta rhythm,
q is the index of alpha rhythm,
r is the index of alpha rhythm,
s is the power of high-frequency beta,
t is the maximum amplitude of the EEG signal,
and f is the maximum range of the autocorrelation function in lead C3. The level of significance of the equation is $p < 0.00002$.

4.2.3 Gaussian Kernel Function

In AI, kernal function are regularly used to characterize similar calculate to learn the relationship among peoples through a kernal trick, and in specific the Gaussian pieces are generally utilized. Their figure a comparability framework by utilizing the Gaussian piece work with an distance of SPD frameworks, there are, where $\sigma > 0$ is the data transfer capacity of the Gaussian bit work. This development characterizes a weighted chart, where the hubs compare to the k subjects {Ri} is the weight grid of the chart.

$$W_{i,j} = \exp\left(-\frac{\mathrm{d}^2(R_i, R_j)}{\sigma}\right),$$

i.e., the log-Euclidean distance, that thinks about a particular calculation of complex. The LEU among R_i and R_j is shown by

$$\mathrm{d}_{\mathrm{leu}}(R_i, R_j) = ||\log(R_i) - \log(R_j)||_F,$$
$$\mathrm{d}_{\mathrm{ck}}(R_i, R_j) = ||(R_i)_{low} - (R_j)_{low}||_F,$$
$$\mathrm{d}_{\mathrm{eu}}(R_i, R_j) = ||R_i - R_j||_F.$$

4.3 Classification

Is a strategy wherein set of free factors called highlights are taken care of to a calculation alongside the relating class of those highlights? The calculation learns those highlights and change its boundaries so that it can observe classes of highlights which are obscure. For order reason, machine learning classification recorded beneath are utilized.

For classification purpose, many different methods are proposed and listed below:

- K-nearest neighbor (KNN) [11]
- Linear regression (LR) [8, 14]
- Linear dimension [1]
- RBF nu-support vector machine (RBF nu-SVM) [8, 16]
- Decision tree classifier (DT classifier) [7]
- Time frequency analysis and statistics [5]
- Conventional method [6, 12]
- Hierarchical extreme learning machine classification [9]
- Perceptual decision method [13]
- neuropsychological method [10]
- Neurofeedback training (NET) [17]

4.4 Data Set

In [1], multiple dataset has been taken to check the IQ level of students, incorporating different datasets collected by various sensors for a marvel of interest that increased more valuable data than any individual dataset does, in light of the fact that the diverse datasets can give proportional information. They attempted 20 understudies from their underlying significant length of planning to be requested in three systems of learning designs. An understudy was then presented to a preliminary of knowledge with growing degrees of difficulty. The time length to address the test taken for understudies as of late assembled in four chief designs submitted by the Kolb test are explored [2].

Offspring of ages 5, 7, 12, 14, 16, and 18 [15]. The answer time was assessed and then repeat of answer time for individual question was resolved. They have seen similar instances of theta EEG change at a year and tracked down a judicious association with verbal and nonverbal scholarly capacities assessed at 2, 3, and 7 years of age [3]. A total of 33 individuals with 15 females were age- and sex-facilitated with 33 controls [10]. The subjects (human) are picked so much that they have no arrangement of encounters of frontal cortex wounds or any psyche issues [9]. Kids with learning inadequacies and sound control kids during the rest of the stages in which the eyes maybe opened or closed [4].

Healthy children of 7–10 years, right-handed, with normal or helped to-conventional visual sharpness took interest for observation. All of them were drawn nearer to keep a healthy life framework with an 8-h night rest during 48 h prior to the investigation [5]. An aggregate of 50 solid understudies 18 to 19 years. Tests involved 25 male and 25 female individuals [6].

All subjects were correctly given, had common or re-examined vision, and had no arrangement of encounters of brain illness [8]. Acquired from 492 strong subjects still with eyes open with relaxed up fixation on an all-inclusive cross-hair on a dull background [12].

Data collected from 14 subjects is used in this study [7]. The data set contained 77 male and female individuals aged from 19 to 26 years [14]. The chosen members with an ID who appeared in any event one EEG seizure in nonstop 24-h EEG write down records with excellent signal processing [16]. A total of 40 grown-ups were determined to have ADHD and age- and sex-coordinated with 40 healthy people [11].

Techniques: A total of 30 good health members were dazed and chosen rapidly and appointed into two gatherings. All members were right-handed, with no experience, and were given to take NFT before [17]. Examination 1 contained 25 right-handed subjects (14 female) somewhere in the range of 18 and 38 years old. Investigation 2 involved 14 right-handed subjects (three female) somewhere in the range of 18 and 24 years old (M = 20.6) [13].

4.4.1 Pre-Preparing

EEG was filtered somewhere in the range of 0.5 and 45 Hz utilizing a band-pass FIR channel. Later taking away that standard dependent on the mean of every information channel, basic ancient artifacts were removed out with the assistance of EEGLAB, ADJUST through independent component

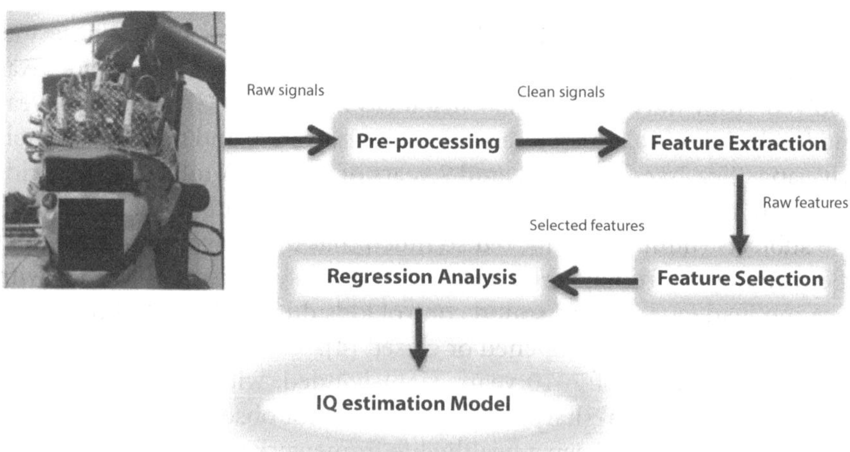

(a)

Figure 4.4(a) The entire producer for IQ estimation [8].

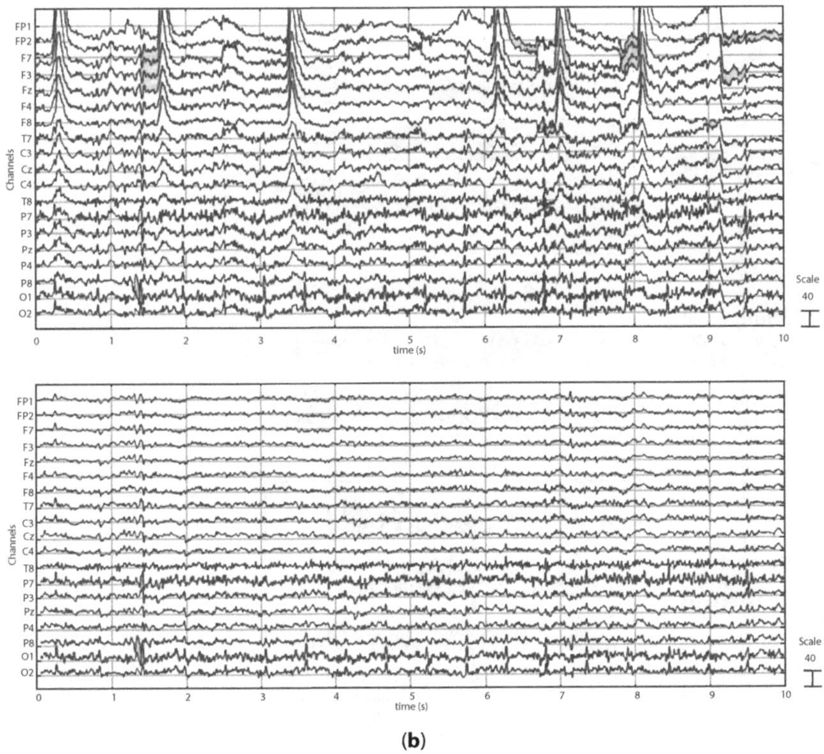

Figure 4.4(b) One example to show before and after preparing [8].

analysis (ICA) in view of runica algorithm under MATLAB is depicted in Figures 4.4(a, b).

4.4.2 EEG Data Producer

Members were situated easily in a dark room with a low degree of natural temperature. EEG information recording was persistently done during a 6-min resting eyes-shut condition with a 128-channel framework. Inside Net Station, EEG information was gathered by utilizing a 1,000-Hz testing rate and carefully separated. All terminal impedances were held under 50 kω. Information was referred to terminal Cz. The progressing EEG recording was constantly checked to guarantee information quality and to avoid watchfulness in all members.

In MATLAB, EEG signals were in this way down-sampled to 250 Hz and changed to the normal reference. An individual component's examination decayed the deliberate EEG signals toward individual segments and those

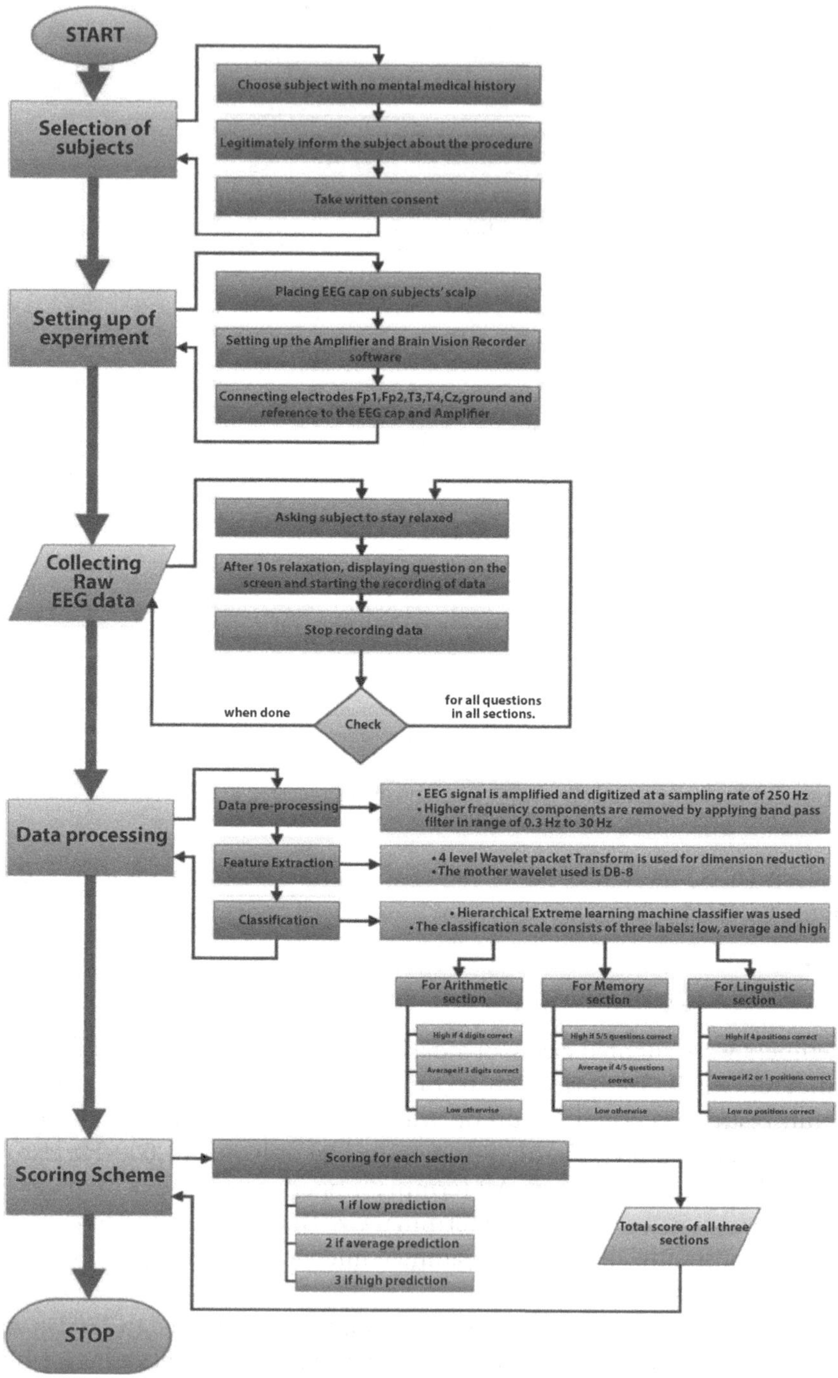

Figure 4.5 Flowchart of EEG test [9].

parts define eye expression. All EEG information were physically assessed, and channels consisting of muscle or some other neuroscience antiquity are disposed. A wavelet change was utilized to acquire time-recurrence area computed. The EEG was done in the four groups, and the overall force of the four groups was determined in measurable investigation as these lists were dependable to distinguish EEG attributes in members with ADHD.

The cognitive test comprises two stages: the retention stage and the memory phase. Then, the subject is permitted to unwind. After relaxation time, in memory stage, other numbers are shown successively to the subject and those people are approached to react yes or no in light of whether individual numbers were available in the recently shown set of numbers. The EEG information is gathered during both the retention stage and also in the memory stage as shown in Figure 4.5.

4.5 Information Obtained by EEG Signals

EEG information is gathered using the EEG Easy cap which comprises cathodes and terminal holders for association. They place of cathode holders on the cap is as per the 5/10 global framework. Students were chosen, and they do not have a set of experiences of brain operations or other mind issues. Each student is approached to tell composed assent prior to stepping through the exam. Subjects are told truly. The chosen subjects are of different years of age. The two segments of the information obtaining measure are as per the following [8]:

1) Structure: This part clarifies the primer undertaking needed to associate the equipment components and run the product while.
2) Explore: The test segment clarifies the genuine interaction recording of information.

4.5.1 System Structure

The subject is situated easily in a quiet room without outside commotion or aggravation. The subject is approached to remain quiet with negligible development during the examination to forestall pointless adjustments in the wave designs because of actual developments. An EEG cap was put on the testing person's scalp. The cathodes were put by the 5–10 worldwide framework. Each of these terminals is then joined to the EEG Easy cap. High-chloride gel is utilized prior to associating the terminals on the scalp to escalate the signs. The terminals are additionally associated with the intensifier which is thus associated with the PC. Cerebrum Vision recorder and Brain Vision analyzer were the two programming bundles utilized. Waves were written down by utilizing the recorder though the pre-preparing of signs is finished utilizing an analyser. Choosing a cathode is a significant undertaking engaged with EEG information assortment. It is fitting to choose just the pertinent anodes for the wanted application. Wrong selection of terminals may result in undesirable results in the EEG signals [17].

4.5.2 Numerical Examination

The explained statistics of neuropsychological execution are determined. A intellectual performance of CP coordinated with controls were looked at utilizing Mann–Whitney *U*-tests, contingent upon a scattered of all information. Numerical investigations of a brain psychological information were performed utilizing SPSS version, PSS statistical, and so on [10].

4.5.3 EEG Circumference

The captured information decreases with the expanding irregularity level of EEG foundation. Furthermore, then, they looked at two gatherings of the subjects with ordinary EEG foundations and the subjects with unusual EEG foundations. The recognition execution in subjects with an ordinary foundation are fundamentally better than those with an unusual one. The chi-square test shows that the patients in the two gatherings were homogeneous concerning the variables EEG release design, IQ level, and seizure understandable [18, 19].

4.6 Discussion

In this examination, all of these papers are disclosed about how to test a scholarly degree of people groups as per their ages. In some investigations, they have clarified that the intelligence level can be determined by doing some tests that were given to subjects and people who have ordinary brain system with no problem. The objective of all these is to discover the degrees of intelligence of individuals by checking their IQ level and by giving some unique tests to each individual. Furthermore, they will take perusing of the level of intelligence by giving some activities like social, sports, test, etc.

4.6.1 Comparison Between IQ Levels With Different Methods

These papers show a comparative analysis of IQ level using different methods of classification techniques in various patients. In graph 1, the accuracy of different methods can be improved from 30.93% to 100% by the combination of different classifications. The support vector machine classification (SVM) method has the highest 100% accuracy compared to others as depicted in Table 4.1.

Various techniques such as K-nearest neighbor (KNN), linear regression (LR), linear dimension [1], BF nu-support vector machine (BF nu-SVM), decision tree classifier (DT classifier), time frequency analysis and statistics, etc., are used to find the intellectual level of patients and each level as their own test process. The application of each technique was also described. This paper evaluates the techniques and methods in testing

Table 4.1 Comparison between IQ levels with different methods.

Technique	Key reference	Merits	Demerits	Application
Linear dimension	1	First paper which successfully demonstrated both resting state and test-based fMRI	Accuracy rate is not mentioned.	We nonlinearly merge rest-state and task-based FC associations to search for an unrivalled depiction.
Machine learning classifier	7	Analyze the patient experiencing scholarly formative problem utilizing many AI characterization	Complicated due to the use of more classifications	Intellectual developmental disorder problem is solved.
Linear dimension	8	Very useful in cognitive testing; does not need conscious learning	Accuracy is low.	It solves many cognitive problem processes.
Hierarchical extreme learning machine classification	9	It uses direct method to classify EEG signals to find the intelligence level.	In old IQ test find intelligence indirectly	Intelligence level is detected.
Support vector machine (SVM)	16, 8	High accuracy	All data are recorded at home without video.	Testing of intelligence level is completed.

intellectual levels by using different processes like tests, games, etc. By considering the table above, we can say that straight measurement classifier is the most utilized technique to discover the intelligence level or degrees of people groups, yet it contrasted with the exactness of SVM. It is low now and then, other than strategy, the cycle likewise assumes a significant position in figuring the intelligence level.

4.7 Conclusion

This chapter shows a different comparable study method to check the intellectual level of patients using different methods of classification. The IQ level consists of different cognitive, behavior, development, memory power, etc. Based on these, different classification methods are used to find the intellectual level. By analyzing this studies may consider that picking in excess of two arrangement dependent on AI or deep learning give more exactness contrasted with other process. Ongoing papers give data on how to realize the level of intelligence levels of understudies depending on their age, yet, in addition, we can get the level of intelligence of people groups whose age is over 60 and still dynamic and they were not influenced by any psychological problem. From these studies, we understood the different methodologies used to measure the intelligence level test dependent on this. We are going to introduce a novel way to deal with recognizing the intelligence level depending on age in a further research.

References

1. Li Xiao, Julia M. Stephen, Tony W. Wilson, Vince D. Calhoun, and Yu-Ping Wang.," Alternating Diffusion Map Based Fusion of Multimodal Brain Connectivity Networks for IQ Prediction," *IEEE,* PP. 0018-9294 (c) 2018.
2. Felisa M. Cordova, *, Hernan Díaz M.a , Fernando Cifuentes a , Lucio Cañeteb, Ferdi Palominos.," Identifying problem solving strategies for learning styles in engineering students subjected to intelligence test and EEG monitoring," ELSEVIER, PP.CC BY-NC-ND license 2015. https://creativecommons.org/licenses/by-nc-nd/4.0/
3. E. J. H. Jones 1*, A. Goodwin2, E. Orekhova3,8, T. Charman 2, G. Dawson 4, S. J. Webb5,6,7 & M. H. Johnson1.," Infant EEG theta modulation predicts childhood intelligence," *Sci. Rep.*, PP. 10:11232,2020.
4. Lutz Jänckea, Mohamad Yassin Sakab, Omer Badawoodc and Nsreen Alhamadid.," Resting-state electroencephalogram in learning-disabled

children: power and connectivity analyses," *Clin. Neurosci.*, received 16 October 2018 accepted 24 October 2018.

5. Vadim Grubov, Vladimir Maksimenko, Vladimir Makarov.," Features of brain activity in children during cognitive tasks of different types," *IEEE*, PP. 2019.
6. M. M. Anoor, K.H . K. Anwar, M. S. A. M. Ali2 and A. H. Jahidin., "Classification OF STUDENTS' IQ LEVEL USING EEG INTELLIGENCE CLASSIFIER-based MODEL," Research Article, PP.17th October 2017. https://www.ajol.info/index.php/jfas/article/view/165392
7. Talha Anwar.," A Machine Learning approach for Recognizing Intellectual Development Disorder using EEG," *IEEE*, PP. 2020.
8. Shabnam Firooz, Seyed Kamaledin Setarehdan.," IQ estimation by means of EEG-fNIRS recordings during a logical mathematical intelligence test Shabnam Firooz, Seyed Kamaledin Setare," ELSEVIER, Received 10 January 2019; Received in revised form 20 May 2019; https://www.sciencedirect.com/science/article/abs/pii/S0010482519301738?via%3Dihub
9. Shashikant T. Kadam, Vineet M.N. Dhaimodker, Milind M. Patil, Damodar Reddy Edla*, Venkatanareshbabu kuppili.,"EIQ: EEG based IQ test using wavelet packet transform and hierarchical extreme learning machine" ELSEVIER, Received 16 January 2019; Received in revised form 15 April 2019; Accepted 16 April 2019, https://www.sciencedirect.com/science/article/abs/pii/S0165027019301190?via%3Dihub
10. Olga Laporta-Hoyos, Kerstin Pannek, Júlia Ballester-Plané, Lee B. Reid, Élida Vázquez, Ignacio Delgado, Leire Zubiaurre-Elorza, Alfons Macaya, Pilar Póo, Mar Meléndez-Plumed, Carme Junqué, Roslyn Boyd, Roser Pueyo ., "White matter integrity in dyskinetic cerebral palsy: Relationship with intelligence quotient and executive function," *NeuroImage: Clin.*, Received date: 13 January 2017 Revised date: 5 May 2017 Accepted date: 8 May 2017. https://www.sciencedirect.com/science/article/pii/S2213158217301109?via%3Dihub
11. Hui Li1,2 , Qihua Zhao1,2 , Fang Huang1,2 , Qingjiu Cao1,2 , Qiujin Qian1,2, Stuart J. Johnstone3 , Yufeng Wang1,2 , Changming Wang4 , and Li Sun1,2.," Increased Beta Activity Links to Impaired Emotional Control in ADHD Adults With High IQ," Article,2017 https://journals.sagepub.com/home/jad
12. asileios C. Pezoulas 1 *, Michalis Zervakis 1 , Sifis Michelogiannis 2 and Manousos A. Klados 3.," Resting-State Functional Connectivity and Network Analysis of Cerebellum with Respect to IQ and Gender," *research*, 26 April 2017. https://www.frontiersin.org/articles/10.3389/fnhum.2017.00189/full
13. Roger Ratcliff, Per Sederberg, Troy Smith, Russ Childers.," A Single Trial Analysis of EEG in Recognition Memory: Tracking the Neural Correlates of Memory Strength," *Sci. Direct*, Accepted date: 29 September 2016. https://www.sciencedirect.com/science/article/abs/pii/S0028393216303633?via%3Dihub
14. E. P. Stankova* and I. Y. Myshkin.," Association between Individual EEG Characteristics and the Level of Intelligence," *Cortex*, Received July 25, 2016; in final form, September 5, 2016.

15. Marjolein M. L. J. Z. Vandenbosch1 | Dennis van 't Ent1 | Dorret I. Boomsma1 | Andrey P. Anokhin2 | Dirk J. A. Smit.," EEG-based age-prediction models as stable and heritable indicators of brain maturational level in children and adolescent, "*Research*, PP. 9 December 2018. https://onlinelibrary.wiley.com/journal/10970193
16. Shetty, Shruthi H., *et al.* "Supervised Machine Learning: Algorithms and Applications." *Fundamentals and Methods of Machine and Deep Learning: Algorithms, Tools and Applications* (2022): 1-16.
17. Karkera, Thejas, and Chandra Singh. "Autonomous bot using machine learning and computer vision." *SN Comput. Sci.*, 2 (2021): 1-9.
18. Lei Wanga,∗, Xi Longa,b, Ronald M. Aarts a,b, Johannes P. van Dijkc,d, Johan B.A.M. Arends a,c.,"EEG-based seizure detection in patients with intellectual disability: Which EEG and clinical factors are important?," ELSEVIER, PP. 6 December 2018, https://www.journals.elsevier.com/biomedical-signal-processing-and-control
19. TingYing Wei1, DaWei Chang1, YouDe Liu1, ChenWei Liu1, ChungPing Young1, ShengFu Liang1,2 and FuZen Shaw3.," Portable wireless neurofeedback system of EEG alpha rhythm enhances memory," *Research*, PP,2017, https://creativecommons.org/licenses/by/4.0/

5

Machine Learning Methods in Radio Frequency and Microwave Domain

Shanthi P.[1] and Adish K.[2*]

[1]*Department of Electronics and Telecommunication Engineering, Radio Frequency and Microwave Engineering, RV College of Engineering, Bengaluru, Karnataka, India*

[2]*Department of Electronics and Telecommunication Engineering, RV College of Engineering, Bengaluru, Karnataka, India*

Abstract

Computing power has greatly increased in the previous few years due to rapid advancements in semiconductor technology. Machine learning (ML) methods have attracted a slew of new applications because of this significant boost to computing. Many researchers working on the design and optimization of electronic circuits are now shifting toward the ML-based approach to synthesize the circuits. These ML-based approaches have gained significant importance because of the aid they provide as they can be deployed at various levels, from design to modeling to testing of components. Complex or nonlinear problems can be easily and efficiently solved using the ML approach. Thus, it is much suited for the automation of radio frequency (RF) circuits where the input-output relation is complex. Furthermore, the employment of ML-based techniques in RF electronic design automation (EDA) tools boosts the performance of such tools. The chapter presents a comprehensive review on the recent research advancements and ML techniques used for the optimization of RF circuits.

Keywords: Machine learning, automation, RF circuit design

**Corresponding author*: shanthip@rvce.edu.in

Chandra Singh, Rathishchandra R. Gatti, K.V.S.S.S.S. Sairam, Manjunatha Badiger, Naveen Kumar S. and Varun Saxena (eds.) Modeling and Optimization of Signals Using Machine Learning Techniques, (75–96)

5.1 Introduction

Radio frequency (RF) circuits are an important component of today's electronics. These devices are now used in a wide range of applications, including automotive, medical, healthcare, and electronics from a security perspective. Radio frequency devices/components are being used in more than 60% of all radio frequency manufactured each year. Machine learning (ML) alternatives can greatly reduce the progress period and still enhance efficiency.

The flow of digital IC design, where many techniques were created and developed, is very different from the flow of RF circuit design.

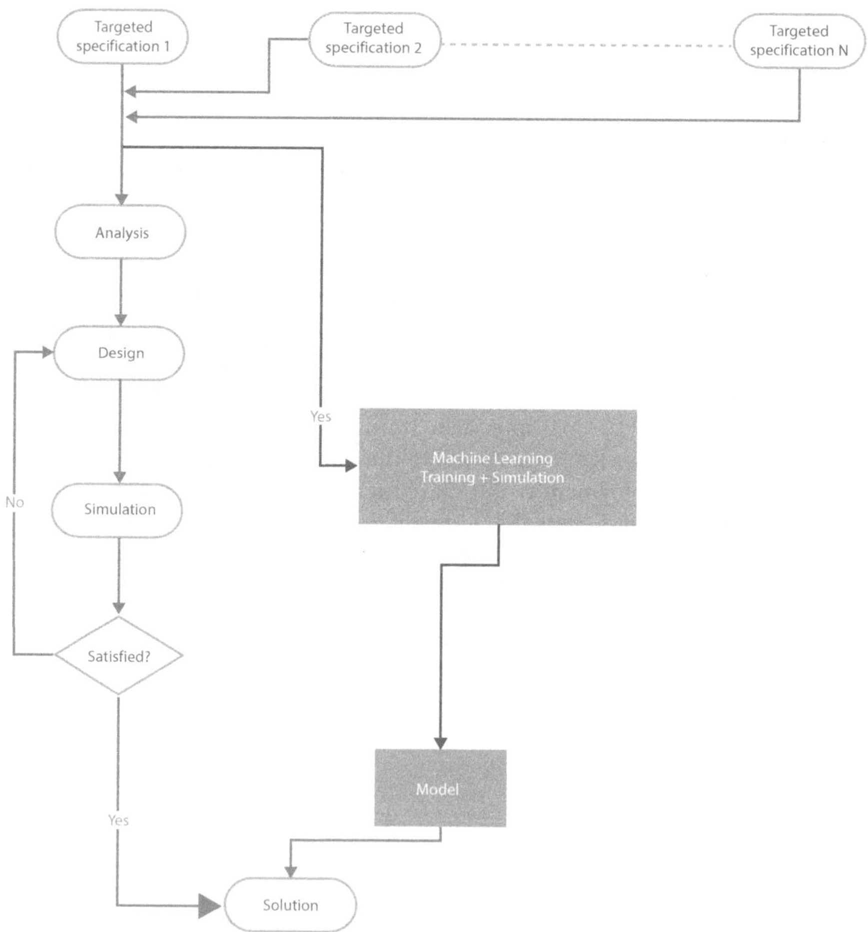

Figure 5.1 ML design flowchart.

The nonlinear behavior of RF circuits and the processing of RF signals are the growing difficulties seen in modern applications, and technologies in nanometer combination add issues and put more pressure to RF IC designers. As a result, ML is the focus of this extensive study, and nowadays, ML continues to change the society in a variety of ways. Machine learning is also opening up new horizons for how computationally smart tools for RF circuit design can help design engineers work more efficiently. Using the traditional flow methodologies, the design engineer reiterates the process for each specification. Figure 5.1 shows the conventional flow of ML-based design methodologies.

While a designer's expertise, skills, and intuition are crucial, the lack of formalization severely restricts knowledge sharing and reuse. The ML-based architecture, on the other hand, generates solutions quickly. In Section 5.2, the basics of machine learning are discussed, including the different types of models with various kinds of supervision techniques. Section 5.3 discusses existing and new methods for modeling and synthesis of RF circuits using machine learning techniques. Finally, the conclusions are drawn in Section 5.4.

5.2 Background on Machine Learning

Machine learning is based on the concept of artificial intelligence (AI). Unlike AI, which focuses on creating expert systems, machine (or statistical) learning focuses on the given statistics of data [1]. Bayes' article on probability [2] established the theoretic basis for ML and served as the basis for primitive machine learning practices like the Markov Chains and the naïve Bayes. The first neural network machine was proposed back in 1951, but artificial neural network (ANNs) did not gain popularity until Frank Rosenblatt's perceptron [3] and backpropagation [4], which were published in 1958 and 1986, respectively. Later, several other breakthroughs have been developed, and there are now many new techniques for developing machine learning systems to solve regression problems. The goal is to categorize the given data. For example, we can take e-mails which are classified as "spam mail" or "no-spam mail" by an e-mail spam filter. Regression systems, on the other hand, attempt to define variables that are dependent as functions of the data. All ML systems must be able to generalize the new data while avoiding over-fitting of the provided data for training.

When a machine learns the variance of data instead of studying the processes that are producing the set of data, it is said to be overfitting [1, 5, 6].

Another important feature in ML systems is the sum and type of control. The data used to train the method in supervised learning must contain the optimal solution, referred to as a marker. The mark may be categorical (for sorting problems) or constant-valued (for problems with continuous values) in regression problems. Linear regression, polynomial regression, support vector machines (SVMs), decision trees, and ANNs are examples of supervised learning algorithms. The data provided here in unsupervised learning are not labeled, and the algorithms of ML group the data to a point based on their characteristics. Unsupervised learning includes clustering, visualization, reduction in dimension, and anomaly detection. The K-means algorithm and the principal component analysis algorithms, as well as their variants, are popular unsupervised learning algorithms. Supervised classification is illustrated in Figure 5.2(a) by linear regression (which, despite its name, is a classifier). Polynomial regression is illustrated in Figure 5.2(b), and the K-means algorithm is mainly used for the clustering of data, as shown in Figure 5.2(c).

Some semi-supervised algorithms are also there where the data used to train the complete system are not properly labeled, and the system mixed both the supervised and unsupervised learning algorithms by itself. Deep neural networks built on auto encoders have been trained by unsupervised learning, and then are tuned using techniques of supervised learning [7].

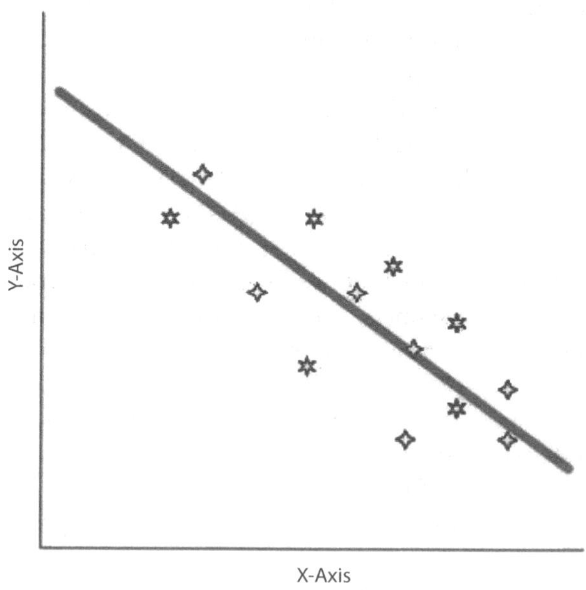

Figure 5.2 (a) Linear regression.

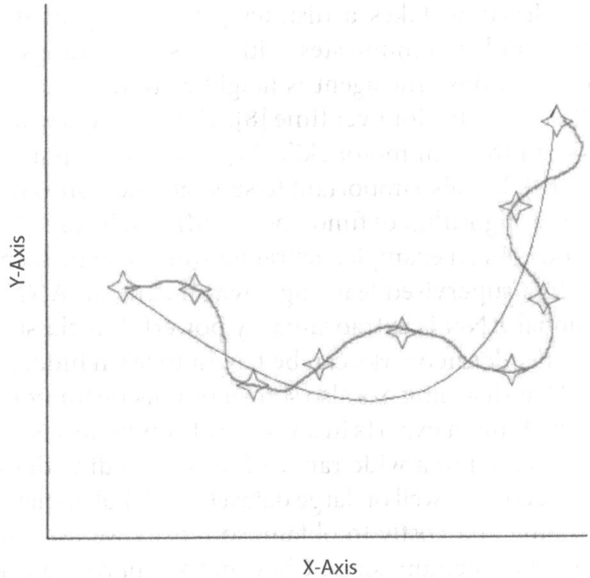

Figure 5.2 (b) Polynomial regression.

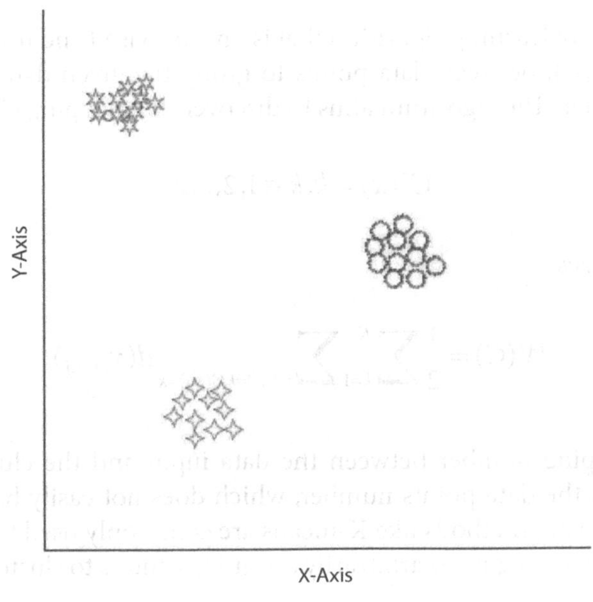

Figure 5.2 (c) K-means.

Reinforcement learning takes a distinct path. An agent in a real-time system watches and communicates with its surroundings by choosing and performing actions. The agent is taught a strategy that improves the desired result of the behavior over time [8]. These devices will then be used on robots to train them in motor skills [9] or to play sophisticated games such as chess [10]. It is also important to separate the results from the algorithm as the same algorithm of fundamental ML can be used in any or both of these methods. As an example, neural networks can be seen with any of the methods. In a supervised learning environment, an ANN in the form of a convolutional ANN is extraordinarily powerful in classifying images, whereas auto-encoder networks can be taught to learn hidden space without guidance. Deep learning has also shown promising improvement in the field of defeating human experts in a variety of competitions [11]. Machine learning is well-suited to a wide range of uses, including designing. Most algorithms work equally well on large datasets [12], but sometimes the data can be challenging and costly to obtain, so mostly we use datasets which range from small to medium sizes. Choosing an appropriate approach for the targeted application is a very crucial decision. The choices are numerous. Some electronic design automation (EDA) approaches are briefly listed in the following sub-sections.

5.2.1 Clustering

This is a type of learning algorithm that is unsupervised and uses a distance metric, d(xi,xj), between data points to group the given data into a pre-defined cluster. The algorithm aims to discover the mapping C.

$$C^*(x) = k, k \in 1,2,..,K \tag{5.1}$$

that minimizes

$$W(C) = \frac{1}{2}\sum\nolimits_{k=1}^{K}\sum\nolimits_{C(x_i)=kC(x_j)=k} d(x_i, x_j) \tag{5.2}$$

The mapping number between the data input and the clusters grows rapidly with the data points number, which does not easily become tractable. As a result, methods like K-means are commonly used to solve clustering. K-means begins by arbitrarily assigning centers to clusters (or using certain spreading criteria), then it iterates the two steps below until no further change is possible:

- Determine the teaching points that are closest to each center according to the other centers.
- The middle of each cluster can be modified repeatedly and becomes the mean of the data points that are there in that cluster.

This technique is mostly useful for reducing the volume of data that needs to be stored without sacrificing too much detail. The clustering algorithm is used [13] for minimizing the data needed to train an SVM classifier for the IC fault diagnostics, while the fuzzy K-means is used [14] to group the elements during size optimization to apply the time-consuming simulation (Monte Carlo) only for a few preliminary resolutions.

Though clustering can save money, calculating how many clusters to use without missing information can be challenging. Clustering is also affected by the distance metric and function scaling.

5.2.2 Principal Component Analysis

Principal component analysis is also another type of unsupervised learning algorithm like the clustering algorithm; it is majorly used to minimize data without sacrificing content. It is a linear matrix operation that maximizes the variance by transforming the function space into a latent space. Taking the covariance of the results, S is defined in Equation 5.3.

$$S = \frac{1}{N}\sum_{n=1}^{N}(x_n - \overline{x})(x_n - \overline{x})^T \tag{5.3}$$

The eigenvectors, which correspond to the corresponding high eigen values, are the additional principal components. Data are defined with fewer features where only the components with higher variance are held. Pessoa *et al.* [15] optimized a voltage-controlled oscillator and an amplifier by using a principal component analysis to reduce the number of configuration variables. This algorithm is a linear matrix operator that cannot accommodate nonlinear data; however, by using the kernel trick, which is shown in the article [16], it can be employed for nonlinear data.

5.2.3 Naïve Bayes Algorithms

Naïve Bayes algorithms are fast and scalable with training and scoring. The classifier computes the posterior using Bayes law, assuming that the features are independent, as shown.

$$p(G/x) = \frac{p(G)\prod_{j=1}^{m} p(x_j/G)}{\sum_{g \in \mathcal{G}} p(g)p(x/g)} \tag{5.4}$$

The decision rule can be formalized only with prior probability; thus, the denominator will be constant for any given function vector.

$$\hat{G} = \underset{G \in \mathcal{G}}{\arg\max}\left(p(G)\prod_{j=1}^{m} p(x_j/G)\right) \tag{5.5}$$

These classifiers are simple to comprehend and construct. They are simple to train and do not take a lot of data to achieve useful results. Even though naïve Bayes is a functional solution in many implementations, it is based on the concept of function freedom, which is not applicable in most real-life cases. For fault detection, Watson and Gupta [18] used naïve Bayes as their go-to option.

5.2.4 Support Vector Machines

A support vector machine is a learning algorithm that is supervised and used for data separation in which the margin is maximized between classes by making it fit the boundary conditions. H(x), which is a function of space, expands the decision space to increase the classifier's accuracy and mostly changes to the nonlinear boundaries in the original space. The margin is maximized when we have the case of a non-separable group.

$$min \,||\,\beta\,||\,\text{subject to}\begin{cases} y_i(h(x_i)^T\beta) \geq 1 - \xi_i \forall_i \\ \xi_i \geq 0, \sum \xi_i \geq \text{constant.} \end{cases} \tag{5.6}$$

The solution for the following problem consequently is found by maximizing the dual of the first equation. The next equation gives the associated decision boundary.

$$L_D = \sum_{i=1}^{N} \alpha_i - \frac{1}{2}\sum_{i=1}^{N}\sum_{i'=1}^{N} \alpha_i \alpha_{i'} y_i y_{i'} \langle h(x_i), h(x_{i'}) \rangle$$

$$f(x) = \sum_{i'=1}^{N} \alpha_i \alpha_{i'} y_i y_{i'} \langle h(x_i), h(x_i) \rangle + \beta_0 \tag{5.7}$$

The above equation result with nonzero coefficients on which the restrictions are exactly satisfied. As a result, the linear grouping of certain data points at the class's corner, also known as SV or support vector, describes the boundary. If the data are linearly separable, the SVM easily finds the best linear separator; for nonlinear data, the idealogy discussed previously helps the SVM to separate in the higher dimensions. Extracting information from various constraints is challenging, making tuning and selecting the appropriate kernel to use difficult tasks. Support vectors often perform poorly where the data dimension approaches the number, which is the numerical equivalent of the total points. Support vectors that define unfeasible areas of the space of solutions during sizing optimization were used in a study by Dumesnil [21] to prevent wasteful circuit simulation.

5.2.5 Artificial Neural Networks

In recent years, ANNs and deep learning have gained popularity in various fields; wherever large amounts of data are being used, ANNs can be implemented. The perceptron is its most fundamental component, consisting of a layer of threshold units that compute a biased sum of the inputs along with z and applies it to a nonlinear activation function:

$$\begin{aligned} h_w(x) &= \text{activation} \\ &= \text{activation}\,(W^T \cdot x) \\ &= \text{activation}\,(w_1 \cdot x_1 + w_2 \cdot x_2 + \cdots w_n \cdot x_n) \end{aligned} \tag{5.8}$$

Here, x denotes the values of the input, and w denotes the linear threshold unit's weights vector. This layer predicts a weight for every x, and therefore, can be used to train by reinforcing the relationship that the weights have that leads to correcting the forecast.

Table 5.1 Summary of machine learning techniques used in different papers.

Reference number	Component optimized	Machine learning methods used	Assistance provided
[17]	Radio frequency and microwave components and MESFET	ANN	Microwave designs using neural network-based computer aided design (CAD)
[18]	Radio frequency and microwave components, HEMT and MESFETS	ANN	Development of modeling of nonlinear microwave devices review
[19]	Radio frequency-coplanar waveguide (CPW) components	ANN (EM-based)	For accurate performance estimations, efficient modeling of CPW components
[20]	Radio frequency-uniplanar compact-photonic band gap (UC-PBG) Rectangular waveguide	ANN (radial basis function or RBF-MLP)	An efficient way of modeling radio frequency systems for nonlinear microwave applications
[21]	Radio frequency-MESFET	ANN (WNN-MLP)	Large-signal nonlinear power transistors and circuits can be designed faster.
[22]	Linear radio frequency amplifier synthesis	Integrating adaptive population generation, naïve Bayes classification, Gaussian process, and differential evolution	High-frequency amplifiers can be built more quickly and efficiently.

(Continued)

Table 5.1 Summary of machine learning techniques used in different papers. (*Continued*)

Reference number	Component optimized	Machine learning methods used	Assistance provided
[23]	Radio frequency circuit's synthesis	Genetic algorithm (GA) + ANN (MLP)	Efficient RF circuit synthesis using GA-assisted ANN
[25]	Slotted waveguide antenna	ANN	Eliminating the need for time-consuming simulations can help speed up the design process.
[26]	Antenna (stacked patch)	Gaussian process regression (Kriging)	Reduces the number of necessary simulations by 80%
[27]	E-shape antenna	Linear regression	The optimum results are obtained in simulation

An ANN is a perceptron, consisting of one or more layers, which can be trained by the training algorithm [4]. In newer applications, ANNs are replacing whole processing pipelines and can create efficient end-to-end ML systems. The ANN is a very adaptable construct. Different functions can be implemented in the same network using these multi-faceted methods. When using ANNs, the real price is the total number of parameters that can be changed. Not like SVMs, in which the resolutions are the convex function's best, the ANN's weight optimization often contributes to local cost function optima. As a result, initialization is an essential aspect of teaching. The ANNs are also used extensively for modeling, synthesis, schematic generation, and fault checking. A summary of machine learning techniques is depicted in Table 5.1.

5.3 ML in RF Circuit Modeling and Synthesis

For RF and microwave modeling and architecture, neural networks have been used, with ANN-based algorithms being used for faster design speeds. As a result, relative to more costly traditional methods, a more reliable answer for the whole device can be achieved in less time. From theory to reality, ANNs for radio frequency and microwave modeling is explained at length [17]. The paper's authors noted that the machine learning approach, namely neural networks, is a very promising replacement for orthodox methods, which have limited range and accuracy. The authors discussed many examples of neural networks being used in modeling. According to many scientists, neural networks are tempting substitutes to traditional approaches such as numerical modeling methods that are computationally intensive, observational methods that can be difficult to achieve with new systems, and empirical modeling solutions, which have a restricted range and precision. They show how neural networks can be used for modeling signal propagation delays in a VLSI network on printed circuit boards (PCBs), CPW discontinuities, and MESFETs. Interconnected networks in PCBs, CPW discontinuities, and MESFETs are all examples of previous research. Finally, they demonstrate how to optimize microwave circuits using CPW models. The same authors provided a thorough review of modeling problems and ANN-based nonlinear modeling methods, including the modeling of transistors and complex neural network circuits [18]. Microwave examples are used to demonstrate the simulation methods discussed. Electromagnetic (EM) simulations are another approach for modeling CPW circuit components using ANN [19]. Individual EM-based ANN models are used in modeling CPW transmission lines, short and open circuit stubs, step-in spacing discontinuities, and T-junctions. Several EM simulations with meaningful input/output relationships were used to train the models, and their performance was directly affected. The backpropagation algorithms are used to build a feed-forward network with multiple layers, one being input, one which is covered, and one that gives output. Without using costly EM simulations, the built models are used to build a CPW-folded double-stub filter and a 50- or 3-dB power-divider circuit. The system suggested here can also be used for other aspects of microwave/RF architecture.

An ANN-based EM approach requires a lengthy period of training for effective modeling, otherwise, performance will suffer. The effective resilient backpropagation (RPROP) algorithm used in the training process [20] provides a solution for modeling RF devices with a RBF/MLP modular framework. A well-known strategy "divide and rule," with the suggested structure seen in Figure 5.3, is used by the authors. The complex architecture problem is grouped into sub-problems and distributed throughout the modular structure's neural networks. They say that by using a modular structure, an EM-based ANN can be more effective. The RBF structure components are organized to take advantage of the RBF and MLP neural networks' local and global approximation characteristics, where the radial network's limited approach and the MLP network's universal approach increase the modular structure's generalization ability. The developed method is demonstrated using a patch antenna with a PBG substrate and a UC-PBG rectangular waveguide. In comparison to using RBF and MLP separately, combining them (modular model) results in a significant generalization potential that is not affected by the number of hidden neurons in the network.

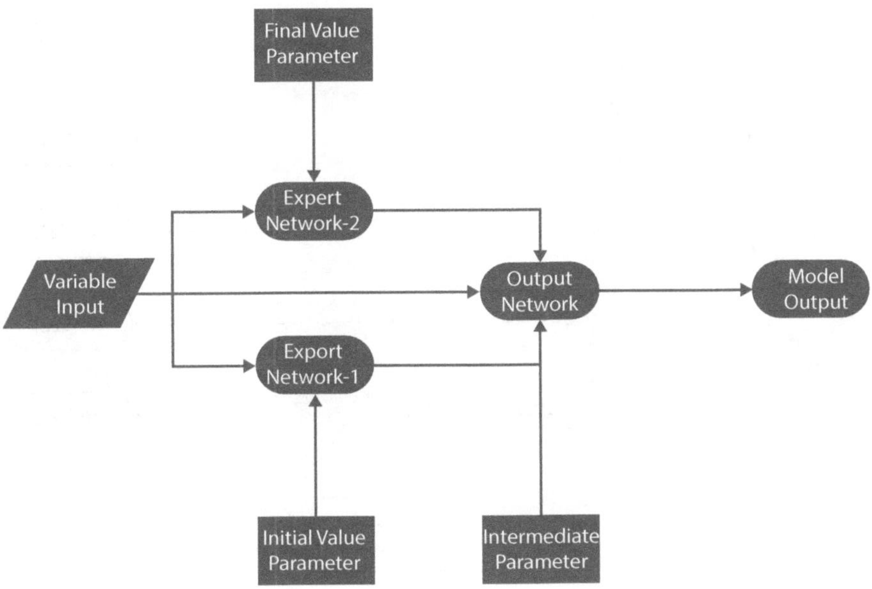

Figure 5.3 The proposed framework [20].

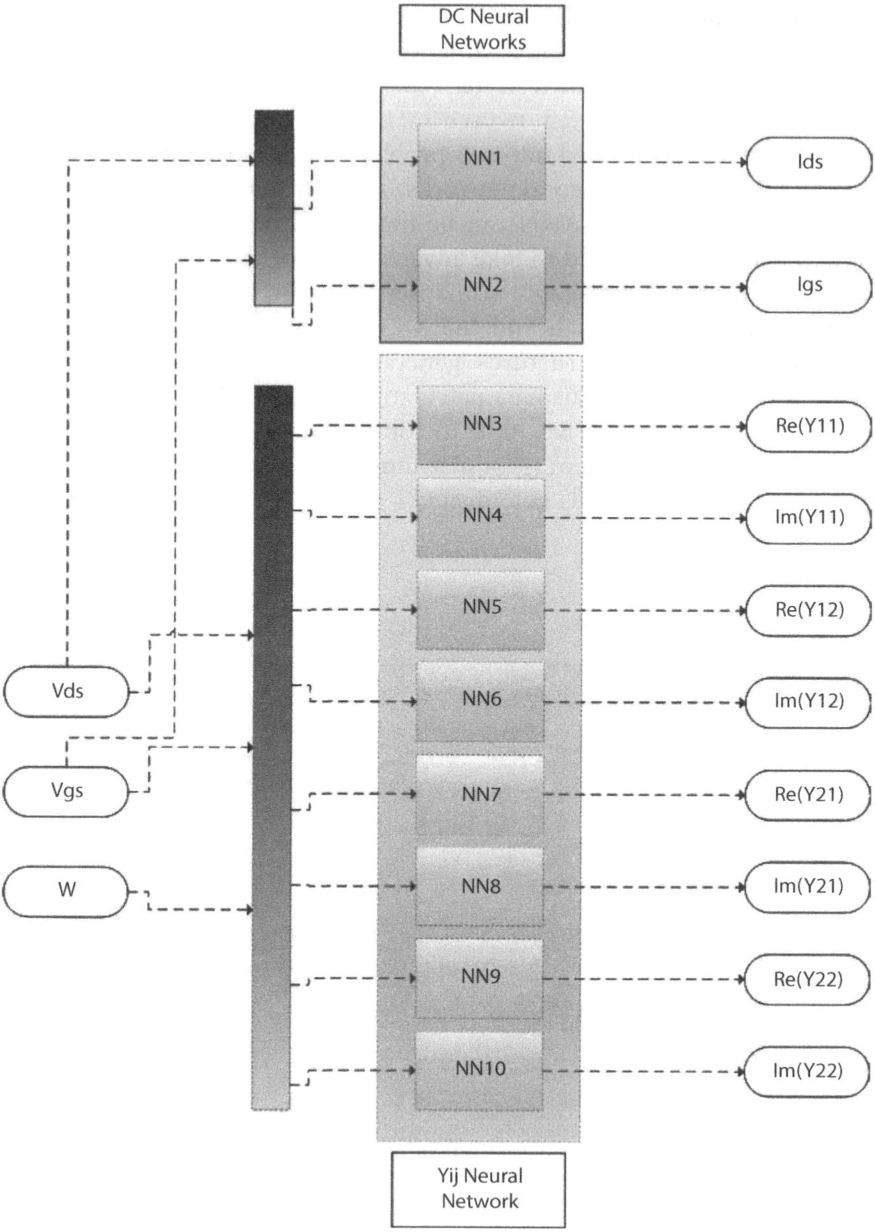

Figure 5.4 Volterra-ANN device model [21].

An article [21] prefers wavelet neural networks to basic MLP and Gaussian radial basis feature networks. The transistor modeling example is shown in Figure 5.4, in which 10 neural networks are used to match the lumped identical circuits exactly. Four neural networks were used in the circuit modeling example. The real and imaginary portions of the input and output voltages are the five inputs. This method of modeling allows for input and output loading to be considered. Learning was carried out on measurement points of number 2625, with promising effects on new data. More general neural network-based models also could solve the issues with the commonly used lumped electrical circuit models. These models are computationally effective and reliable, resulting in a very complicated model parameter extraction process with the need for a precise circuit layout.

The EMLDE was presented [22] for millimeter wave frequency RF amplifiers. The proposed ABGPDE algorithm helps to solve low-dimensional but more complicated issues and results in costly optimization problems, which reformulates the problem into a different hierarchical structure.

The EMLDE system benefited from the parallel computing approach as well. The EMLDE will produce results that are equivalent to those obtained by directly applying a global optimization algorithm using EM simulations, which is the better framework with respect to solution efficiency and less computational cost. The algorithm is shown in Figure 5.5.

They demonstrated that a 100-GHz, three-stage, linear amplifier with more than a 10-dB gain and a 20-GHz bandwidth (BW) could be synthesized in a clock time of 25 hours (nine times speedup). As a result, high-optimization ability and efficiency were achieved.

This procedure is repeated until all specification criteria have been addressed. This method was used on a traditional cascade low noise amplifier (LNA) circuit. For training, 235 valid LNA designs were produced at random. Six variables were used to target 10 template parameters. The models were found to accurately forecast the LNA's activity within 5% of the time. During the mapping of circuit performances to circuit sizing, neural networks can also be used in succession [23]. Figure 5.6 depicts the proposed synthesis approach as a block diagram. In contrast to previous methods, the ANNs' inputs are the performances, and every design constraint is obtained by using a specific ANN. The network is given the optimal performance dataset as the input and output constraints were selected.

This ANN's learning mechanism is regulated by a GA, which means that the algorithm chooses which style to use, MLP or RBF, as well as the magnitude and design constraint of the output. When the ANN is complete, the yield is fed into a second ANN, which is tasked with determining a

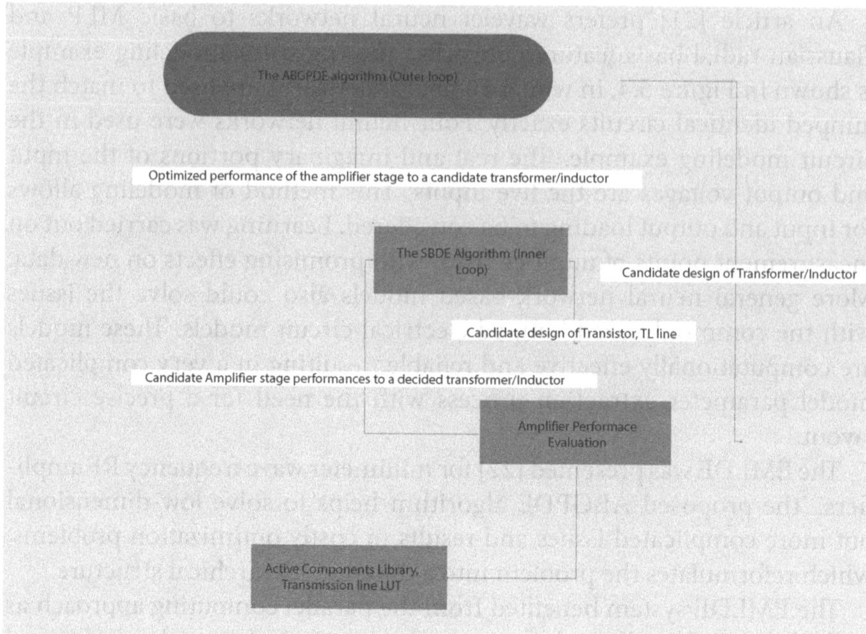

Figure 5.5 The framework of the EMLDE method [22].

second design parameter based on the performance parameters and the first design parameter.

Now regarding antenna design and fabrication, there are multiple approaches to get the required parameters for the design. The particle swarm optimization method (PSOA), an evolutionary algorithm, was used [24] to build a multiband patch antenna with ANNs. The ANN creates a mapping function after the PSOA determines the antenna's geometrical parameters, which are connected to the frequency and the BW of the antenna. Figure 5.7 depicts the flow chart of the optimizer with four input parameters. The findings revealed that the design process can be improved by eliminating the need for time-consuming simulations, resulting in a considerable reduction in computing load. The designed antenna has also been built and tested, and the measured and simulated findings [24] are very similar.

The E-shaped antenna design using differential evolution (DE) and the Kriging method is described [25]. The feed position, slot position, patch width (W), slot width (Ws), slot length (Ls), and patch length (L), are all optimized. It is claimed that comparable findings may be reached using alternative optimization approaches while decreasing the number of

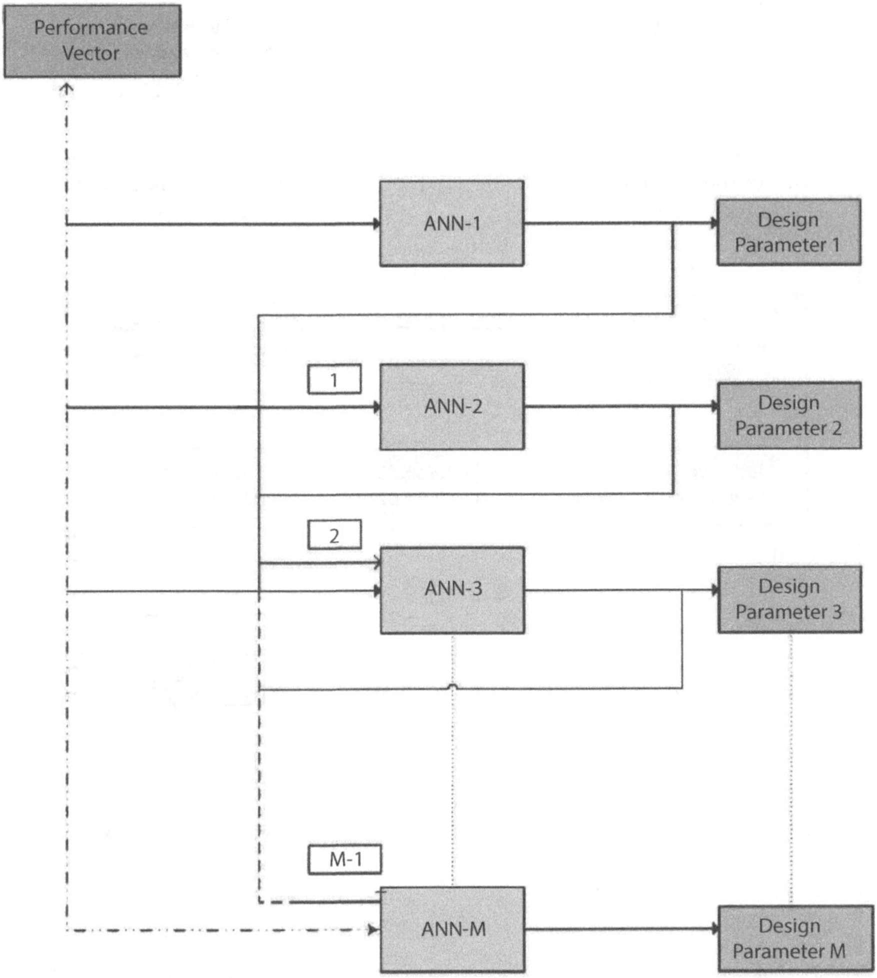

Figure 5.6 ANN array methodology as a block diagram [23].

required simulations by 80%. At frequencies of 5.0 GHz and 5.5 GHz, the magnitude of the S11 parameter should be kept to a minimum. After running the suggested approach five times, optimal solutions were discovered, and the model demonstrated excellent prediction accuracy. Comparison is made with optimization approaches, such as self-adaptive differential evolution [26], and wind-driven optimization [27]; combining machine learning with evolutionary algorithms has been shown to give a quicker convergence rate with equal solution quality. Differential evolution, which is self-adaptive [26], and optimization, which is wind-driven [27], were

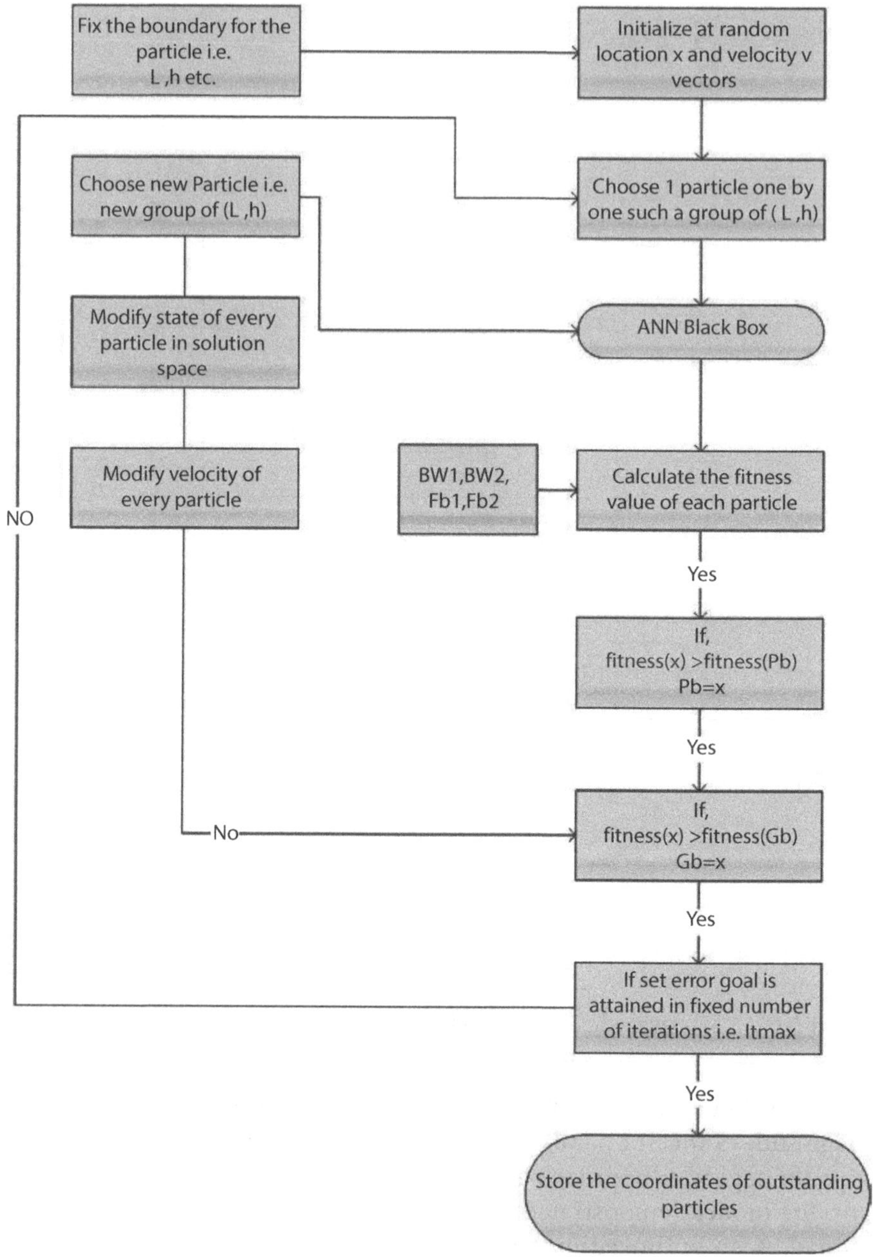

Figure 5.7 Flowchart of the suggested improved optimizer [24].

demonstrated to achieve the same optimization goals while decreasing the number of simulations required by 82.3% and 77.9%, respectively. The construction of microstrip antenna with an ultra-wide band using a combination of regression and an evolutionary algorithm was discussed [28]. The BW, the return loss (RL), and the central frequency are estimated using an ML approach in conjunction with an evolutionary algorithm (CFD). The regression in ML techniques allows a curve to be fitted using a discrete collection of known data points that will be the antenna parameters collected from earlier simulations performed. The other antenna parameters remain constant and consider design constraints (BW > 9.0 GHz, RL < -20 dB, and CFD = 0.37 Hz). A prototype method is utilized to discover the optimum values for 'Ws' and 'Ls'. Using 170 datasets, it was demonstrated that the best outcomes were obtained when meeting certain criteria [29].

5.4 Conclusion

Machine learning approaches are being used in a variety of applications, where their improved learning power makes them ideal for solving any problem, which is not linear in nature. Machine learning techniques have positively affected IC architecture at various design stages, from the initial simulation of devices to the final testing of the integrated chips. Machine learning modeling attempts aim at producing precise models at various stages of design and replace the simulator with these models, particularly in RF applications, to reduce human effort and design time. Creating technology-independent models will also allow them to be used in other technologies to solve a problem. Given the growing popularity of ANNs, researchers can revisit their use in RF IC modeling, where uncertainty and reliability issues have yet to be completely tackled. Furthermore, the ability to model CMOS devices technology-agnostic through ANNs would certainly lead to the inclusion of RF ICs in EDA tools. Machine learning synthesis also makes it possible for optimization tools to work in a large, fast, and accurate space. With a precise model of the RF circuits, in a time-conserving way, engineering can now do all things that were previously thought to be unfeasible by making more complex yet optimized designs, which are well-tailored for the application they want, opening a lot of opportunities in the future of circuit design.

References

1. K. P. Murphy, *Machine learning: a probabilistic perspective (adaptive computation and machine learning series)*, MIT Press., 2012.
2. M. Bayes and M. Price, "An Essay towards Solving a Problem in the Doctrine of Chances. By the Late Rev. Mr. Bayes, F. R. S. Communicated by Mr. Price, in a Letter to John Canton, "*Philos. Trans.,* (1683- 1775), vol. 53, pp. 370–418,1763.
3. F. Rosenblatt, "The perceptron: A probabilistic model for information storage and organization in the brain," *Psychol. Rev.*, vol. 65, no. 6, pp. 386–408, 1958.
4. D. E. Rumelhart, G. E. Hinton and R. J. Williams, "Learning representations by back-propagating errors," *Nature*, vol. 323, no. 6088, pp. 533- 536,1986.
5. J. Hastie, Trevor, Tibshirani, Robert, Friedman, *The Elements of Statistical Learning The Elements of Statistical Learning Data Mining, Inference, and Prediction*, Second Edition,2009.
6. C. M. Bishop, *Pattern Recognition and Machine Learning*, ser. Information science and statistics. Springer, 2007, vol. 16, no.4.
7. I. Goodfellow, Y. Bengio and A. Courville, *Deep Learning*. MIT Press, 2016.
8. R. S. Sutton and A. G. Barto, *Reinforcement Learning*, Second Edition, 2018.
9. J.Peters and S.Schaal, Reinforcement learning of motor skills with policy gradients," *Neural Netw.*, vol. 21, no. 4, pp. 682–697, 2008.
10. D.Silverand *et al.* "Mastering the game of Go without human knowledge," *Nature*, vol.550 ,no. 7676, pp. 354–359,2017.
11. V. Mnih and *et al.*, "Human-level control through deep reinforcement learning," *Nature*, vol. 518, no. 7540, pp. 529–533,2015.
12. Z. H. Zhou and *et al.*, "Big data opportunities and challenges: Discussions from data analytics perspectives [Discussion Forum]," *IEEE Comput. Intell. Mag.*, vol. 9, no. 4, pp.62- 74, 2014.
13. A. Zhang, C. Chen and B. Jiang, "Analog circuit fault diagnosis based UCISVM," *Neurocomputing*, vol. 173, pp. 1752–1760,2016.
14. A. Canelas and *et al.*, "FUZYE: A Fuzzy C-Means Analog IC Yield Optimization using Evolutionary-based Algorithms," *IEEE Trans. Comput. Aided Des. Integr. Circuits Syst.*, vol. 0070, no. 3,2018.
15. T. Pessoa and *et al.*, "Enhanced analog and RF IC sizing methodology using PCA and NSGA-II optimization kernel," in *Proceedings of the Design, Automation and Test in Europe Conference and Exhibition (DATE)*, 2018, pp.660–665.
16. Q.-J. Zhang, K. C. Gupta and V. K. Devabhaktuni, "Artificial neural networks for RF and microwave design-from theory to practice," *IEEE Trans. Microwave Theory Tech.*, vol. 51, no. 4, pp. 1339–1350,2003.

17. V. K. Devabhaktuni and *et al.*, "Neural networks for microwave modeling: Model development issues and nonlinear modeling techniques," *Int. J. RF Microwave Comput. Aided Eng.*: Co-sponsored by the Center for Advanced Manufacturing and Packaging of Microwave, Optical, and Digital Electronics (CAMP mode) at the University of Colorado at Boulder, vol. 11, no. 1, pp. 4–21,2001.
18. P. M. Watson and K. C. Gupta, "Design and optimization of CPW circuits using EM-ANN models for CPW components," *IEEE Trans. Microwave Theory Tech.*, vol. 45, no. 12, pp. 2515–2523, 1997.M. G. Passos, P. d. F. Silva and H. C. Fernandes, "A RBF/MLP modular neural network for microwave device modeling," *Int. J. Comput. Sci. Netw. Secur.*, vol. 6, no. 5A, pp. 81–86, 2006.
19. Y. Harkouss and *et al.*, "The use of artificial neural networks in nonlinear microwave devices and circuits modeling: An application to telecommunication system design (invited article)," *Int. J. RF Microwave Comput. Aided Eng.*, vol. 9, no. 3, pp. 198–215,1999.
20. B. Liu, N. Deferm, D. Zhao, P. Reynaert and G. G. E. Gielen, "An Efficient High-Frequency Linear RF Amplifier Synthesis Method Based on Evolutionary Computation and MachineLearning Techniques," in *IEEE Trans. Comput. Aided Des. Integr. Circuits Syst.*, vol. 31, no. 7, pp. 981-993, July 2012, doi:10.1109/TCAD.2012.2187207.
21. E. Dumesnil, F. Nabki and M. Boukadoum, "RF-LNA circuit synthesis using an array ofartificial neural networks with constrained inputs," in *IEEE International Symposium on Circuits and Systems (ISCAS)*, 2015, pp.573–576.
22. Badiger, Manjunatha, and Jose Alex Mathew. "Tomato plant leaf disease segmentation and multiclass disease detection using hybrid optimization enabled deep learning." *J. Biotechnol.*, 374 (2023): 101-113.
23. Badiger, Manjunatha, *et al.* "Leaf and skin disease detection using image processing." *Global Transitions Proc.*, 3.1 (2022): 272-278.
24. Manjunatha Badiger, Dr Jose Alex Mathew. "A Survey on Low Power SBCs for Optimized Implementation of Deep Learning Models." *J. Surv. Fish. Sci.*, 10.2S (2023): 4173-4189.
25. EnginAfacan, Nuno Lourenço, Ricardo Martins, Günhan Dündar. "Review: Machine learning techniques in analog/RF integrated circuit design, synthesis, layout, and test", *Integration*, 2021.
26. S. K. Jain, "Bandwidth Enhancement Of Patch Antennas Using Neural Network Dependent modified Optimizer," *Int. J. Microwave Wireless Technolog.*, vol. 8, no. 7, pp. 1111–1119, 2016.
27. M. D. Gregory and Z. Bayraktar and D. H. Werner, "Fast Optimization of Electromagnetic Design Problems Using the Covariance Matrix Adaptation Evolutionary Strategy," *IEEE Trans. Antennas Propag.*, vol. 59, no. 4, pp. 1275–1285,2011.

28. Bayraktar and M. Komurcu and J. A. Bossard and D. H. Werner, "The Wind DrivenOptimization Technique and its Application in Electromagnetics," *IEEE Trans. Antennas Propag.*, vol. 61, no. 5, pp. 2745–2757, 2013.
29. C. R. Silva and S. R. Martins, "An Adaptive Evolutionary Algorithm for UWB Microstrip Antennas Optimization Using a Machine Learning Technique.," *Microw. Opt. Technol. Lett.*, vol. 55, no. 8, pp. 1864–1868, 2013.

6

A Survey: Emotion Detection Using Facial Reorganization Using Convolutional Neural Network (CNN) and Viola-Jones Algorithm

Vaibhav C. Gandhi[1]*, Dwij Kishor Siyal[2], Shivam Pankajkumar Patel[2] and Arya Vipesh Shah[2]

[1]*Department of Computer Engineering, Madhuben and Bhanubhai Patel Institute of Technology (MBIT), (Constitute College of CVM University), Gujarat, India*
[2]*TY CSE Department, School of Engineering & Technology, Navrachana University – Vadodara, India*

Abstract
According to a World Health Organization (WHO) report in 2015, one in five people in India suffer from depression in their lifetime, which accounts for almost 300 million people. Most cases in India are not even reported, so the number is likely to be greater than this. With the increase in everyday stress from a job or daily life, or in the case of students, academic stress, people have not been giving much care to their mental health, which in time leads to depression. According to WHO data from 2016, India's suicide rate was 16.5 per 100,000 individuals, the highest in Southeast Asia. Suicide is the main cause of death in the 15–19-year-old age group in India; something was clearly wrong with our perception of mental health and the problems related to it. Keeping this in mind, we came up with an idea where we created an application that detects user's emotions and suggests a binaural beats playlist, which would help them calm their emotions.

Keywords: Emotion detection, CNN, face recognition, depression, Viola Jones method, deep learning, classification, images

**Corresponding author*: Vaibhavgandhi2424@gmail.com

Chandra Singh, Rathishchandra R. Gatti, K.V.S.S.S.S. Sairam, Manjunatha Badiger, Naveen Kumar S. and Varun Saxena (eds.) Modeling and Optimization of Signals Using Machine Learning Techniques, (97–114) © 2024 Scrivener Publishing LLC

6.1 Introduction

6.1.1 Purpose

Mental health can be a challenging area to tackle, and seeking professional help can be the only sure way to tackle such issues. However, not everyone gets the help they need, which can lead to unexpected issues. During these tough times of COVID-19, people who are already suffering from mental health issues have not been able to seek help for their problems. With no adequate help and lifestyle changes, such as working from home and spending 90% of the time in their homes, not all are able to cope with their issues [1].

To help people tackle some of their mental health issues and help them focus on their lives, we came up with the idea where users can have their moods detected and then they are given binaural beats' suggestions, which are scientifically proven to help alter people's mental states. A flow diagram for this application is depicted in Figure 6.1.

6.1.2 Process Flow

- ❖ The user opens the application on the login screen and enters their credentials.
- ❖ After entering valid credentials, the user is taken to the home screen. From that screen, the user chooses from four options, i.e. "Detect," "Depressed," "Relaxed," and "Focus."

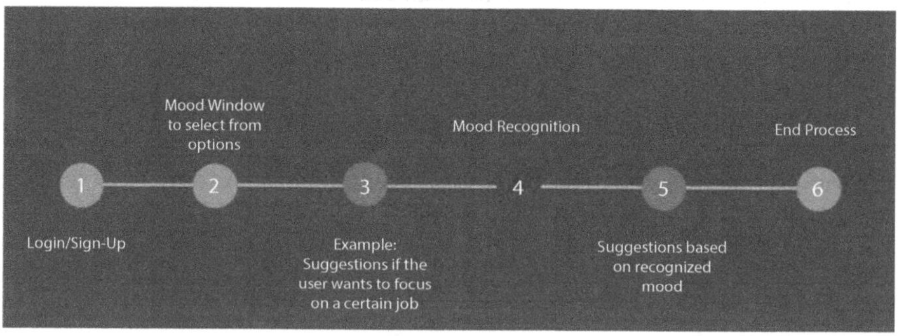

Figure 6.1 Application flow diagram.

- ❖ If the user selects "Detect," the user will be taken to the camera window in which the application will ask permission from the user for camera access. If the user allows it, the user's face will be detected through the camera, and an emotion will be recognized.
- ❖ After emotion recognition, the user will be taken to the suggestion window where the application will show a list of binaural beats specific to that emotion to enhance their mood.
- ❖ The user will be taken to YouTube, on their selected binaural beat, then the application will be closed.

6.2 Review of Literature

Author(s)/Year	**Research Paper: Prudhvi Raj Dachapally/2017**
Topic	Facial Emotion Detection Using Convolutional Neural Networks and Representational Autoencoder Units
Objectives	The study has been carried out to present the sentiment analysis model for detecting facial emotions.
Methods	Representational Autoencoder Units (RAUs), Convolutional Neural Networks (CNNs)
Findings	It was found that the performance of the CNNs was much better than RAUs in terms of accuracy.
Applications	The study helps improve the quality of emotion recognition through cameras/images.

- ❖ Studying this research paper, we learned that there was another method for emotion detection which is RAU.
- ❖ This paper shows a background that is theorized in both RAU and CNNs.
- ❖ It also shows the comparison of accuracy and training models using both methods.
- ❖ In conclusion, CNN was much better than RAU; hence, we chose CNN for our emotion detection model.

Author(s)/Year	**Blog: Nishank Sharma**
Topic	How to Do Facial Emotion Recognition Using a CNN?
Objectives	To figure out how to perform facial emotion recognition using a CNN.
Methods	CNNs
Findings	The winning model was accurate 71.1% of the time.
Applications	The blog helps improve the quality of emotion recognition through the camera.

- Studying this blog, we learned how to train our CNN model for emotion recognition, identify the different types of filters, and obtain better accuracy [5].
- The highest accuracy of the model is 71.1%.

Author(s)/Year	**Blog: Daphne Cornelisse**
Topic	An Intuitive Guide to Convolutional Neural Networks
Objectives	We will look at CNNs and how the brain architecture influences them at a high level.
Methods	CNNs
Findings	CNNs can be used to classify and recognize images.
Applications	This blog aims to increase the accuracy of camera-based emotion identification.

- Studying this blog, we learned about humans perceiving through the eyes and processing through the primary cortex of the brain.
- We also learned about the history of CNN and how it came to be, thanks to D. H. Hubel and T. N. Wiese's study in the 1950s and 1960s [6].
- The first CNN paper, LeNet-5, was published in 1998 by Bengio, Le Cun, Bottou, and Haffner. Architecture, categorization, feature extraction, and training were also covered.

Author(s)/Year	Blog: ujjwalkarn
Topic	An Intuitive Explanation of Convolutional Neural Networks
Objectives	To see if CNN could be used with the images dataset.
Methods	CNNs
Findings	CNNs have two main features: feature extraction and classification.
Applications	The blog helps improve the quality of emotion recognition through cameras/images.

- We learned how CNN is used to recognize handwritten digits and object identification (functioning of CNN) while reading this blog [7].
- We also learned about the basic concepts of CNN and its training.

6.3 Report on Present Investigation

6.3.1 Analysis of the Model

6.3.1.1 Emotion Recognition

We found several datasets for emotion recognition, but facial emotion recognition (FER)-2013 is among the popular ones. Hence, we used FER-2013 to train the model. The FER dataset contains low-quality grayscale images with a size of 48 * 48 pixels, which are not reflective of real-world settings. As a result, we amassed our collection of 2,156 photos. The labels "glad," "sad," "neutral," "disgusted," "surprised," "fearful," and "angry" are applied to each image. The data are evenly distributed throughout the seven emotions [8, 9].

What is the definition of a confusion matrix? The forecast outcomes of the classification issue are summarized in a confusion matrix. The results of correct and incorrect predictions are totaled and then split down by class using count values. The confusion matrix is depicted below in Figure 6.2.

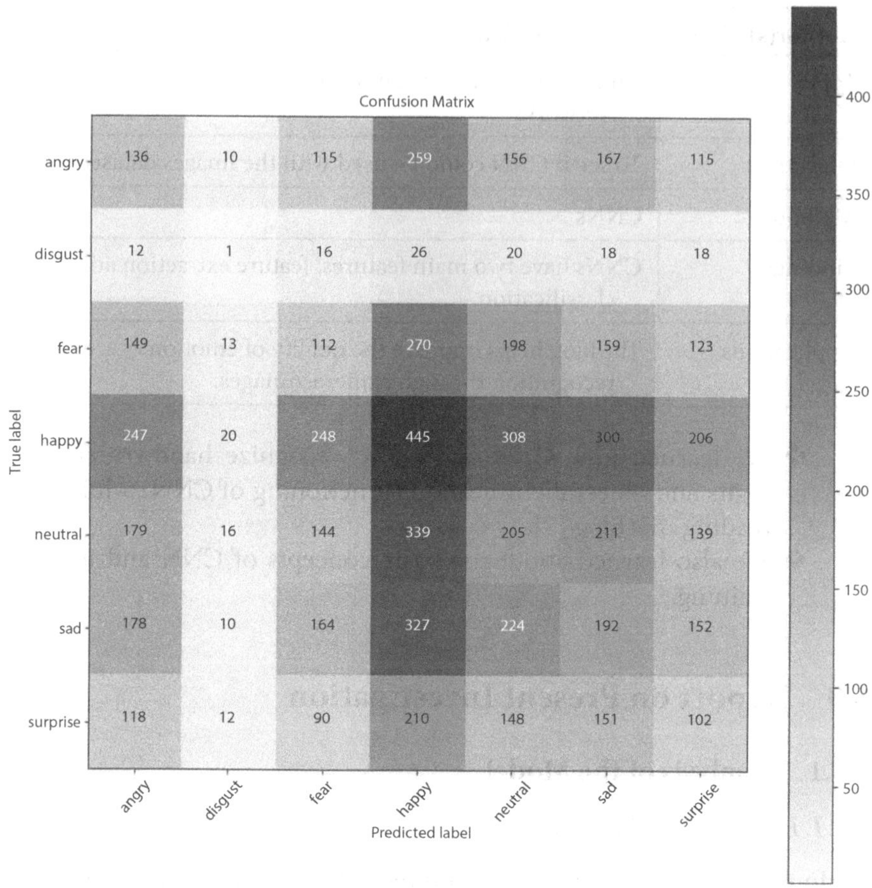

Figure 6.2 Confusion matrix.

6.4 Algorithms

6.4.1 CNN

- A CNN is a deep learning method that takes an image as input and allocates importance (trainable weights and biases) to different aspects/objects in the image, allowing them to be distinguished.
- CNNs are one of the most common types of neural networks used for image detection and recognition. Scene recognition, target tracking, and facial recognition are just a few of the applications for CNNs [1, 2].

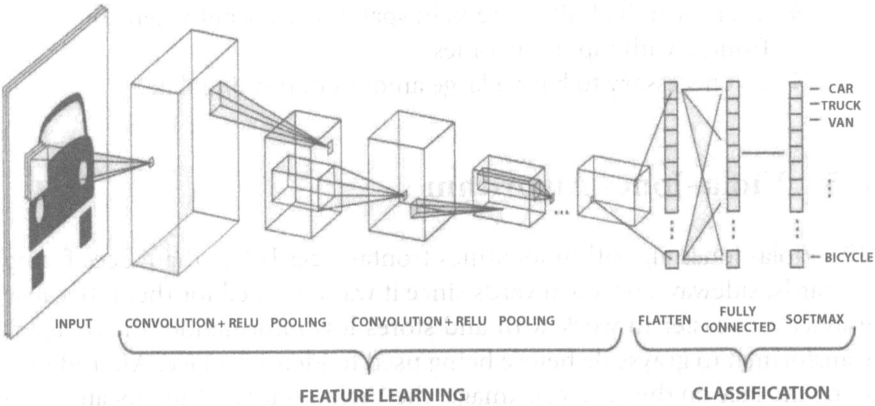

Figure 6.3 Image input feature learning and classification.

- An image is fed into a CNN for it to be categorized and processed into one of the categories, such as dog, cat, lion, or tiger, among others. An image is interpreted by the machine as an array of pixels, the size of which is determined by the resolution of the image. It will appear as h * w * d, where h = height, w = width, and d = depth, depending on the image resolution. An RGB image, for example, is a 6 * 6 * 3 matrix array, and a grayscale image is a 4 * 4 * 1 matrix array.
- Each input image will go through a number of convolution layers, pooling, totally linked layers, and other procedures and filters before being processed by a CNN (also known as kernels). Image input feature learning and classification is depicted in Figure 6.3.
- The soft-max function can then be used to categorize an object having between 0 and 1 probabilistic values using probabilistic values between 0 and 1 [3, 4].

6.4.2 Advantages

- ❖ In picture recognition challenges, high accuracy is required.
- ❖ It detects the significant elements of the given data without the need for human intervention.

6.4.3 Disadvantages

- ❖ The position and orientation of an object are not encoded by CNN.

- There is an inability to remain spatially invariant when confronted with input outcomes.
- It is necessary to have a large amount of training data.

6.5 Viola–Jones Algorithm

The Viola–Jones algorithm identifies frontal faces better than faces facing upwards, sideways, or downwards since it was designed for them. Because grayscale is easier to work with and stores less information, the image is transformed to grayscale before being used to identify a face. After identifying the face on the grayscale image, the Viola–Jones method locates it on the color image [5–7]. It is used for face detection. There are two stages in the Viola–Jones algorithm: training and detection.

6.5.1 Training

- We are in the process of teaching the computer to recognize these traits. We are feeding it data, and then teaching it to make predictions based on it. As a result, the algorithm effectively establishes a minimum standard to decide if anything can be counted as a function [8]. The algorithm reduces the image to 24 * 24 pixels and looks for traits that have been previously taught to be able to see features in their various and differing ways, a large amount of facial image data is needed. That is why we must provide the algorithm with a large amount of facial image data in order for it to be qualified. Viola–Jones fed 4,960 images to their algorithm.

6.5.2 Detection

- Viola–Jones creates a box (as seen on the right) and searches inside for a face. Basically, it is hunting for Haar-like elements [10, 11].
- The characteristics below imagine a box with two sides, one light and the other dark, which show how the model decides what the characteristic is in the image is. As in the edge of a brow, one side may be lighter than the other at times. It is likely that the center of the box is shinier than the rest, giving it a nose-like look. Face-like traits (Haar-like features) are

detected in smaller stages by a variety of boxes, and the data from both of those boxes combined aids the algorithm in determining the location of the face [12, 13]. Viola and Jones discovered three sorts of Haar-like characteristics in their research: edge features, line features, and four-sided features. These characteristics aid the model's understanding of the image. Consider how a black-and-white representation of a table's edge would appear. It will have a black-and-white effect because one side will be lighter than the other [14, 15].

6.6 Diagram

6.6.1 Working Diagram for Systems

The first window the user encounters is the login menu, which will help the user to log into the application working diagram for the system depicted in Figure 6.4. If the user does not have the account, then using the same screen, the user can sign up and then log into the account. After entering valid credentials, the user is taken to the home screen, where the user can choose from four options, i.e. "Detect," "Depressed," "Relaxed," and "Focus." In the 3 options: Detect, Depressed, and Relaxed, the application directly shows the suggestions to the user, and the user can ultimately skip the emotion recognition part. If the user selects the "Detect" option, the user will be taken to the camera window in which, through the camera, the user's face will be detected, and an emotion will be recognized. After emotion recognition, the user will be taken to the suggestion window where

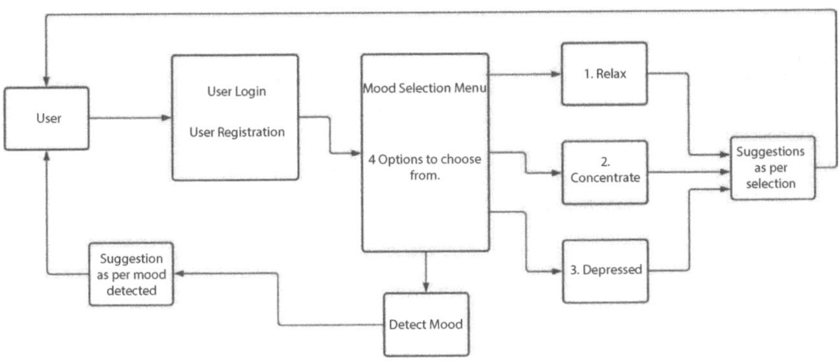

Figure 6.4 Working diagram for the system.

the application will show a list of binaural beats specific to that emotion to enhance the user's mood. The user will be then routed to YouTube, where the user can listen to the specific playlist [16].

6.6.2 The Application's Use Case Diagram

- Figure 6.5 shows how the user can interact with the application and get the best output. The user has to sign in first into the application where the credentials of the user are identified. If the credentials match up in the system, the user is redirected to the "home screen." If the credentials do not match up, then an error will be shown to the user, and the user will be redirected to the login screen. After signing into the application, the user is directed to the home screen, where the user can select from the four options. The user can directly select the mood and get suggestions. The other option is that the user can get his/her mood detected and

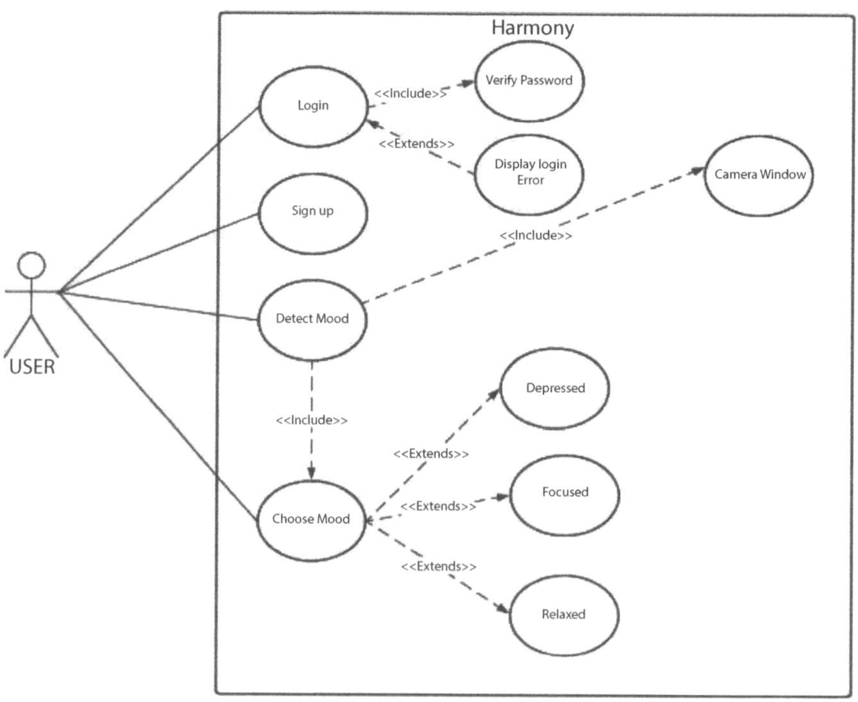

Figure 6.5 The application's use case diagram.

then get the suggestions. If the user selects the mood-detection option, the user is directed to the camera screen. Here, using the first algorithm, i.e. the Viola–Jones algorithm, the face of the user is detected. Then, using the second algorithm, i.e. the CNN, the emotion of the user is recognized. After the emotion is recognized, the user is given suggestions on suitable binaural beats playlist based on the emotion. The playlist is directly linked to the YouTube page, so that the user can directly listen to the playlist. Results are depicted in Figures 6.6, 6.7, 6.8, 6.9, 6.10.

6.7 Results and Discussion

Sign in/Log in:
It will help the user to login / sign in to the application.

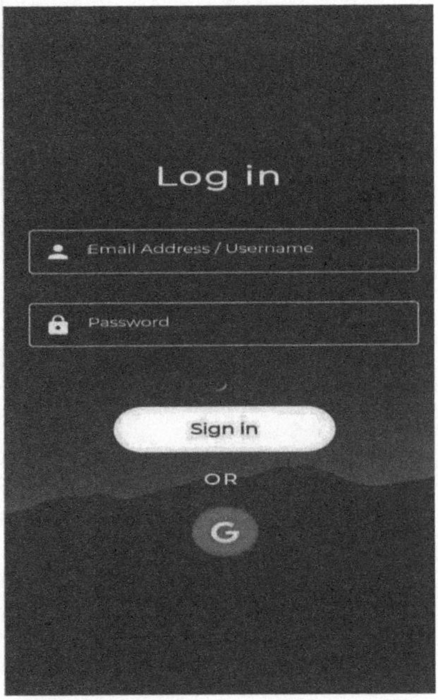

Figure 6.6 Log in page.

Home Screen:
Here, the user can choose from three options, where the user can directly get the suggestions, or , using the 4th option, the user's mood can be detected and then the user can get the suggestions based on the mood detected.

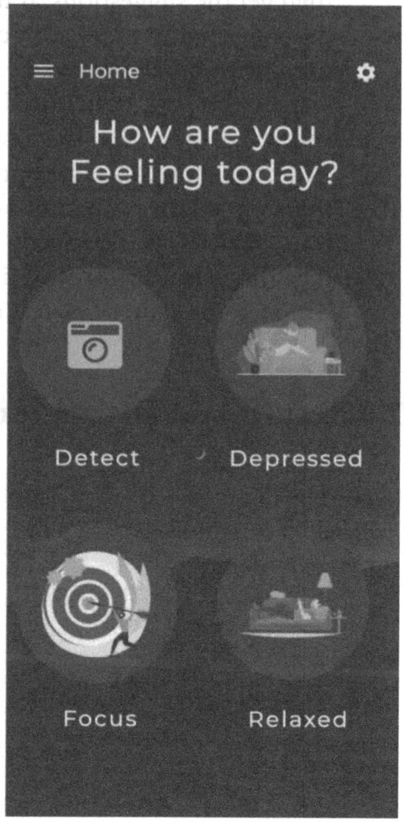

Figure 6.7 Emotion detection of the user.

Suggestion Screen: It suggests binaural beats for relaxation.	

Figure 6.8 Music suggestions for the user.

Emotion Detection:

Here, the user can directly get his/her mood detected using the front camera, and then get suggestions based on the mood detected.

Figure 6.9 User expression detection.

Suggestion Screen After Detection:

- ❖ The user above was smiling; hence, the "Happy Emotion" was triggered. Therefore, the user is suggested binaural beats based on the emotion detected.
- ❖ Similarly, the user can get suggestions based on different moods as well.

Figure 6.10 Music is played based on the emotion.

6.8 Limitations and Future Scope

Through this application, we can detect emotions from a user's facial expressions. While a person might be sad, but shows happy facial expressions, the application will recognize that the person is happy. Thus, it will show suggestions of happy binaural beats. In the future, if we can detect the true emotions of a person using the camera, the improved model can be attached to the application and can be used to detect emotions more accurately.

6.9 Summary and Conclusion

In this research, we describe an emotion-recognition method, from beginning to end. On various benchmarks, we show that our novel deep architecture, along with our massive, in-house acquired data, outperforms competitive commercial and academic algorithms. We summarize that CNN for emotion detection and the Viola–Jones algorithm for face recognition are the best algorithms for this purpose when compared to other algorithms. This application will help with predicting people's moods and can also help make them feel better.

References

1. Sarah Guido and Andreas C. Muller "*Introduction to Machine Learning with Python: A Guide for Data Scientists*".
2. Stuart Russell and Peter Norvig, "*Artificial Intelligence: A modern Approach*", 3rd Edition, Prentice Hall.
3. Elaine Rich and Kevin Knight, "*Artificial Intelligence*", Tata McGraw Hill.
4. Trivedi M.C,"*A Classical Approach to Artificial Intelligence*", Khanna Publishing House, Delhi.
5. Shetty, Shruthi H., *et al.* "Supervised Machine Learning: Algorithms and Applications." *Fundamentals and Methods of Machine and Deep Learning: Algorithms, Tools and Applications* (2022): 1-16.
6. Soans, R. V., Hegde, A., Singh, C., & Kumar, A. (2017, October). Object tracking robot using adaptive color thresholding. In *2017 2nd international conference on communication and electronics systems (ICCES)* (pp. 790-793). IEEE.
7. Gatti, Rathishchandra R., *et al.*, eds. *Self-Powered Cyber Physical Systems.* John Wiley & Sons, 2023.
8. Badiger, Manjunatha, and Jose Alex Mathew. "Tomato plant leaf disease segmentation and multiclass disease detection using hybrid optimization enabled deep learning." *J. Biotechnol.* 374 (2023): 101-113.
9. Badiger, Manjunatha, *et al.* "Leaf and skin disease detection using image processing." *Glob. Transit. Proc.*, 3.1 (2022): 272-278.
10. Manjunatha Badiger, Dr Jose Alex Mathew. "A Survey on Low Power SBCs for Optimized Implementation of Deep Learning Models." *J. Surv. Fish. Sci.* 10.2S (2023): 4173-4189.
11. Prudhvi Raj Dachapally, School of Informatics and Computing, Indiana University:"Facial Emotion Detection Using Convolutional Neural Networks and Representational Auto encoder Units"
12. Rai Bellipady, S., Shetty, S.M. & Airbail, H. Quality of Information Analysis in WSN: An Application in BASN. *J. Ambient Intell. Human Comput.* (2019). https://doi.org/10.1007/s12652-019-01362-7
13. Prasad, B. Krishna, and Keerthana Rajeev. "A performance evaluation of convolution neural networks for kinship discernment: An application in digital forensics." *Intell. Decis. Technol.* 16.2 (2022): 379-386.

14. Airbail, Harisha, *et al.* "Deep learning-based approach for malware classification." *Int. J. Intell. Def. Support Syst.* 6.2 (2021): 61-80.
15. Harisha, A., Amarnath Mishra, and Chandra Singh, eds. *Advancements in Cybercrime Investigation and Digital Forensics.* CRC Press, 2023.
16. Karkera, T., & Singh, C., Autonomous bot using machine learning and computer vision. *SN Comput. Sci.*, 2, 4, 251, 2021.

7

Power Quality Events Classification Using Digital Signal Processing and Machine Learning Techniques

E. Fantin Irudaya Raj[1]* and M. Balaji[2]

[1]*Department of Electrical and Electronics Engineering, Dr. Sivanthi Aditanar College of Engineering, Tamil Nadu, India*
[2]*Sri Sivasubramania Nadar College of Engineering, Tamil Nadu, India*

Abstract
Good quality electrical power is mandatory for the correct operation of electronic equipment in industrial, commercial, and residential appliances. Due to the arrival of many electronic devices and many nonlinear loads connected to the consumer side, faults occur in the power system, and lightning-like natural events cause various power quality (PQ) problems in the modern power system. It will lead to various issues like the malfunction of the protective devices in the substation, permanent damage to the sensitive equipment in home appliances, and faulty operation of medical equipment. There are numerous techniques adopted for mitigating these power quality problems by using flexible alternating current transmission system (FACTS) devices, active and passive filters, adopting various standards prescribed by IEEE, IEC, ANSI, etc. in manufacturing industries. The first step in solving the PQ issue is identifying and classifying PQ events more accurately. For this purpose, the digital signal processing technique and machine learning methodologies are adopted. In this process, a feature extraction technique is employed to extract features from the raw PQ data; feature selection and feature optimization are also carried out. Finally, the machine learning tool acts as a classifier that classifies the actual PQ event present in the power system. In this, the feature extraction technique must be more accurate for the efficient functioning of the classifier. We discussed feature extraction based on the wavelet transformation (WT) in the present work in more detail. Feature selection and optimization play

**Corresponding author*: fantinraj@gmail.com

Chandra Singh, Rathishchandra R. Gatti, K.V.S.S.S.S. Sairam, Manjunatha Badiger, Naveen Kumar S. and Varun Saxena (eds.) Modeling and Optimization of Signals Using Machine Learning Techniques, (115–144) © 2024 Scrivener Publishing LLC

a vital role in optimizing the data sets to the classifier after the feature extraction. In this regard, we are using a genetic algorithm (GA) for this purpose. Then, after machine learning, models like the probabilistic neural network (PNN), back propagation neural (BPN) network, and support vector machine (SVM) are used as a classifier to classify the various PQ events. The performance parameters to identify the efficiency of these classifiers are also discussed. Finally, based on the performance parameters, the most suitable classifier for PQ event classification is determined.

Keywords: Power quality (PQ), machine learning, probabilistic neural network (PNN), digital signal processing, wavelet transformation (WT), back propagation neural (BPN) network, support vector machine (SVM)

7.1 Introduction

Any dysfunctional occurrence on a power system that manifests itself as a deviation in a current or voltage that interferes with the regular electrical or electronic equipment operation is referred to as a power quality (PQ) problem [1]. Harmonics, voltage sag and swell, overvoltage, undervoltage, voltage flicker, and transients are a few critical PQ problems that generally arise in the power system. The PQ problems affect the various sensitive equipment on the consumer side and supply side. The primary sources behind the PQ problems are natural events like lightning, faults occurring on the supply side, large nonlinear loads connected on the consumer side, and power electronic devices [2].

Four significant concerns arise due to these PQ problems. They are the following: (1) Modern equipment, such as power electronic devices and embedded-based controllers, are more sensitive to changes in power quality than older equipment. (2) The overall system efficiency and good power factor maintenance are increased by using filter arrangements, using shunt capacitors for power factor correction, controlling the voltage of power electronic devices, and adopting adjustable-speed motor drives in the power system. As a result, the harmonic levels on power systems are increasing, which will have a long-term influence on power system operations. (3) Consumers are more aware of the difficulties with power quality. Consumers are becoming more aware of switching transients, power outages, swells, and sags, and they are demanding providers to enhance the quality of power supplied. (4) In today's world, so many components are linked together in a network. Because unified processes are used, every element in the network is more critical. A small failure of a single component has far more significant effects [3].

There are numerous techniques and devices adopted to mitigate these PQ problems. Many international standard organizations, like the International Electrotechnical Commission (IEC), the American National Standard Institute (ANSI), the National Electricity Regulator (NER), and the Institute of Electrical and Electronics Engineers (IEEE), prescribe various standards and define PQ problems precisely [4]. The basics behind identifying PQ problems play a vital role in eliminating those problems from the system. The advancement in digital signal processing techniques and machine learning (ML) approaches provide a new opportunity to identify and classify PQ problems more efficiently. The precise identification of PQ problems enables devices like the flexible alternating current transmission system (FACTS) to act accordingly and eradicate those problems in the power system [5].

The rest of the chapter is organized as follows: Section 7.2 discusses the methodology adopted for PQ event classification in the present work. Various PQ problems and their important sources and causes are discussed in Section 7.3. Section 7.4 explains feature extraction through wavelet transformation (WT) techniques from a distorted voltage signal of a power system. Feature selection and optimization using a genetic algorithm (GA) are discussed in Section 7.5. The classification of PQ problems based on the features extracted from the voltage signal using machine learning techniques like the support vector machine (SVM), the back propagation neural (BPN) network, and the probabilistic neural network (PNN) are covered in Section 7.6. This section further discusses various performance parameters of the classifier. Finally, Section 7.7 provides the summary and conclusion of the present work.

7.2 Methodology for the Identification of PQ Events

Numerous works are listed in the literature to identify PQ problems that occur in the power system accurately and precisely. In the current work, we adopted the digital signal processing methodology for feature extraction and machine learning techniques to classify the PQ problems. Figure 7.1 shows the flowchart of the methodology adopted to classify PQ events in the present work.

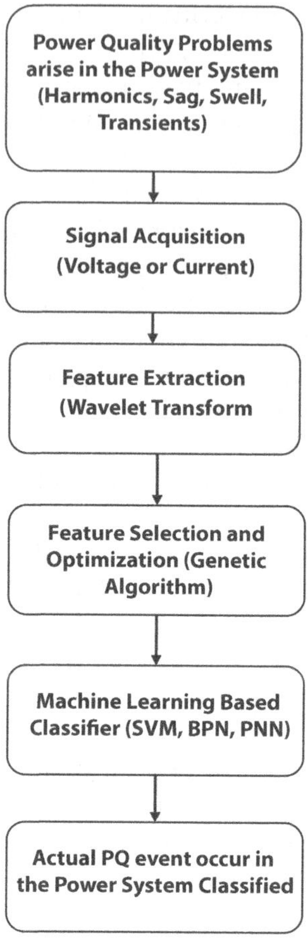

Figure 7.1 Methodology of the proposed work.

7.3 Power Quality Problems Arising in the Modern Power System

As mentioned in the introduction, PQ problems create more problems for the consumer side's sensitive equipment. The nonlinear loads, supply-side faults, and power electronics devices are the major sources of PQ problems. In this section, we are going to see the definitions and various sources of important PQ problems. The PQ problems can be classified as short-duration voltage variation, long-duration voltage variation, transients, and harmonics. Sag and swell fall under short-duration voltage variation. It may happen with either current or voltage waveforms.

Overvoltage and undervoltage are examples of the long-duration voltage variation. Transients are classified as impulsive transient and oscillatory transient. All these voltage sags are present in AC and DC systems, but harmonics are present in the AC system alone; it can be further differentiated into inter-harmonics.

7.3.1 Sag

Sag is a type of short-duration voltage variation. It prevails in the system in less than 1 minute. IEEE defines it as the decrement between 0.1 and 0.9 per unit in the root mean square (RMS) value of the current or voltage for the duration of 10 milliseconds to 1 minute. The same phenomenon is defined as "dip" by IEC. Voltage sags are mainly due to system faults. However, it can also be due to the energization of heavy loads and the starting of huge motors on the consumer side [6]. The RMS waveform of the voltage sag event is shown in Figure 7.2.

7.3.2 Swell

Swell is also another type of short-duration voltage variation. It can also prevail in the system in less than 1 minute. IEEE defines it as the increment between 1.1 and 1.9 per unit in the RMS value of the current or voltage for the duration of 10 milliseconds to less than 1 minute. The primary source behind the voltage swell is supply-side faults, switching off large loads and motors on the consumer side, and de-energization of capacitor banks [7]. The RMS value of the voltage swell is shown in Figure 7.3.

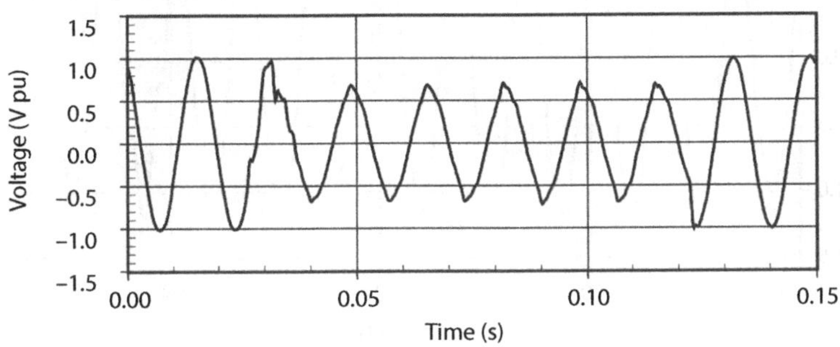

Figure 7.2 RMS waveform of the voltage sag event.

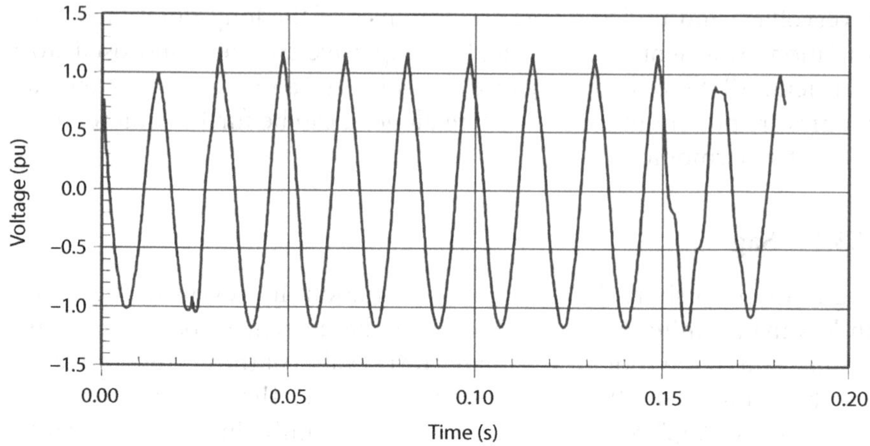

Figure 7.3 RMS waveform for the voltage swell event.

7.3.3 Overvoltage

Overvoltage is a type of long-duration voltage variation. It will prevail in the system for more than 1 minute. IEEE defines it as the increment of RMS voltage in the range greater than 1.1 per unit for a duration of more than 1 minute. It usually happens due to the energization of the capacitor bank and switching off large loads [8]. Figure 7.4 shows the RMS waveform of the overvoltage event.

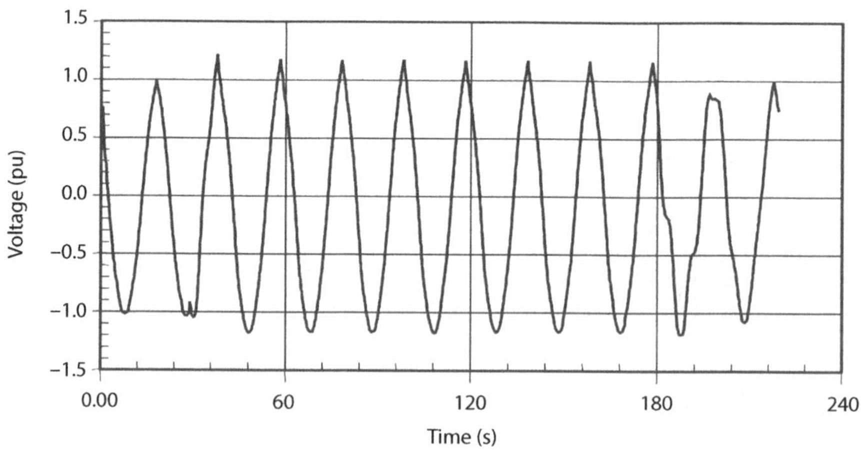

Figure 7.4 RMS waveform of the overvoltage event.

7.3.4 Undervoltage

Undervoltage is another type of long-duration voltage variation. It can also prevail in the system for more than 1 minute. IEEE defines it as the decrement of RMS voltage in the range less than 0.9 per unit for the duration of more than 1 minute. It happens mainly due to turning on large loads or big motors [9]. It also may occur due to the de-energization of the capacitor banks. The overloaded circuits also result in undervoltage. If undervoltage persists for a very long time, then it may lead to sustained interruption. Figure 7.5 shows the RMS waveform of the undervoltage event in the power system.

7.3.5 Impulsive Transient

Typically, impulsive transient overvoltage occurs due to the unavoidable effects of lightning or capacitor switching operations in the power system. It is a transient response of a system or circuit during a sudden state transition. It is defined, as per IEEE, as the sudden and immediate non-power frequency alteration in the steady-state condition of nominal voltage [10]. This type of transient is unidirectional in nature. Further, it initiates and induces oscillatory transients in the power system. Figure 7.6 depicts the impulsive transient overvoltage that occurs due to lightning.

7.3.6 Oscillatory Transient

As per IEEE, the oscillatory transient is well-defined as the sudden non-power frequency alteration in the steady-state condition of the voltage or current that is bidirectional in nature. Usually, the oscillatory transient is a power system response to the impulsive transient [11]. It may also occur due

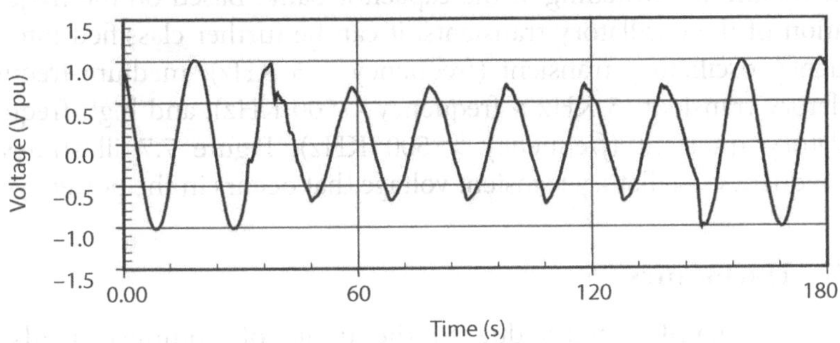

Figure 7.5 RMS waveform of the undervoltage event.

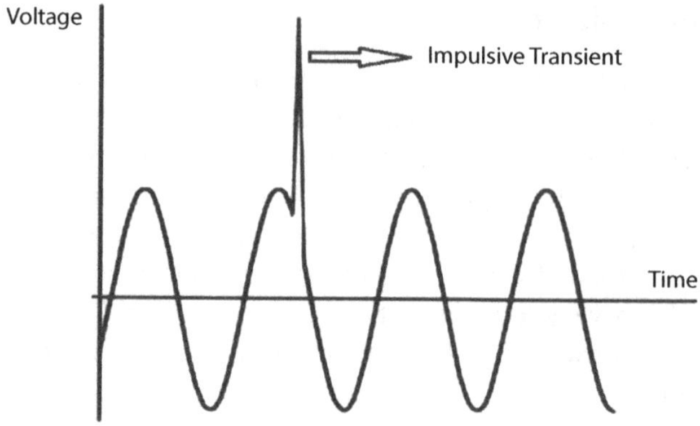

Figure 7.6 Impulsive transient overvoltage due to lightning.

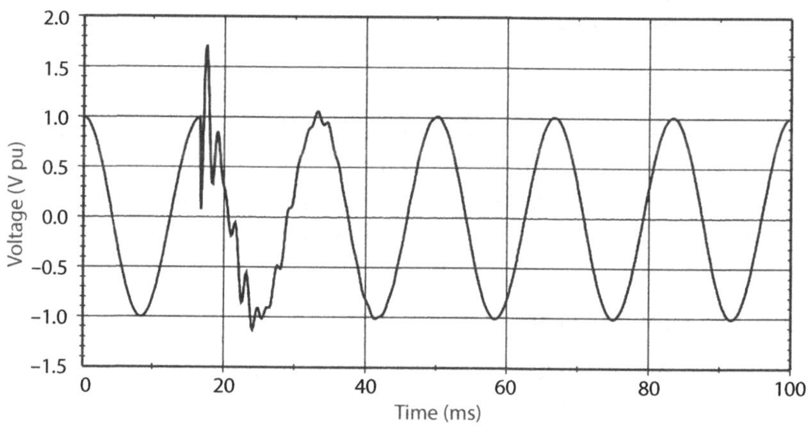

Figure 7.7 Low-frequency oscillatory transient voltage.

to the on and off switching of the capacitor bank. Based on the frequency variation of the oscillatory transients, it can be further classified into low-frequency oscillatory transient (frequency < 5 KHz), medium-frequency oscillatory transient (5 KHz > frequency < 500 KHz), and high-frequency oscillatory transient (frequency > 500 KHz). Figure 7.7 illustrates the low-frequency oscillatory transient voltage that occurs in the power system.

7.3.7 Harmonics

Harmonics mainly occurs due to the usage of nonlinear loads and solid-state drives on the consumer side. As per IEEE, it is defined as the

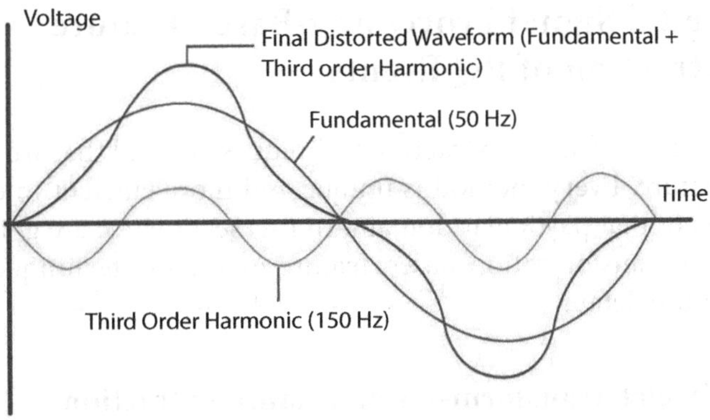

Figure 7.8 Harmonics and the resultant waveform.

sinusoidal currents or voltages having frequencies that are integer multiples of the fundamental frequency (50 Hz or 60 Hz). Due to the presence of harmonics in the power system, overheating issues will arise. Furthermore, in an induction motor, cogging torque will occur, and the transformer core saturation will also happen due to harmonics. IEEE standard 519–1992 explains harmonics and the various ways to quantify harmonics in the power system [12]. It also provides detailed guidelines to the manufacturer to manufacture the devices with proper shielding and filtering arrangements to eradicate the harmonics from the system. The harmonics that do not have an integer multiple of the fundamental frequency are called inter-harmonics. Figure 7.8 shows the fundamental component, the third-order harmonic component, and the resultant distorted voltage waveforms.

From the above subsections, we can easily understand the basic concepts of important PQ problems that normally occur in the modern power system in detail. We can easily figure out that these problems differ from the other with minute variations with respect to time. Some separate methodologies and devices are used to mitigate these problems individually, so accurate prediction and classification of PQ events are mandatory for the precise operation of those devices. In this context, digital signal processing tools and machine learning approaches are employed to identify and classify the exact PQ event present in the system. The following section expresses the feature extraction techniques used to extract useful information from the distorted signal in the power system that leads to PQ problems.

7.4 Digital Signal Processing-Based Feature Extraction of PQ Events

There are so many feature extraction techniques [13–16] that are explained in the literature. Every method is unique and more efficient in extracting the features of that particular domain. In this section, we are going to discuss wavelet transformation-based feature extraction techniques for PQ events in more detail.

7.4.1 Wavelet Transform-Based Feature Extraction

In signal analysis, the continuous wavelet transform (CWT) is particularly effective. It is used in a variety of engineering domains because of its ability to investigate local signal discontinuities. The key advantage of this approach is that it has a variable window size, which is comprehensive for slow frequencies and narrow for rapid ones. As a result, an excellent time-frequency resolution is achieved throughout all frequency ranges [17, 18]. Because the majority of power system disturbances are non-stationary transients, wavelets are ideal for PQ research. Typically, a wavelet, $\psi \in L^2(R)$, is a function with an average of zero.

$$\int_{-\infty}^{+\infty} \psi(t)dt = 0 \tag{7.1}$$

The CWT of a signal $x(t)$ is defined as,

$$CWT_{\psi}x(a,b) = \frac{1}{\sqrt{a}} \int_{-\infty}^{+\infty} x(t)\psi^{*}\left(\frac{t-b}{a}\right)dt \tag{7.2}$$

Where $\psi(t)$ is a mother wavelet, the * signifies the complex conjugate, "a" is a scaling parameter, and "b" is a translation parameter. The length of the wavelet and oscillation frequency are determined by the scaling parameter "a", while the shifting position is determined by the translation parameter "b". Through sampling, the continuous wavelet function is converted to a discrete function. This will be used in many practical applications. In Equation 7.2, the scaling and translation parameters are replaced by a_0^m and $nb_0a_0^m$, respectively.

$$\psi_{m,n}(t) = a_0^{-\frac{m}{2}} \psi(a_0^{-m} t - nb_0) \tag{7.3}$$

Where "m" and "n" indicate the frequency and the time localization, respectively. The basis for multiresolution analysis and dyadic orthonormal WT is provided by choosing $a_0 = 2$ and $b_0 = 1$.

7.4.2 Multiresolution Analysis

The signal is dismantled and rebuilt at various resolution levels using scaling functions and wavelets as building blocks. This method's main purpose is to construct signal representations at different resolution levels. Each level of this analysis has two filters: a high pass filter (HPF) and a low pass filter (LPF) [19, 20]. Function space can be divided into a series of low-frequency and high orthogonal frequency spaces using a set of orthogonal wavelet functions $\psi(t)$ and scaling operations $\phi(t)$. The decomposed signal's $[f(t)$'s$]$ time-frequency picture is a scaled and translated version of the wavelet utilized in the multiresolution analysis. The original signal $f(t)$ can be decomposed into the following components:

$$f(t) = \sum_k c_j(k)\phi_{jk}(t) + \sum_{j=1}^{k} \sum_k d_j(k)\psi_{jk}(t) \tag{7.4}$$

Where j is known as the level of the wavelet decomposition. The weighted sequences of the projection of space of $f(t)$ are c_j and d_j. Figure 7.9 shows the decomposition of four levels using this analysis.

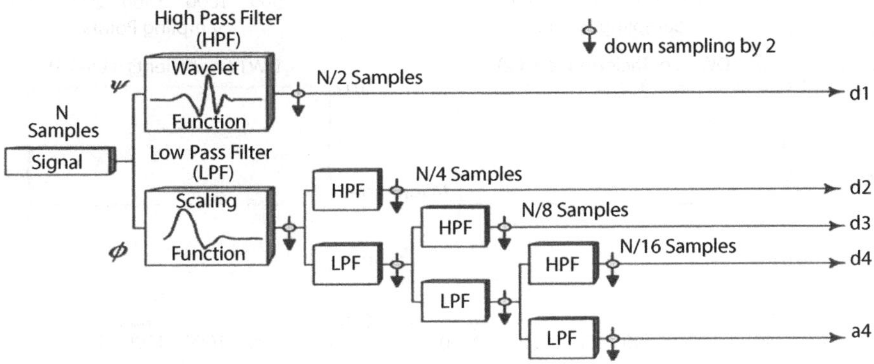

Figure 7.9 Multiresolution decomposition of the signal.

7.4.3 Future Generation and Extraction

Extraction of futures is crucial in signal processing procedures since training a network can be challenging with rough and big data sets. Discrete wavelet transform (DWT) is a fantastic strategy for enhancing disruption and reducing data to coefficients that correlate to the signal's closeness to the wavelet on a specific region and time scale. This study considers four different types of PQ problems. Harmonic distortions, transients, voltage sag, and voltage swell are the four types. In MATLAB, the PQ signals for these four different classes are created using customizable models with various parametric values. The PQ distorted signals were sampled at a sampling frequency of 12.8 kHz for 10 cycles in the range of 256 points per cycle.

The experimented disturbance signals were analyzed using the fourth-order Daubechies wavelet (db4), which is more similar to power disturbance signals. To make sure that PQ disturbance features are retrieved in low- and high-frequency spectra, with 13 levels of multiresolution analysis, the sampled distorted PQ signals (voltage or current) were decomposed.

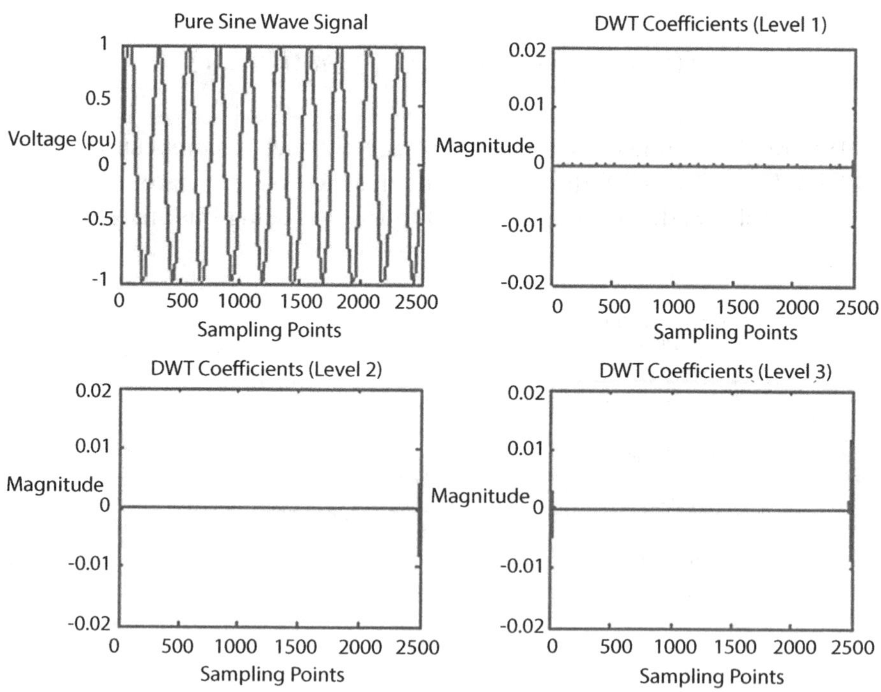

Figure 7.10 Plot of DWT coefficients of a pure sine wave.

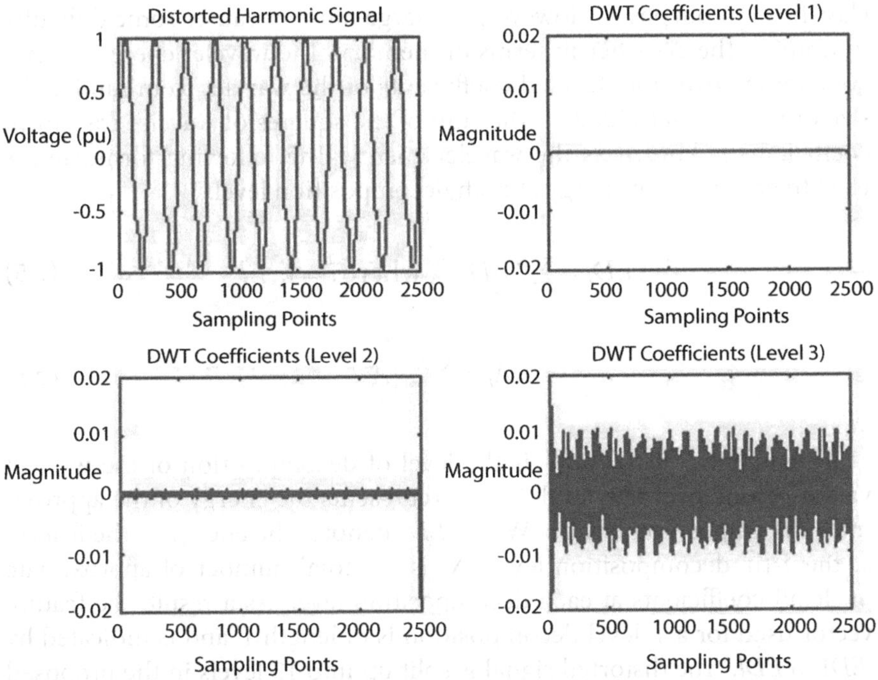

Figure 7.11 Plot of DWT coefficients of a harmonics waveform.

For each disturbance signal, the 13-level decomposition yielded 13 detail coefficients and one approximation. Therefore, it is sufficient to use only the 13-level decomposition's detail coefficients (D1 to D13). Thirteen feature sets (a single set has 2560 coefficients) from a single PQ distorted signal are obtained [21]. All these feature sets with their coefficients are used for classification; however, the number of these large datasets with detail coefficients is too large for training the classifier. As a result, the size of the data set must be reduced. Figure 7.10 shows a pure sine wave plot with a plot of levels 1, 2, and 3 of DWT coefficients. The plot of the distorted harmonic signal with the plot of levels 1, 2, and 3 of DWT coefficients is shown in Figure 7.11. Similarly, other PQ problems also had different levels of DWT coefficients.

7.4.4 Wavelet Energy

As we saw in the previous subsection, 13 different feature sets are acquired for only one PQ distorted signal. Each set has 2560 coefficients. This huge number of feature sets and coefficients are further used for classification purposes. However, this large number causes some difficulty in training the classifier in terms of memory. The wavelet energy criteria are adopted over the detailed coefficients of the wavelet to minimize the size of the data set [22, 23]. The sum of the squares of wavelet transform coefficients is known as the wavelet energy. The following formulae are used to compute the energy at each decomposition level.

$$ED_i = \sum_{j=1}^{N} \left|D_{ij}\right|^2, \text{ where } i = 1,\ldots l \tag{7.5}$$

$$EA_l = \sum_{j=1}^{N} \left|A_{lj}\right|^2 \tag{7.6}$$

In equation 7.5, $i = 1,\ldots l$ is the level of decomposition of the wavelet varying from level 1 to level l. EA_l represents the energy of the approximate decomposition level l. While ED_i denotes the energy of the feature at the i-th decomposition level. N is the total number of approximate or detail coefficients at each decomposition level. As a result, the feature vector used for a l level decomposition is of length l and is indicated by $ED1$ to EDl. The distorted signal is split up into 13 levels in the proposed approach, resulting in a feature vector dimension of 13. The wavelet energy distributions of the PQ problems, such as voltage sag and harmonics, are shown in Figures 7.12 and 7.13, respectively. The remaining PQ problems will also have different wavelet energy distributions. The differentiation of features is based on the differences in these energy patterns. These distinguishing characteristics are critical to the classifier's performance.

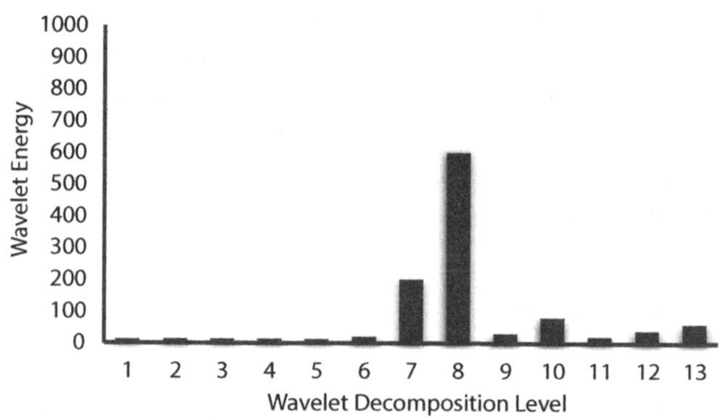

Figure 7.12 Wavelet energy distribution level of a voltage sag.

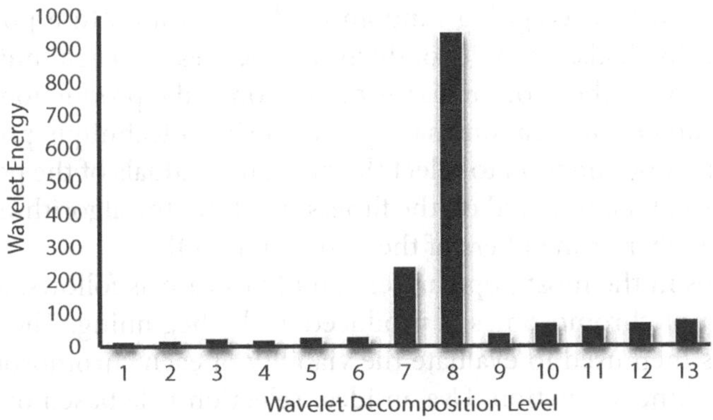

Figure 7.13 Wavelet energy distribution level of voltage harmonics.

7.5 Feature Selection and Optimization

Feature selection and optimization are other crucial steps after feature extraction. In feature extraction, a greater number of features are getting extracted from the PQ event. As a result, an enormous amount of data is directly given as an input to the classifier, and there might be a chance for misclassification to occur [24, 25]. Feature selection and optimization steps are highly needed to avoid this situation. In this step, a various number of features that are extracted are converted into a minimum and required level of precise datasets. Feature selection and optimization can be carried out by any optimization algorithm. There are so many optimization algorithms proposed in the literature [26–29]. However, nature-inspired algorithms are more efficient in this process. There are so many naturally inspired algorithms available like GA, particle swarm optimization, ant colony optimization, and whale optimization. In the present work, we are using GA for this purpose.

7.5.1 Genetic Algorithm

A GA is a search-based technique used in machine learning to solve optimization problems. This algorithm is significant because it solves tough issues that would otherwise take an enormous amount of time to solve. A GA can estimate known solutions and simulate the evolution of complex systems. It has a lot of parallel processing capabilities and does not necessitate the use of derived data. It can solve issues involving continuous functions, multi-objective problems, and discrete functions, among others.

A GA begins by assigning random or default values to a population/group of individuals. Then, it conducts a fitness test on each individual in the population. It then continues the evaluation and reproduction until the requisite number of iterations have passed, using a technique provided in the reproduction function to select the fittest individuals of the population to reproduce. Finally, based on the fitness function, the algorithm presents the best member or members of the population [30].

The steps in the most popular version of GAs are as follows: a random population of chromosomes is produced at the beginning. The objective function is then used to evaluate the viability of each chromosome using a specified fitness function. Use an ideal selection rule based on the idea that the better one's fitness is more likely the one to be chosen. This fitness evaluation stage permits maintaining the best chromosomes and rejecting the poorest. The next stage is to reproduce the population after the selection procedure is accomplished. Genetic operators such as crossover and mutation are used to achieve this. Finally, until a user-defined convergence condition is met, the process is repeated [31].

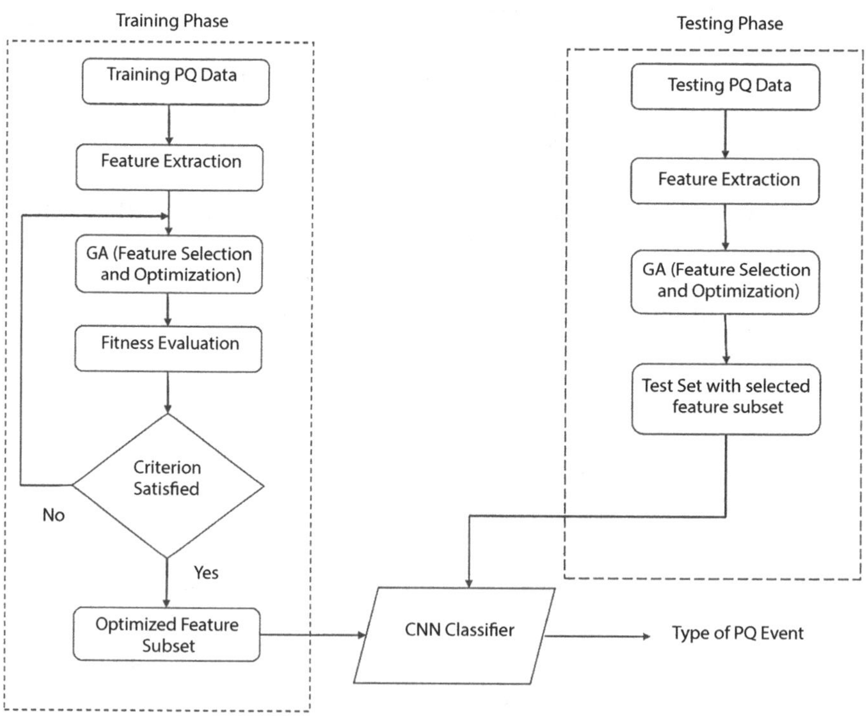

Figure 7.14 GA-based feature optimization and selection for SVM classifier.

Figure 7.14 shows the flowchart of GA-based feature selection and optimization for SVM. Similarly, the GA can be used along with other classifiers like the probabilistic neural network (PNN) and the back propagation neural (BPN) network to classify the PQ events more precisely.

7.6 Machine Learning-Based Classification of PQ Disturbances

Machine learning (ML) has received a lot of attention recently. The fact that it provides a unified framework for incorporating intelligent decision-making into various fields is one of the key driving elements behind the ML boom [32–35]. It is a field of research that looks at how to turn empirical data into useful models using computational techniques. This field is developed from artificial intelligence (AI) and traditional statistics. Many researchers defined ML as a subset of AI. These ML approaches can be more useful in the classification of PQ events. It will provide a more precise classification compared with any other traditional method. A few of the important ML approaches used in PQ event classification will be discussed in this section.

7.6.1 Support Vector Machine Classifier

Both linear and nonlinear inputs can be classified using the SVM approach. It starts by mapping each data point through an n-dimensional attribute vector, with 'n' denoting the total number of attributes. The hyperplane that divides these data components into two groups is then found, with the minimum separation for both categories maximized and classification errors minimized [36]. The length of both the selection hyperplane and the class's neighboring occurrence defines the minimum closeness for a grouped class. Each feature vector is initially plotted as a point in an n-dimensional space, with each feature becoming the value of a specific coordinate. First, we must find the hyperplane that separates the two categories by the greatest amount of distance to finish the categorization. Figure 7.15 shows an SVM classifier in a basic workflow with a hyperplane.

In nonlinearly separable scenarios, the SVM uses the kernel function, $K(x_i,x_j)$, and nonlinearly moves the training points to a high-dimensional spatial domain, where linear separation may be possible. The sigmoid, polynomial, linear, and Gaussian radial basis function (GRBF) kernel functions of SVMs are as follows:

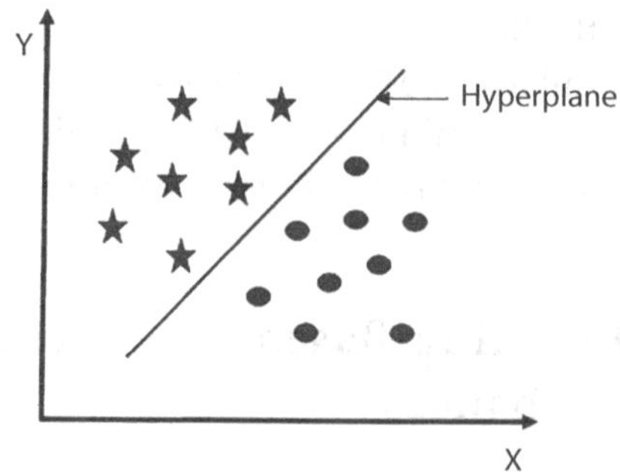

Figure 7.15 The SVM workflow (the hyperplane maximizes the separation of two different classes).

$$K(x_i, x_j) = e^{\left(\frac{-\|x_i - x_j\|^2}{2\sigma^2}\right)} \tag{7.7}$$

Where σ is the Gaussian radial basis kernel function parameter. After this function is adopted, the quadratic programming problem is also getting changed. The decision function is getting changed as shown in Equation 7.8 after the training.

$$f(x) = sgn\left(\sum_{i=1}^{l} y\alpha^{*} k(x, x_i) + b^{*}\right) \tag{7.8}$$

The kernel parameter and the penalty parameter, often known as hyper-parameters, can be used to adjust the performance of the SVM. These variables have an impact on the total number of support vectors and the SVM's maximizing margin.

7.6.2 Artificial Neural Network Classifier

An artificial neural network (ANN) is another type of machine learning algorithm influenced by the working of the neurons in the human brain. Many axon connections connect nerve cells in the biological brain, forming a graph-like pattern. These connections can be rearranged to help in storage, processing, and data adaptation. These approaches could also be visualized as a collection of distributed nodes. Because of the interconnection, one node's output becomes the input for future processing on another node. Nodes are frequently grouped into a multilayered array based on the

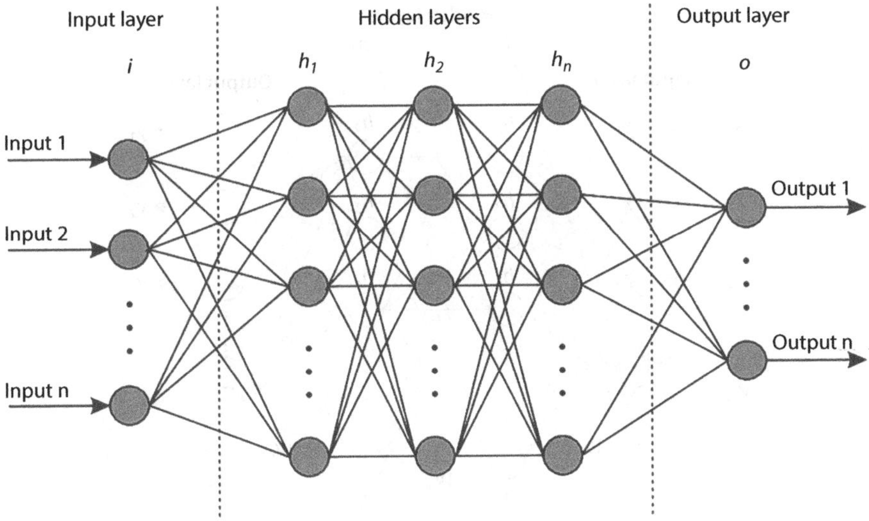

Figure 7.16 Artificial neural network with three hidden layers.

changes they perform. In addition to the source and destination nodes, an ANN design can comprise one or more hidden layers. Nodes and edges are given weights to adjust communication signal quality, which can then be enhanced or lowered through repeated training. These ANNs can make quick and accurate decisions based on the PQ test data and then change the edge values, node, and matrix. Figure 7.16 depicts an ANN with many interconnected nodes with three hidden layers.

There are different types of ANN configurations adopted in the PQ event classification process. They are the BPN network and the PNN.

7.6.2.1 Back-Propagation Neural Network

To accurately classify PQ events, a two-layer feed-forward ANN with a BPN network for error correction and learning is used. Figure 7.17 shows the BPN network adopted for PQ event classification. After completing numerous different trials, the structure of the BPN network and the training parameters were chosen to achieve the optimum performance, the number and size of the hidden layers, the kind of activation functions, and the learning rate [37]. The number of hidden neurons is set to 10. In the hidden-layer neurons, the employed transfer function is "tan-sigmoid."

Only one neuron is used in the output layer to determine the disturbance class. The output layer's transfer function is "purelin" and the output

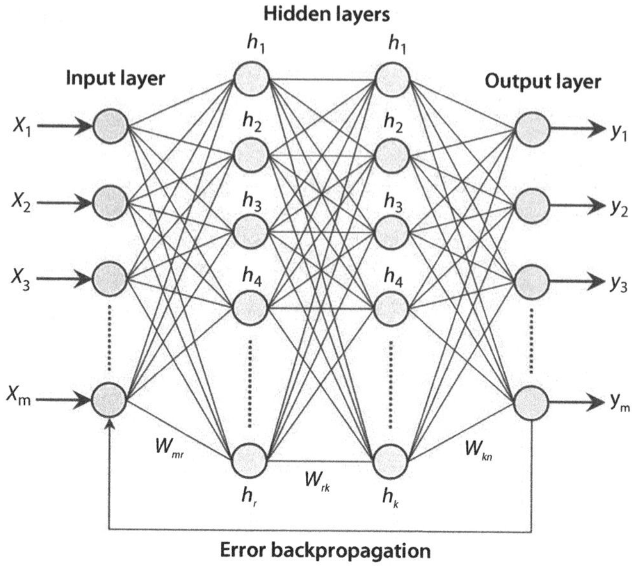

Figure 7.17 Back propagation neural network.

must specify the classes from 1 to 10. To increase the speed of the training process, the training function, Levinberg-Marquardt ("TrainLM"), is utilized. A back-propagation algorithm is used to train the network. The error measure is calculated from Equation 7.9, and the weights are updated as in Equation 7.10.

$$E = \sum_{j=1}^{Q} \sum_{k=1}^{M} (d_{kj} - x_{kj})^2 \tag{7.9}$$

$$\Delta w_{k+1} = -\eta \nabla wE + \alpha \Delta wK \tag{7.10}$$

Where α is the momentum constant, Q is the total number of training samples, d_{kj} is the target value, M is the number of output neurons, η is the learning rate, and ∇wE is the gradient of E with respect to w.

The gradient-descent technique was employed in a group mode. The learning rate has a significant impact on a gradient-descent method's performance. The training might well fluctuate backward and forward if the η was too high. Similarly, it will take an extended time to achieve convergence if the η was too low. Therefore, an adaptive learning rate is employed, which aims to retain the step size as large as feasible while avoiding oscillation to address this issue. The learning rate is adjusted in response to

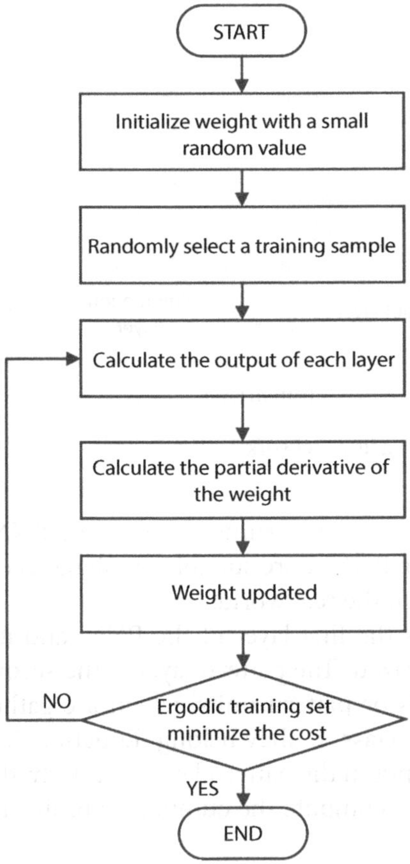

Figure 7.18 Process flow of back-propagation training algorithm.

the local error surface's complexity. This study employed a learning rate of 0.1. The flowchart of the back-propagation training algorithm is shown in Figure 7.18.

7.6.2.2 Probabilistic Neural Network

In classification applications, PNNs provide a scalable alternative to traditional back-propagation neural networks, eliminating the requirement for enormous forward and backward calculations associated with conventional neural networks [38]. It can also be adopted in PQ event classification. Moreover, PNNs can work with smaller sets of training data. Figure 7.19 shows the simple PNN used for classification. Compared with

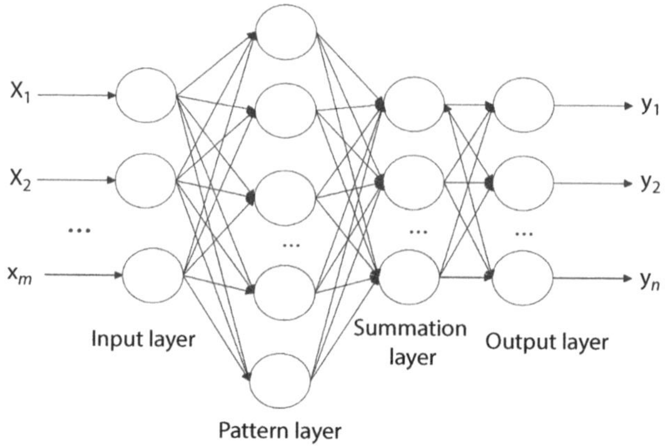

Figure 7.19 Probabilistic neural network.

other ANN approaches, the learning speed of the PNN is very high. This quality will make the PNN more suitable for signal classification and fault diagnosis problems in the real world.

The input layer is the first layer of the PNN, and it receives the input vectors to be categorized. The pattern layer is the second layer of PNN. It is also called kernels or pattern units, which are gathered in M groups depending upon the class k_i they belong. Equation 7.11 is used to determine each unit's output in the pattern layer. Similarly, the summation layer uses Equation 7.12 to compute the cumulative probability of belonging to a particular class.

$$\phi_{ij}(x) = \frac{1}{(2\pi)^d \sigma^d} e^{\left[-\frac{(x-x_{ij})^T (x-x_{ij})}{2\sigma^2}\right]} \tag{7.11}$$

Where $i = 1,\ldots,N$ and $j = 1,\ldots,M$.

M is the total number of classes in the training samples. N is the number of neurons in the kernel. d is the dimension of data in the sample space, σ is the smooth parameter, and x_{ij} is jth mode ith implicit vector.

$$f_{iN_i}(x) = \frac{1}{N_i} \Sigma_{j=1}^{N_i} \phi_{ij}(x) \tag{7.12}$$

Each class in the summation layer corresponds to a unit, and summation layer units are only connected to units in the pattern layer with their own classes. As a result, the summation layer units' input is only stacked with its own output in pattern layer units of the same class. The decision-making layer's output of a PNN is simple threshold identification. It is intended to choose a single neuron as the complete system's output. The output layer's neurons are a type of competitive neuron that operates in a one-to-one fault pattern. The number of neurons in the output layer is determined by the number of fault categories in the training samples, which is the same as the number of fault classes in the training samples. Equation 7.13 expresses the input of the output layer neurons.

$$\rho(x) = arg\,\max[\alpha_i f_{iN_I}(x)] \tag{7.13}$$

Where $\rho(x)$ is the obtained estimated class through PNN and α_i is known as the class i prior probability. The following are the steps of the learning algorithm of PNN:

Step 1: All the training samples x are stated as vectors. $x = x_1, x_2, \ldots, x_d$
Step 2: The input layer's neurons get the initial training sample vector. The weights between both the units in the pattern and input layers are then set $[x_1 = w_1]$.
Step 3: Repeat the previous step for the remaining units $[x_m = w_m]$, where m is equal to $1,2,3,\ldots,n$.
Step 4: The pattern layer and input layer are connected after training. However, the pattern and accumulation layers are sparsely connected in the PNN.

After the training of the PNN, the classification of PQ events can be carried out using the following three steps:

Step 1: The input layer receives a test sample. The following nonlinear function is used to calculate each unit of the pattern layer:

$$f_w(x) = \frac{1}{2\pi^{\left(\frac{d}{2}\right)}\sigma^d} e^{-\frac{(x-x_{ij})^T-(x-x_{ij})}{2\sigma^2}} \tag{7.14}$$

Step 2: The neural cells in the summation and pattern layer that are connected and have their estimated results $f_w(x)$ are added together. As a

result, each summation layer unit receives a signal. These layers are connected to the pattern model layer. The strength of a signal received in the summation layer is equal to the test sample's probability.
Step 3: The desired category of test samples is the largest estimated result after calculating the summation layer results.

7.6.3 Performance Prediction of the ML Classifiers

The ML classifier's performance in PQ event classification is predicted based on precision, accuracy, specificity, and sensitivity. Table 7.1 shows these parameters and the formula used to calculate them. In this, TP means true positive prediction. It means that the PQ problem is correctly classified. On the other hand, TN means true negative; it denotes the accurate prediction of non-PQ events classified. If the classifier wrongly predicts the PQ event in a sense, it is denoted as FP (false prediction) [39, 40]. Similarly, false negative (FN) means wrong classification or prediction of the non-PQ event.

The classification precision, accuracy, specificity, and sensitivity of the different classifiers with a different number of samples tested are shown in Table 7.2. Similarly, Figures 7.20–7.23 show the ML classifiers (used in PQ event classification) and their performance comparison graphically.

From the above performance parameters calculation and comparison, we can conclude that the PNN-based classifier exhibits overall accuracy in the range of 90.7%, which is comparatively higher than the BPN

Table 7.1 Performance parameters of the ML classifiers.

Sl. no	Parameter	Formula
1	Precision (P)	$P = \frac{TP}{TP + FP}$
2	Accuracy (Acc)	$Acc = \frac{TP + TN}{TP + TN + FP + FN}$
3	Specificity (SP)	$SP = \frac{TN}{TN + FP}$
4	Sensitivity (Se)	$Se = \frac{TP}{TP + FN}$

Table 7.2 A comparison of performance parameters of the different ML classifiers with varying test samples.

No. of samples	SVM (values in %)				BPN (values in %)				PNN (values in %)			
	P	*Acc*	*Sp*	*Se*	*P*	*Acc*	*Sp*	*Se*	*P*	*Acc*	*Sp*	*Se*
100	82.5	83.5	83.6	83.1	86.5	85.6	85.6	86.1	88.3	88.4	89.2	89.2
200	86.3	84	82.1	84.2	82.5	87	86.7	85.1	86.2	87	88.7	88.3
300	85.1	82	83	86.1	84.9	88	87.6	84.7	83.4	86	89.6	88.6
400	88.2	80	84	82.6	88.1	86	87.4	86.4	88.6	89.2	90.3	88.7
500	81	86.3	86.2	83.4	86.2	89	88.1	87.3	90	88	91.6	88.6
600	86.2	84	88	88	88	90	83.2	89.3	92.5	91.1	90	89.7
700	88	86	81.3	87.2	89.2	89	89.6	90.2	91	92.3	92.3	90.2
800	92	90	88	88.3	93.2	91	90.2	92.1	93.5	96	96	92.6
900	91	92	89.5	90.7	94	94	93.1	90.7	94	95	95	91.6
1000	92	91.6	90.3	88.1	93	93	94	93.2	96	94.7	95.7	94.7
Overall	87.2	85.9	85.6	86.1	88.5	88.5	89.2	88.5	90.3	90.77	91.8	90.2

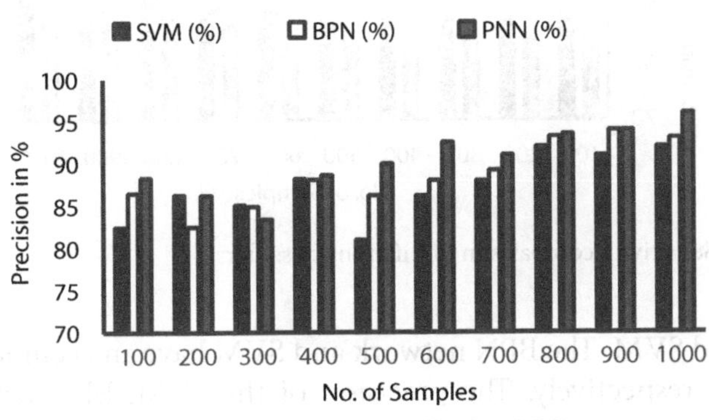

Figure 7.20 A comparison of the classifier's precision with a different number of samples.

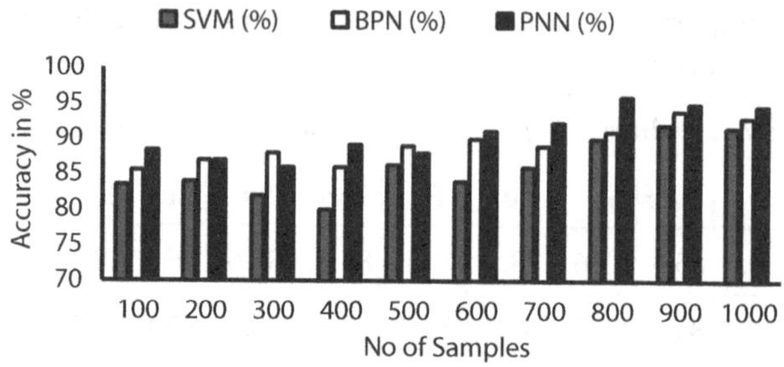

Figure 7.21 A comparison of the accuracy of the different classifiers with various number of test samples.

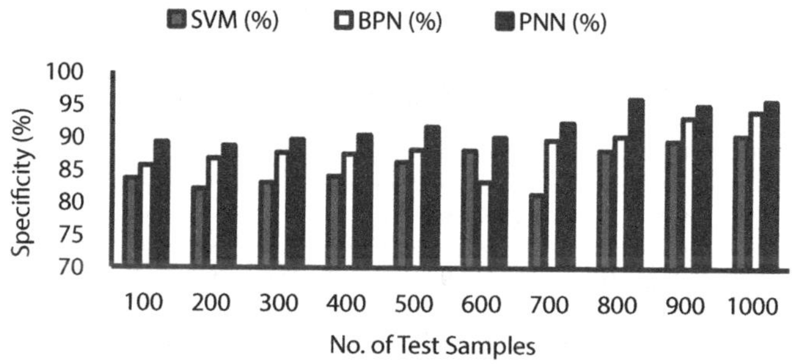

Figure 7.22 Specificity comparison of different classifiers.

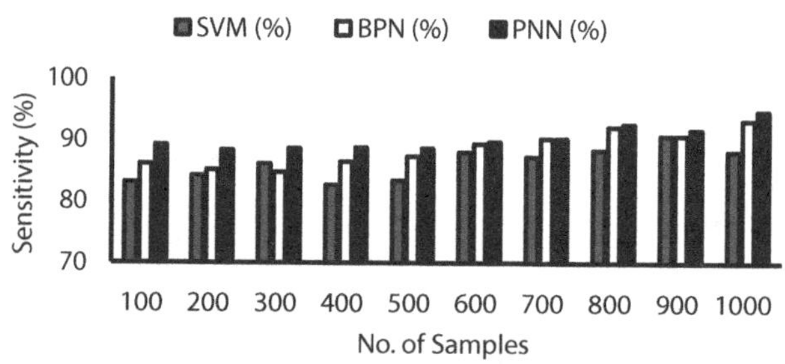

Figure 7.23 Sensitivity comparison of different classifiers.

network and SVM. The BPN network and SVM have an accuracy of 88.5% and 85.9%, respectively. The precision of the SVM, BPN network, and PNN classifiers are 87.2%, 88.5%, and 90.3%, respectively. At the precision level, even the PNN is good to compare with other classifiers. The specificity and sensitivity of the PNN classifier are in the range of 91.8%

and 90.2%, respectively. Compared with SVM and BPN network's specificity and sensitivity, PNN has higher values. Thus, the PNN classifier efficiently classifies the PQ events occurring in the power system and is more suited for a real-time environment.

7.7 Summary and Conclusion

Power quality problems are one of the critical problems in the modern-day power system. The various important PQ problems occurring in the power system with their sources and their impacts are also discussed. The most important sources of PQ problems are nonlinear loads and power electronics devices. Nowadays, most energy suppliers impose a penalty on the consumers who are responsible for these PQ problems. So, these problems need to be eradicated from the system. Many devices like filters, FACTS, and voltage regulators are developed in recent times to overcome or minimize this issue. But all these devices needed a quick and precise identification of PQ problems that occur in the system. Many conventional systems prevail in the present world to identify PQ events. Using digital signal processing techniques and machine learning techniques to identify and classify the PQ events attract more interest in recent times. The present work discussed the wavelet transformation-based feature extraction and nature-inspired algorithms like the GA for feature selection and feature optimization. The feature extracted and optimized is then classified using machine learning approaches like the BPN network, SVM, and PNN. The performance of the classifier is measured in terms of precision, accuracy, specificity, and sensitivity. The PNN classifier exhibits an excellent overall percentage of different performance parameters compared with other classifiers. From all these things, we can conclude that the PNN classifier is more suited for PQ problems classification in the real modern world.

References

1. Stones, J., & Collinson, A. (2001). Power quality. *Power Eng. J., 15*(2), 58-64.
2. Chattopadhyay, S., Mitra, M., & Sengupta, S. (2011). Electric power quality. In *Electric power quality,* (pp. 5-12). Springer, Dordrecht.
3. Bollen, M. H., & Gu, I. Y. (2006). *Signal processing of power quality disturbances,* (Vol. 30). John Wiley & Sons.
4. Reid, W. E. (1996). Power quality issues-standards and guidelines. *IEEE Trans. Ind. Appl., 32*(3), 625-632.

5. Dash, R. L., Tripathy, S., Mohanty, B., & Hota, P. K. (2021, January). Power Quality Improvement of Grid Connected Wind System with STATCOM. In *2021 1st International Conference on Power Electronics and Energy (ICPEE),* (pp. 1-6). IEEE.
6. HassanzadehFard, H., & Jalilian, A. (2021). Optimization of DG units in distribution systems for voltage sag minimization considering various load types. *Iran. J. Sci. Technol. - Trans. Electr. Eng., 45*(2), 685-699.
7. Patil, S., Pawar, S., Mulla, A., & Patil, D. (2021). FC-TBSR Compensator for Reactive Power Compensation and Voltage Swell Mitigation. In *Innovations in Electrical and Electronic Engineering,* (pp. 19-28). Springer, Singapore.
8. Lee, J. Y., Verayiah, R., Ong, K. H., Ramasamy, A., & Marsadek, M. B. (2021). Voltage Oriented Control and Direct Power Control Strategies in solving under and overvoltage conditions for heavy load application on Malaysian Distribution Representative Network. *Electr. Eng., 103*(3), 1597-1612.
9. Appadurai, M., & Raj, E. (2021). Epoxy/silicon carbide (sic) nanocomposites based small scale wind turbines for urban applications. *Int. J. Energy Environ. Eng.*, 1-16.
10. Fitzpatrick, C., Odhner, J. H., & Levis, R. J. (2020). Spectral Signatures of Ground-and excited-state wavepacket interference after impulsive excitation. *J. Phys. Chem. A, 124*(34), 6856-6866.
11. Fantin Irudaya Raj, E., & Appadurai, M. (2021). Minimization of Torque Ripple and Incremental of Power Factor in Switched Reluctance Motor Drive. In *Recent Trends in Communication and Intelligent Systems: Proceedings of ICRTCIS 2020,* (pp. 125-133). Springer Singapore.
12. Sun, J., Wang, G., Du, X., & Wang, H. (2018). A theory for harmonics created by resonance in converter-grid systems. *IEEE Trans. Power Electron., 34*(4), 3025-3029.
13. Singh, U., & Singh, S. N. (2017). Optimal feature selection via NSGA-II for power quality disturbances classification. *IEEE Trans. Ind. Inf., 14*(7), 2994-3002.
14. Liu, H., Hussain, F., Shen, Y., Arif, S., Nazir, A., & Abubakar, M. (2018). Complex power quality disturbances classification via curvelet transform and deep learning. *Electr. Power Syst. Res., 163*, 1-9.
15. Sahani, M., & Dash, P. K. (2018). Automatic power quality events recognition based on Hilbert Huang transform and weighted bidirectional extreme learning machine. *IEEE Trans. Ind. Inf., 14*(9), 3849-3858.
16. Raj, E. F. I., & Balaji, M. (2021). Analysis and classification of faults in switched reluctance motors using deep learning neural networks. *Arab. J. Sci. Eng., 46*(2), 1313-1332.
17. Arrabal-Campos, F. M., Montoya, F. G., Baños, R., Martínez-Lao, J., & Alcayde, A. (2018, May). Simulation of power quality disturbances through the wavelet transform. In *2018 18th International Conference on Harmonics and Quality of Power (ICHQP),* (pp. 1-5). IEEE.

18. Singh, N., Ansari, M. A., Tripathy, M., & Singh, V. P. (2021). Feature Extraction and Classification Techniques for Power Quality Disturbances in Distributed Generation: A Review. *IETE J. Res.*, 1-16.
19. Chamchuen, S., Siritaratiwat, A., Fuangfoo, P., Suthisopapan, P., & Khunkitti, P. (2021). High-Accuracy power quality disturbance classification using the adaptive ABC-PSO as optimal feature selection algorithm. *Energies, 14*(5), 1238.
20. Dekhandji, F. Z., & Rais, M. C. (2021). A Comparative Study of Power Quality Monitoring Using Various Techniques. In *Optimizing and Measuring Smart Grid Operation and Control,* (pp. 259-288). IGI Global.
21. Deivakani, M., Kumar, S. S., Kumar, N. U., Raj, E. F. I., & Ramakrishna, V. (2021, March). VLSI Implementation of Discrete Cosine Transform Approximation Recursive Algorithm. In *J. Phys.: Conf. Ser.,* (Vol. 1817, No. 1, p. 012017). IOP Publishing.
22. Beniwal, R. K., Saini, M. K., Nayyar, A., Qureshi, B., & Aggarwal, A. (2021). A Critical Analysis of Methodologies for Detection and Classification of Power Quality Events in Smart Grid. *IEEE Access.*
23. Ghaffarzadeh, N. (2021). A New Method for Power Quality Events Detection and Classification using Discrete Wavelet Transform and Correlation Coefficients. *Int. J. Ind. Electron. Control Optim.*, 4(1), 47-57.
24. Jain, D., Sangale, D. M. D., & Raj, E. (2020). A Pilot Survey Of Machine Learning Techniques In Smart Grid Operations Of Power Systems. *Eur. J. Mol. Clin. Med., 7*(7), 203-210.
25. Raj, E. F. I. (2016). Available Transfer Capability (ATC) under Deregulated Environment. *J. Power Electron. Power Syst., 6*(2), 85-88.
26. Rodriguez-Guerrero, M. A., Jaen-Cuellar, A. Y., Carranza-Lopez-Padilla, R. D., Osornio-Rios, R. A., Herrera-Ruiz, G., & Romero-Troncoso, R. D. J. (2017). Hybrid approach based on GA and PSO for parameter estimation of a full power quality disturbance parameterized model. *IEEE Trans. Ind. Inf., 14*(3), 1016-1028.
27. Sijini, A. C., Fantin, E., & Ranjit, L. P. (2016). Switched Reluctance Motor for Hybrid Electric Vehicle. *Middle-East J. Sci. Res., 24*(3), 734-739.
28. Mishra, A. K., Das, S. R., Ray, P. K., Mallick, R. K., Mohanty, A., & Mishra, D. K. (2020). PSO-GWO optimized fractional order PID based hybrid shunt active power filter for power quality improvements. *IEEE Access, 8,* 74497-74512.
29. Raj, E. F. I., & Kamaraj, V. (2013, March). Neural network based control for switched reluctance motor drive. In *2013 IEEE international conference on emerging trends in computing, communication and nanotechnology (ICECCN),* (pp. 678-682). IEEE.
30. Mirjalili, S. (2019). Genetic algorithm. In *Evolutionary algorithms and neural networks,* (pp. 43-55). Springer, Cham.
31. Chawda, G. S., Shaik, A. G., Shaik, M., Padmanaban, S., Holm-Nielsen, J. B., Mahela, O. P., & Kaliannan, P. (2020). Comprehensive review on detection

and classification of power quality disturbances in utility grid with renewable energy penetration. *IEEE Access*, *8*, 146807-146830.

32. Ch, G., Jana, S., Majji, S., Kuncha, P., & Tigadi, A. (2021). Diagnosis of COVID-19 using 3D CT scans and vaccination for COVID-19. *World J. Eng.*
33. Agarwal, P., Ch, M. A., Kharate, D. S., Raj, E. F. I., & Balamuralitharan, S. (2021). Parameter Estimation of COVID-19 Second Wave BHRP Transmission Model by Using Principle Component Analysis. *Ann. Rom. Soc. Cell Biol.*, 446-457.
34. Gampala, V., Kumar, M. S., Sushama, C., & Raj, E. F. I. (2020). Deep learning based image processing approaches for image deblurring. *Mater. Today: Proc.*
35. Alimi, O. A., Ouahada, K., & Abu-Mahfouz, A. M. (2020). A review of machine learning approaches to power system security and stability. *IEEE Access*, *8*, 113512-113531.
36. Dekhandji, F. Z., & Rais, M. C. (2021). A Comparative Study of Power Quality Monitoring Using Various Techniques. In *Optimizing and Measuring Smart Grid Operation and Control*, (pp. 259-288). IGI Global.
37. Shaik, N. B., Pedapati, S. R., Taqvi, S. A. A., Othman, A. R., & Dzubir, F. A. A. (2020). A feed-forward back propagation neural network approach to predict the life condition of crude oil pipeline. *Processes*, *8*(6), 661.
38. Ding, J., Xiao, D., & Li, X. (2020). Gear fault diagnosis based on genetic mutation particle swarm optimization VMD and probabilistic neural network algorithm. *IEEE Access*, *8*, 18456-18474.
39. Sun, J., Yan, S., Song, C., & Han, B. (2020). Dual-functional neural network for bilateral hippocampi segmentation and diagnosis of Alzheimer's disease. *Int. J. Comput. Assist. Radiol. Surg.*, *15*(3), 445-455.
40. Shalbaf, A., Bagherzadeh, S., & Maghsoudi, A. (2020). Transfer learning with deep convolutional neural network for automated detection of schizophrenia from EEG signals. *Phys. Eng. Sci. Med.*, *43*(4), 1229-1239.

8

Hybridization of Artificial Neural Network with Spotted Hyena Optimization (SHO) Algorithm for Heart Disease Detection

Shwetha N.[1*], Gangadhar N.[1], Mahesh B. Neelagar[2], Sangeetha N.[1] and Virupaxi Dalal[3]

[1]*Dr. Ambedkar Institution of Technology, Karnataka, India*
[2]*Department of VLSI Design and Embedded Systems, VTU, Belagavi, Karnataka, India*
[3]*Department of ECE, Jain College of Engineering and Research, Belagavi, India*

Abstract
Heart-related illnesses are the leading cause of mortality globally, which causes a high number of deaths in poor- and middle-income nations like India. Large amounts of data are constantly being generated by medical professionals. The generated data can be used to diagnose heart disease in advance, which can efficiently diminish the incidence of various heart-related diseases. Predictions can be done effectively by improving the knowledge identification needed to detect previously unknown patterns. Effective predictions can be made and hidden patterns can be detected by accessing data and concerns collected from healthcare industries. In this work, machine learning technique is used on cardiac disease-related data to try to find out the potential for heart disease before suffering from serious problems. Therefore, an artificial neural network (ANN) is used to predict a coronary illness. Additionally, the spotted hyena optimization (SHO) algorithm is hybridized with ANN to update the weights in an ANN. The implementation is carried out on the MATLAB platform. The proposed method's effectiveness is verified by different patients considering 13 constraints as the dataset. These constraints are evaluated for training and testing each data in the dataset. The efficiency of the proposed approach is shown in comparison with different methods, namely, social learning algorithm (SLO), particle swarm optimization (PSO), gaussian discriminant analysis (GDA), and genetic algorithm (GA).

**Corresponding author*: shwethaec48@gmail.com

Chandra Singh, Rathishchandra R. Gatti, K.V.S.S.S.S. Sairam, Manjunatha Badiger, Naveen Kumar S. and Varun Saxena (eds.) Modeling and Optimization of Signals Using Machine Learning Techniques, (145–172) © 2024 Scrivener Publishing LLC

Keywords: Artificial neural network, heart disease, spotted hyena optimization, cardiovascular disease, weight

8.1 Introduction

The leading cause of death globally is heart disease. Specifically, coronary artery disease (CAD) is a general problem in cardiovascular situations. CAD occurs in the left-peripheral, left-anterior descending and the right coronary arteries are stenotic [1]. Physicians use different methods to diagnose CAD in which angiography is considered the most accurate method. However, this is related to higher expenses and more side effects; so, scholars have been trying to develop other accurate diagnostic methods [2]. CAD causes more severe infirmity and higher mortality than other diseases such as cancer. CAD is a narrowing of coronary arteries. It can cause myocardial infarction, silent ischemia, arrhythmias, angina, unstable angina, sudden death, and heart failure. The term "heart disease" includes different diseases that affect the heart, which is the leading cause of death globally. It also refers to a number of disorders that affect the blood vessels and the heart, as well as the blood that is pumped and distributed throughout the body [3]. Among these diseases are CAD and cardiomyopathy.

Cardiovascular disease (CVD) is one of the main heart diseases which causes disability, severe illness, and death. The highest number of deaths in India and abroad, especially due to a cardiac disease, is compatible with the World Health Organization (WHO). Each year, 17.7 million deaths are because of cardiovascular illnesses, which accounts for 31% of all deaths globally. As a result, it is frequently necessary to consider this mortality rate by appropriately identifying the disease in its early stages [4]. Hence, processing technologies are used to extract the knowledge from databases. The exposed knowledge is frequently used by health administrators to enhance their service quality. Discovered facts can be utilized by medical experts to diminish the extent of contrary drug effects and prescribe therapeutically equivalent alternatives at low costs. In 2005, 53% of deaths were due to chronic diseases, but 29% of deaths were caused by heart diseases. In 2020, CVD was the foremost reason for disability and death in India. Also in India, more than 40.9 million people have been affected by diabetes, and more than 118 million people have been affected by high blood pressure. In 2025, these will increase to 69.9 and 213 million, respectively, unless emergency precautionary measures are taken.

Some of the types of heart problems are CAD, CVD, and cardiomyopathy. A CVD of any kind causes extreme illness, disability, and death. The narrowing of the heart's conduits and coronary corridors lessens the blood and oxygen flow to the heart, resulting in a coronary disease [5]. Blood congealing provides a blockage in the coronary blood vessel, resulting in sudden failure in breathing. If the heart does not receive the needed blood, then chest pains occur. Support vector machine, naïve Bayes, random forest, and other traditional machine learning (ML) tactics help in classifying data. Presently, deep learning is a promising technique being used in various fields including medicine. The artificial neural network (ANN) implemented by deep learning approaches is a data-driven algorithm. The ANN method has the following characteristics: accuracy, speed, nonlinearity, and self-adaptiveness. An ANN uses the activation function, which offers nonlinear mapping among the inputs and outputs. For example, the sigmoid function is the activation function. Therefore, the nonlinear problems of electrocardiogram (ECG) signals classification are solved by the activation function of the ANN.

8.1.1 Objective of the Work

The main objective of this work deals with the ANN and SHO algorithm for heart disease prediction with better accuracy. ANN weight values are optimized by the proposed SHO approach. ANN is designed and implemented for medical diagnosis to show the heart disease data classification. SHO algorithm is utilized with neural networks to achieve a high prediction accuracy.

8.1.2 Scope of the Project

Developing a neural network that integrates hybrid technology and leverages advanced evolutionary computation techniques to ensure it meets the precise specifications demanded by medical applications.

8.2 Literature Survey

Heart disease is the major reason for death worldwide, leading to many deaths in low- and average-income countries, such as India. Medical practitioners constantly generate a great amount of data and such data can be used to detect heart diseases earlier, which can reduce the threat of various heart-related diseases. The decision prediction can be done effectively by improving the knowledge identity needed to detect previously unknown patterns. Efficient forecasting can be done by accessing data collected from healthcare organizations and industries and finding hidden patterns. Chavda *et al.* (2019) [6] utilized the ML scheme on cardiac data and efforts to diagnose the possibility of heart disease before suffering from severe complications. The effectiveness of earlier predictions of cardiac diseases has been implemented and has demonstrated the accuracy of such predictions.

Machine learning is an area where different schemes are developed by many researchers. Lin *et al.* (2019) [7] have proposed artificial intelligence (AI) applications to analyze congenital dyserythropoietic anemia (CDA) and ECG. It involves the Internet of Things (IoT) based front-end hardware, cloud database, smart application with user interface, and AI for cardiac-disease detection. The IoT-based hardware and wearable ECG patches contain Bluetooth modules and analog front-end circuits for the detection of ECG signals. Smart device application users cannot show real-time ECG signals; however, instantly labeled abnormal signals can be achieved in real-time diagnosis. The detected ECG signals are updated in the cloud database, which is utilized to save every ECG signal. It will form big data for AI approaches to diagnose heart diseases. Here, the convolution neural network is utilized, which gives a 94.96% accuracy [12–16].

El-Kased (2018) [8] introduced immuno-analytical evaluation with polyclonal-antibody proficient for representing the troponin-I antigenic determinant and performed the troponin-I proteolytic digestion. For CVD, cardiac troponin-I was an auspicious diagnostic marker. Antigenic determination is dispensed through separating the wreckages by gel electrophoresis. The antigenic determining area acted in the 30–90 amino acid sequence. This strong process is best for a timely diagnosis of diseases, stabilization of patients, and individual treatment.

In developed countries, cancer and CVD are two of the most common diseases. Evidence is gradually showing that anticancer treatment is biologically linked to some detrimental effects on health. Curigliano *et al.* (2020) [9] proposed CVD toxicity in relation to cancer or its treatments and to offer guidelines for the screening, prevention, and monitoring of CVD treatment. CVD can occur as a result of anticancer treatment. Concerns about injury must be weighed contrary to the possible offerings of cancer treatment, including overall survival assistance. The cardio-oncology scope is extensive and involves the process of detection, monitoring, prevention, and treatment. It confirms the safe progress of prospective cancer treatments that reduce side effects.

ECG is used to screen cardiac disorder diagnostic signals by the prevention of CVD. Twelve-lead-based ECG images were utilized for CVD detection systems by Khan *et al.* (2020) [10]. The detection of a single shot to identify CVD utilizes a mobile net v2-based deep neural network (DNN) architecture. It was used to spot cardiac abnormalities, such as anirregular heartbeat, myocardial infarction, normal class, and prior myocardial infarction (MI) history. A 98% accuracy was attained after installation. Experts manually gathered 11,148 regular 12-lead-based ECG pictures via the institute of healthcare and annotated them. The effectiveness of this approach was verified by several cardiologists. On the diagnosis of CAD, Arabasadi *et al.* [11] proposed an accurate hybrid technique. The neural network performance is improved by 10% by the enhancement of weights in a neural network. The weights are trained by a genetic algorithm (GA), which gives enhanced weights in the neural network. Hence, the proposed method achieved better accuracy (93.85%), sensitivity (97%), and specificity (92%) rates on Z-Alizadeh Sani dataset.

8.2.1 Problem Identification

The existence of cardiac alignment is predicted using emotionally supportive networks. Deprived diagnostic outcomes and medical decisions can trigger death. Similarly, not all physicians are reasonable in diagnosing coronary heart disease, in which certainty is a mandatory task. In cases of coronary disease, every moment is precious, emphasizing the critical importance of timely systematic analysis to prevent fatalities. Timely systematic analysis provides a strategic space from death. This outline allows cardiologists to improvise the solution, while people can make mistakes on the side of complex data. The intelligent system quickly adopts an important job in the field of pharmaceuticals and medical amenities under all circumstances and enhances the standard of patient care and health. The need for target techniques to evaluate such a structure is mostly accepted. Given the well-documented nature of pharmaceuticals and medical amenities, well-being is critical; for example, medical master structures and neurological structures must be widely acknowledged in medical practices. In this study, an ML-based neural network approach for CVD was studied.

8.3 Proposed Methodology

An analysis of coronary heart disease considers a significant work where people of all ages and both sexes are significantly affected. Coronary disease poses a severe hazard to people who need to be analyzed efficiently.

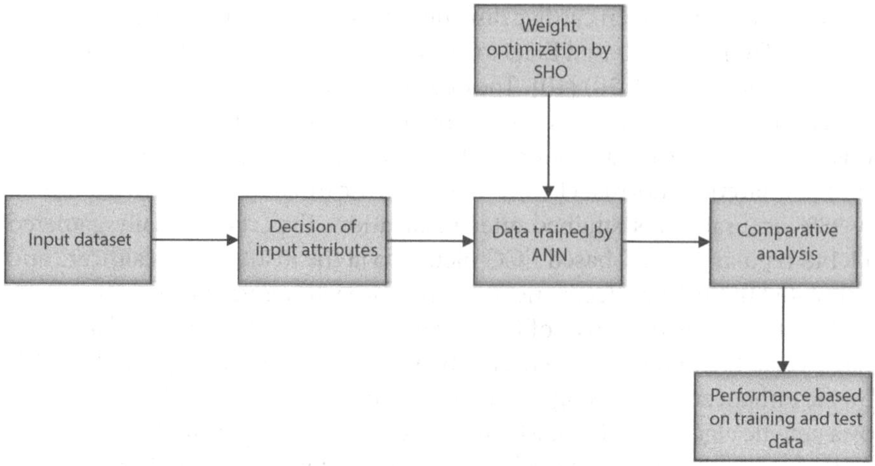

Figure 8.1 Block diagram of the proposed ANN-SHO approach.

This is reasonable for accurately diagnosing coronary heart diseases at an earlier stage to ensure timely treatment and avoid further complications. The principal quality of this task is to incorporate a structure that can predict heart problems in a mechanized manner with extended accuracy. Therefore, the neural network is hybridized with an optimization algorithm to predict the coronary illness. ANN is employed here, and the spotted hyena optimization approach is utilized to upgrade the neural network weights. Figure 8.1 illustrates the proposed method diagrammatically.

The suggested strategy's effectiveness related to the correct diagnosis for coronary diseases is accomplished by boosting the neural network weight that is completed via providing a hereditary computation. A hereditary computation picks up the ideal weight regarding the concealed layers of a neural framework. The given dataset is primarily normalized prior to implementing the computation. By then, the arranged ANN-SHO algorithm is put into the normalized dataset.

8.3.1 Different Kinds of Machine Learning Approaches

Generally, machine learning approaches are categorized as

- Supervised learning,
- Unsupervised learning,
- Semi-supervised learning, and
- Reinforcement learning.

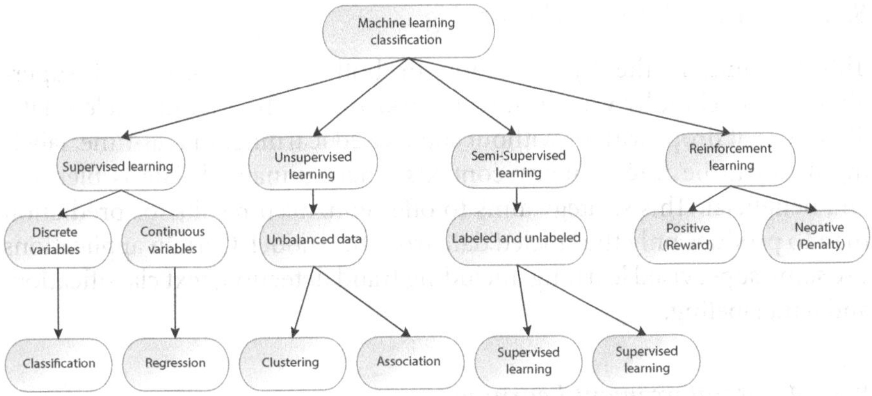

Figure 8.2 Categories of machine learning.

In real-life situations, each type of learning technique solves many problems, which are in the following. The different kinds of machine learning are given in Figure 8.2.

8.3.1.1 *Supervised Learning*

The machine learning function is learned by the supervised learning method that graphs an output/input depending on the sample input/output pairs. The prediction of this method is done by the set of training data and training samples. It is carried out when certain objectives are recognized from a set of specific inputs, which is a task-driven scheme. The tasks of supervised learning are classification and regression, which divide the data and fit the data, respectively. An example of supervised learning is text classification (e.g., predicting a text or a class label, such as a Tweet or a product review).

8.3.1.2 *Unsupervised Learning*

Unsupervised learning investigates unnamed datasets without the need for human intervention; this is known as a data-driven process. It is widely used to isolate the features it creates to identify meaningful trends and structures, exploratory purposes, and grouping in results. General tasks are density estimation, anomaly detection, dimensionality reduction, feature learning, clustering, finding association rules, etc.

8.3.1.3 Semi-Supervised Learning

This learning is the hybridization of both unsupervised and supervised approaches, by way of how it works on unlabeled and labeled data. Therefore, it drops with or without supervised learning. In real-time, labeling data can be rare in many contexts where unnamed data is plentiful and beneficial. This strategy aims to offer enhanced results for prediction than to produce only the labeled data from the model. Certain applications use semi-supervised learning, including fraud detection, text classification, and data labeling.

8.3.1.4 Reinforcement Learning

This learning method allows the software to habitually assess optimal behavior in a specific environment or context to enhance its performance known as an environmentally driven approach. It is based on penalty or reward and its aim is to improve the reward or reduce the risk using insights gained from environmental activists. For AI training, reinforcement is a good tool that can enhance the automation of systems like autonomous driving tasks, robotics, and production and distribution chain logistics, or improve functional performance; however, reinforcement is not intended to be used to solve basic or direct problems. Therefore, to create useful models in different applications filed, various kinds of machine learning methods can be used based on their learning abilities, nature of the model, and target data effect.

8.4 Artificial Neural Network

Artificial neural networks are greatly parallel computing systems, which involve a high number of simple processors with lots of interconnections that are motivated by biological neural networks. An ANN has utilized some principles that are believed to be applied to humans. In the network, the nodes are seen as artificial neurons. These are called ANNs. Inspired by natural neurons, artificial neurons are created as a computational model. Artificial neurological methods are computational standards according to the numerical structures that have architecture and function, contrary to conventional inventions which take place after the cerebrum of a well-developed organism. ANNs or neural systems are also called connective structures, equally conveyed structures, or multipurpose structures, which are then made up of an arrangement of interconnected components that

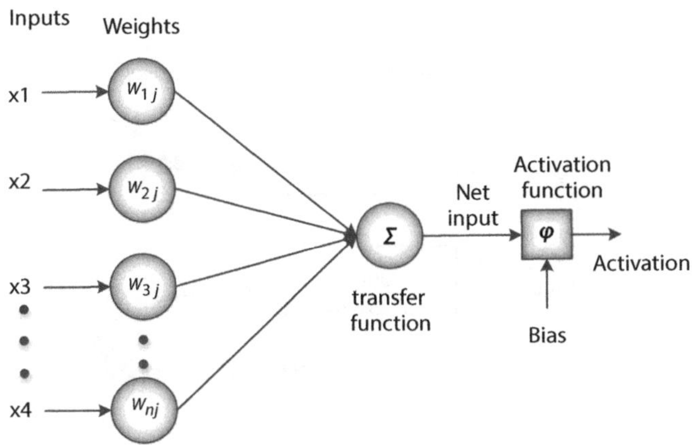

Figure 8.3 Artificial neuron.

function equally well. Neural network (NN) systems can be useful for expanding real-time complications. Their primary advantage is that they can handle more complicated problems than traditional innovations and problems without an algorithmic arrangement that should not be classed in any manner. When using algorithmic approaches to address a problem, the system examines a piece of computation. However, if the system is aware of the exact change to be made, it will be unable to handle the situation. That control realizes how the controller of the system structure sheds light on the dangerous thinking ability of the conventional system to the problems now understood. In general, neural systems are ideal for difficulties that people can grasp, but computers, for the most part, cannot. Examples of these concerns are approval and determination; these necessitate the approval of forms in the data. In the case of a nervous system failure, the computer can retrieve the information closest to the best result or the best information for the loose information. Figure 8.3 shows the artificial neuron structure.

8.4.1 ANN Classification

An ANN consists of input, weights, and output variables. The performance for the network is based on the correlation among the variables of input and output. Generally, an ANN has cells, nodes/units, and a number of processing units named neurons. The ANN process is performed in three layers: input, hidden, and output. In between the input and output layers, the concealed layer is also known as the middle layer. The ANN's different layers are illustrated in Figure 8.4.

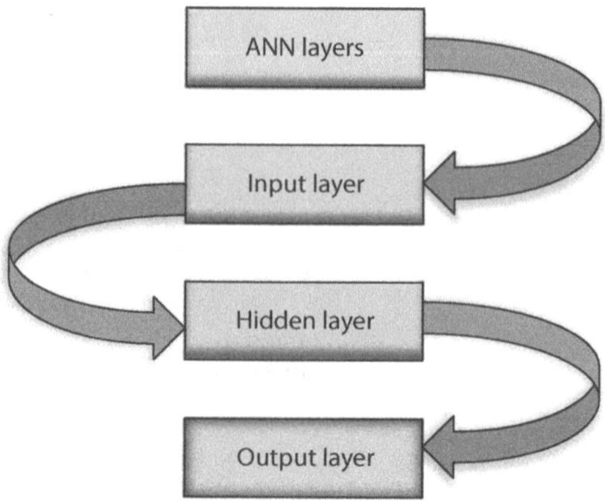

Figure 8.4 Different layers of an ANN.

8.4.1.1 *Input Layer*

The input layer is the first one. The input units in this layer display the source of information provided over the network.

8.4.1.2 *Hidden Layer*

The hidden layer is the second layer, which is mostly dependent on the neural network structures. The inputs are applied to determine the hidden layer performance and their relationship with further hidden layers. Every hidden unit function is established according to the input unit weight and the interaction among them. The hidden unit activation time is determined by the hidden and input unit weights.

8.4.1.3 *Output Layer*

The output layer is the final layer. Every output unit activity depends on every hidden unit weight and the interaction between them. The output layer performance is dependent on the hidden layer activity and output unit weights. Each layer neuron and layer are stated by the designer in the trial-and-error process.

The ANN algorithm is as follows:

Step 1: The input layer data are fed into the hidden layer.
Step 2: From the input layer, input values are used in the hidden layer, and weight values are adjusted and guided to the output layer.

Step 3: The value is changed again by certain weights from the connection among the hidden and output layers.
Step 4: These details are processed, and the output layer provides the final output. Based on the activation function, the output is processed, and an ANN follows trial-and-error means to obtain the optimal solution.

During iteration, the neural network weights are updated in the training phase of NN based on Equation 8.1.

$$\omega_{ji} = \omega_{ji} + l\delta_j a_{ji} \tag{8.1}$$

Where weight associated with i^{th} input to j unit is denoted as ω_{ij}, learning rate is l, and i^{th} input is denoted as x_{ji}. If the output layer units' weight is ω, δ is updated by Equation 8.2. Whereas ω is the hidden layer unit weights, δ is updated by Equation 8.3.

$$\delta_g = O_g(1-O_g)(t_g - O_g) \tag{8.2}$$

$$\delta_h = O_h(1-O_h)\sum_{g \in O} \omega_{gh}\delta_g \tag{8.3}$$

Where the outputs produced by the hidden unit (h) and the output unit (g) are denoted as O_h and O_g, respectively. The target output of unit g is denoted as t_g. The term g is in the range among 1 and the output layer quantity, whereas the term h and the hidden layer quantity. The sigmoid function is expressed in Equation 8.4.

$$F(n) = \frac{1}{1+e^{-n}} \tag{8.4}$$

Figure 8.5 displays the architecture of an ANN, and Figure 8.6 shows the ANN modeling of multiple hidden layers structure.

ANNs are utilized in many fields to process the corresponding information. Different ANN models are established in the real-world and neural networks studied the behaviors of machines and animals. However, ANN techniques are used for data compression, forecasting and pattern recognition, and other engineering applications. To achieve minimum errors, the ANN weights are modified and trained. Here, the ANN weights are optimized by the proposed SHO algorithm to obtain minimum errors.

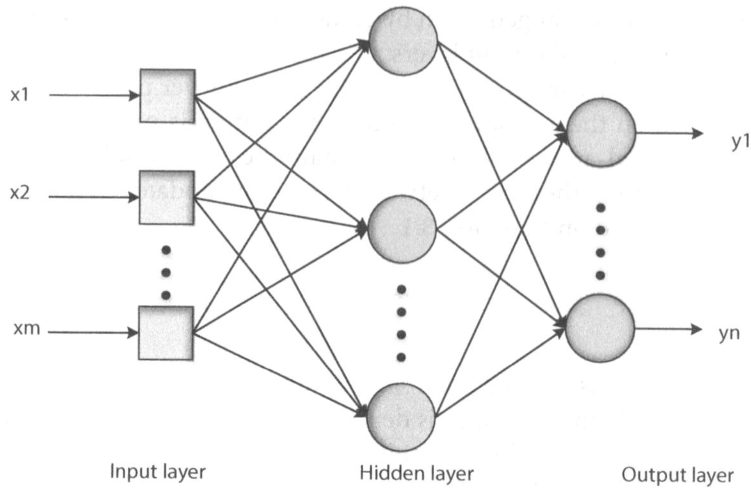

Figure 8.5 ANN architecture.

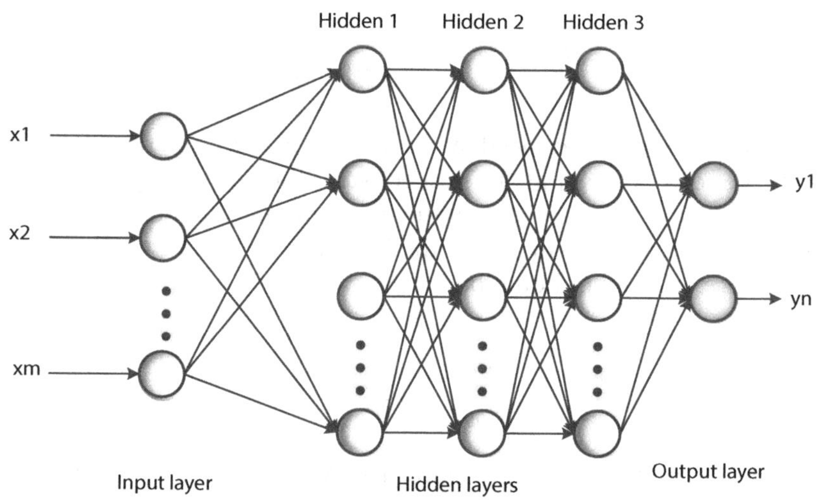

Figure 8.6 Structure of ANN modeling with multiple hidden layers.

8.4.2 Spotted Hyena Optimization

The SHO is the metaheuristic swarm-intelligence optimization algorithm for solving optimization difficulties. It is constructed by the hunting behavior of the spotted hyena (SH). The process of SHO hunting involves three categories:

- At first, considering environmental factors like the availability of resources and the opposition with other animal creatures;
- Secondly, focusing on social likings based on the performance of individuals; and
- Finally, involving the social relations of species.

The inspiration of the algorithm is the social relationship among animals, and it associates this strategy with the SH. SHs are skilled hunters that are bigger than the species of brown and striped aardwolves. They are alson known as 'laughing hyenas' since their sound is similar to a human laugh. SHs are intelligent, complicated, and very social animals with a terrible reputation. For food and territory, they have a good ability to fight until the end. Among the hyenas found, the female members dominate in their clan. When male hyenas leave their family as adults, they join a different clan. In the new family, hyenas get lesser food. A male hyena of the clan will continuously stay with the same individuals for a longer period of time. However, a female SH is permanently present in a stable location. During food searching, SHs create sounds to communicate with each other.

In the algorithm process, SHO is the cohesive cluster among the trusted SHs. SHO considers four important steps as follows:

- Searching,
- Encircling,
- Hunting, and
- Attacking.

8.4.2.1 Searching Behavior

The mechanism of searching defines the capability of exploring an algorithm. Using random numbers that are greater or lower than 1, this method validates the ability. The vector oversees exhibiting the SHO's unpredictable behavior while also avoiding local optima. The SH's prey-searching behavior is displayed in Figure 8.7.

8.4.2.2 Encircling Behavior

The objective of the minimum weight is considered as a better resolution, and each other's search agents appraise its locations with regard to attaining the best solution. Here, the SH can recognize the prey and encircle them. The SHO's hunting behavior is led to the best search agent by a group

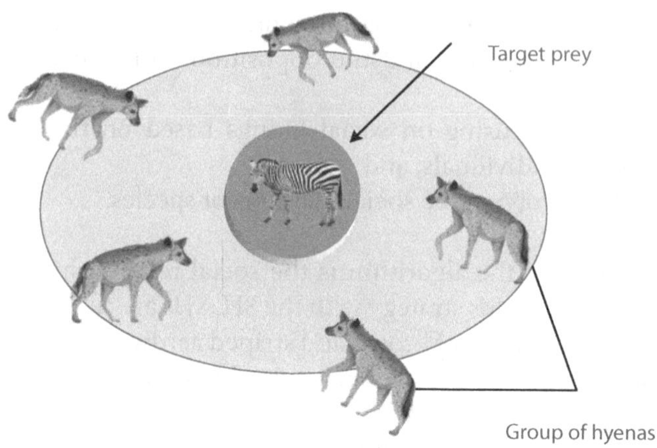

Figure 8.7 Prey searching.

of intimate pals, and better solutions are preserved. The encircling behavior of the SH is stimulated by Equation 8.5, and the search agents' locations are updated by Equation 8.6.

$$\vec{D}_f = |\vec{K}.\vec{P}_p(x) - \overrightarrow{P_f}(x)| \tag{8.5}$$

$$\vec{P}(x+1) = \vec{P}_p(x) - \vec{L}.\vec{D}_f \tag{8.6}$$

Where the distance between the SH and the prey is signified as $\vec{D}_f$. The current iteration and coefficient vectors are signified as x and $\vec{K}$ and $\vec{L}$, respectively. The position vectors of the SH and the prey are represented as $\vec{P}_f$ *and* $\vec{P}_p$, respectively.

The coefficient vectors of $\vec{K}$ *and* $\vec{L}$ are computed in Equations 8.7 and 8.8, respectively.

$$\vec{K} = 2.\vec{c}_1 \tag{8.7}$$

$$\vec{L} = 2.\overrightarrow{f}.\vec{c}_2 - \vec{f} \tag{8.8}$$

$$\vec{f} = 5 - \left(iter \times \frac{5}{max_{iter}} \right) \tag{8.9}$$

Where $iter = 0,1,2,\ldots..,max_{iter}$.

Here, $\vec{f}$ is linearly reduced from five to zero over the iterations and the random vectors, $\vec{c_1}$ *and* $\vec{c_2}$, lies between the 0 to 1 range. There are different numbers of spaces depending on the current position by adjusting the $\vec{K}$ and $\vec{L}$ vector values. The obtained optimal solution is stored thus far, and it forces further search agents to update their status.

8.4.2.3 Hunting Behavior

The hunting behavior and search space are discovered by Equations 8.10 and 8.11.

$$\overrightarrow{D_f} = |\vec{K}.\overrightarrow{C_f} - \overrightarrow{C_n}| \tag{8.10}$$

$$\overrightarrow{C_n} = \overrightarrow{C_f} - \vec{L}.\overrightarrow{D_f} \tag{8.11}$$

$$\overrightarrow{G_f} = \overrightarrow{C_n} + \overrightarrow{C_{n+1}} + \ldots \overrightarrow{C_{n+N}} \tag{8.12}$$

Where $N = count_{nos}(\vec{C}_f, \vec{C}_{f+1}, \vec{C}_{f+2}, \ldots, (\vec{C}_f + M))$.

$$\vec{P}(x+1) = \frac{\vec{C}_f}{N} \tag{8.13}$$

Where the term *nos* represents the total solution that is identical to the optimal solution and the cluster solution is represented as $\vec{C}_f$. Random vectors (M) are in the range of 0.5 to $1.\vec{P}(x+1)$. Maintain a repository of the *N* most favorable optimal values, and iteratively adjust their positions according to the whereabouts of the top-performing search agent.

8.4.2.4 Attacking Behavior

In this phase, the search agent locations are updated based on the greatest solution and the best search agent location. The SHs attack the prey and continuously update their location, which is illustrated in Figure 8.8.

When hyenas search for their prey, they usually besiege the prey. The proposed algorithm balances the exploration and exploitation in the given search space.

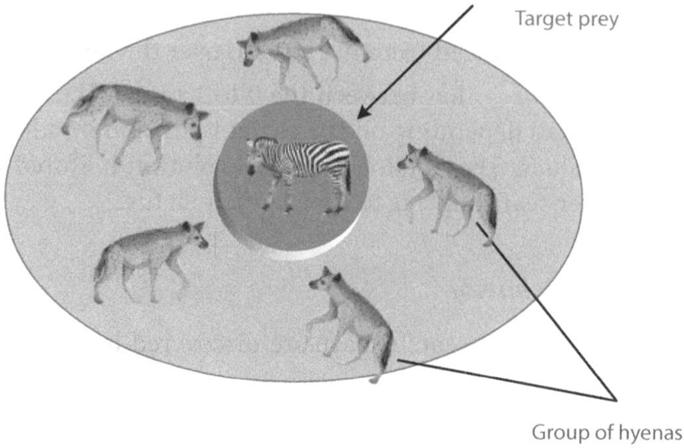

Figure 8.8 Prey attacking.

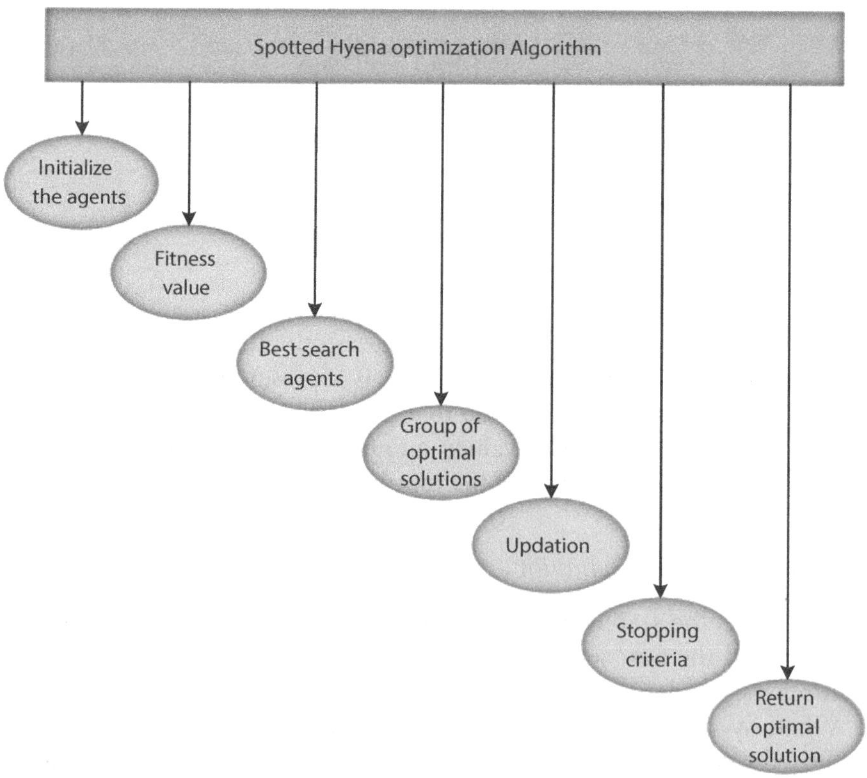

Figure 8.9 Flow chart of SHO.

With reducing $\vec{L}$, partial iteration is applied to the exploration $(\vec{L} > 1)$ and the remaining is dedicated to exploitation $(\vec{L} < 1)$. $\vec{K}$ is present in the range of 0 to 2 to display the SHO algorithm's random performance. This enables the SHO to demonstrate further random behaviors in the optimization process and avoid local optima and choosing exploration. The flow chart of the SHO is depicted in Figure 8.9.

Pseudo code for SHO algorithm

Algorithm 1: SHO

Input: ANN signal
Output: optimal weights
Initialize the SHO population
Initialize f, K, L *and* N
Compute each search agent fitness
Best search agent $\vec{P}_f$
Group of all far optimal solution $\vec{C}_f$
While $(x < max_{iter})$ ***do***
    for every search agent ***do***
        Equation 8.10 is used to update the new search agent
end for
    update f, K, L *and* N
    If search agents go outside the search space and restore it.
    Compute fitness
    Update $\vec{P}_f$ if there is any best solution by Equations 8.6 and 8.7

$$x = x + 1$$

end while
Return P_f

8.4.3 SHO-Based ANN

ANN is trained to discover the optimal weight value. The main problem is discovering the variables for the metaheuristic algorithm and using metaheuristic and appropriate formula to train the ANN.

The ANN training process is formulated by two steps:

- Demonstrating the biases and weights in a proper mode; and
- Describing the objective function.

ANN weights and biases are represented in three ways such as binary, matrix, and vector. Here, considering the vector solution because the SHO algorithm is in a vector. The ANN's final vector is given in Equation 8.14.

$$\vec{V} = \{\vec{\omega}, \vec{\theta}\} \tag{8.14}$$

$$= \{\omega_{1,1}, \omega_{1,2}, \ldots, \omega_{n,\ n}, \quad \theta_1, \theta_2, \ldots, \theta_h\}$$

Where the number of input nodes is denoted as n, from i^{th} node to j^{th} node connected weight is denoted as $\omega_{i,j}$, and j^{th} node bias is represented as θ_j.

The ANN deviation and weights are expressed in the vector form and the ANN fitness function is required to state the objective function. The ANN's objective is to attain the maximum approximation and prediction accuracy for testing and training the samples.

Generally, the neural network presentation is evaluated through the variance among the desired and actual output values of the network. The ANN performance measure is the mean squared error (MSE), which is described in Equation 8.15.

$$MSE = \sum_{i=1}^{m} (a_i^q - d_i^q)^2 \tag{8.15}$$

Here, the number of outputs, desired output, actual output, and training sample are denoted as $m, d, a\ and\ q$, respectively. The objective function of the SHO is minimizing the MSE in the ANN. Hence, the objective is evaluated based on Equation 8.16.

$$obj_{fun} = \min\,(MSE) \tag{8.16}$$

8.4.4 Benefits of SHO in ANN

There are some reasons why using the SHO algorithm to prepare neural networks might be beneficial. When it comes to determining the weight (and inclination) for systems with fixed networks, hereditary computations are particularly good at efficiently searching across large and complicated areas to find practically global optima. SHO calculations may be used to prepare a variety of systems to build up a selected stable framework with just modest changes to the calculations. The ANN engineering is prepared using SHO. Force and propensity will be used to boost learning and speed. Node predispositions may generally be changed during preparation, but they can be adjusted to a specific incentive at the time of system installation. An energy rate allows the system to possibly skirt around local minima to help avoid plunging into one.

8.5 Software Implementation Requirements

SHO algorithm does not have the bias option, and the input data consideration is very limited because it is difficult to analyze huge datasets.

PC Configuration

- CPU- Intel Pentium IV 3GHz
- 20GB hard disk
- 2.9GHz 2GB SDRAM
- Keyboard, mouse, and other peripherals

Software Used

- Operating system: Windows 2007/x5
- Development platform: MATLAB 7.0 /higher version
- Graphical user interface (GUI)

Reasons for Using MATLAB

For technical computations, MATLAB is a better-quality language. It combines computational, visual, and easy ways of programming. Use the MATLAB background wherever solutions and problems are articulated in acquainted mathematical code, typically in the data acquisition, math computation, simulation, modeling, prototyping, development of algorithm, data analysis, exploration, scientific, and engineering graphics.

8.5.1 Results and Discussion

According to experts who have already created the dataset, there are a total of 13 parameters in the input test as per the University of California Irvine (UCI) repository. For heart disease detection, the dataset is first normalized based on the min–max method. Data normalization is a must-need because the dataset has classes, zero values, variables, and binary values.

Figure 8.10 displays the GUI menu module. The GUI menu is arranged for the sequential output. For the sequential output, the button list has been organized. For the input data, the heart disease dataset is first normalized into zero and one. After that, the data is used to train the suggested ANN-SHO method for detection. The tested data was used to compare to the trained data in order to determine whether or not the learning was complete. Normalizing the datasets is required to move the process to saturation mode.

Figure 8.11 displays the training data, and the normalized dataset with target is shown in Figure 8.12. In the dataset, the row indicates the considered patients as Patient 1, Patient 2, and so on. Individual patient parameters are represented in the columns as sex, age, chest pain type, etc. Those are age in years, sex (one = men; zero = women), nature of chest pain (one = typical type angina; two = typical type angina; three = non-angina pain; four = asymptomatic), blood sugar in fasting (one represents > 120 mg/dl; zero represents < 120 mg/dl), rest ECG – resting electrographic results (zero = normal; value one: having ST-T wave abnormality; value two = showing credible or certain left ventricular hypertrophy), Exang – exercise induced angina (one = yes; zero = no), slope – slope of the peak exercise ST segment (one = unsloping; two = flat; three = down sloping), CA – number

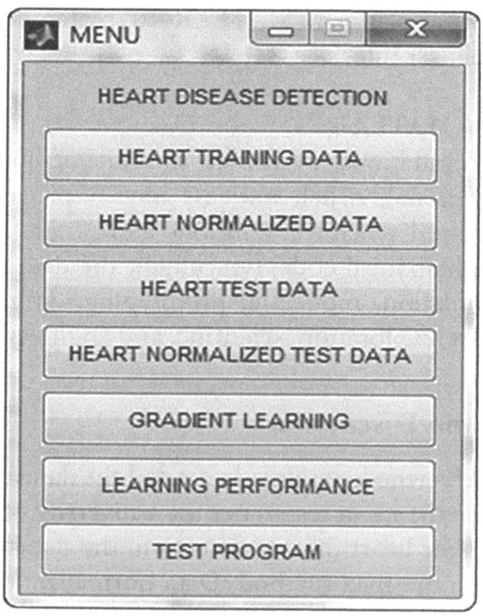

Figure 8.10 Overall GUI modules.

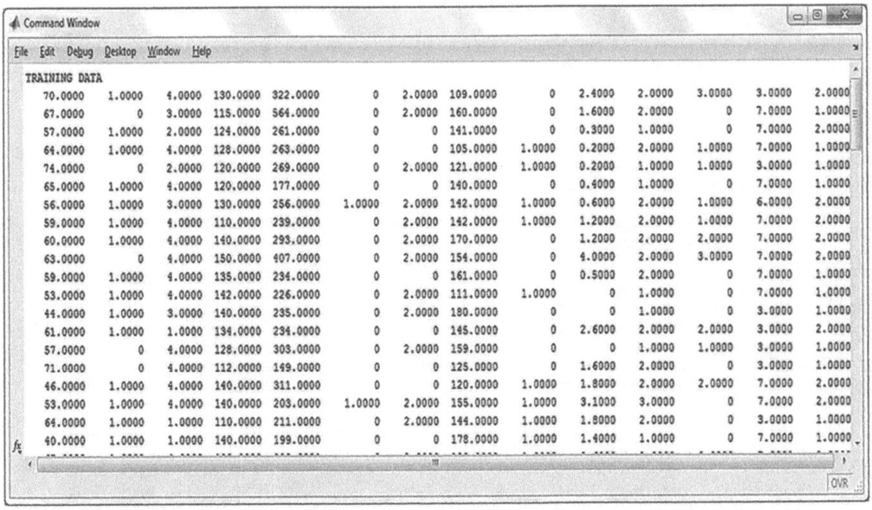

70.0000	1.0000	4.0000	130.0000	322.0000	0	2.0000	109.0000	0	2.4000	2.0000	3.0000	3.0000	2.0000
67.0000	0	3.0000	115.0000	564.0000	0	2.0000	160.0000	0	1.6000	2.0000	0	7.0000	1.0000
57.0000	1.0000	2.0000	124.0000	261.0000	0	0	141.0000	0	0.3000	1.0000	0	7.0000	2.0000
64.0000	1.0000	4.0000	128.0000	263.0000	0	0	105.0000	1.0000	0.2000	2.0000	1.0000	7.0000	1.0000
74.0000	0	2.0000	120.0000	269.0000	0	2.0000	121.0000	1.0000	0.2000	1.0000	1.0000	3.0000	1.0000
65.0000	1.0000	4.0000	120.0000	177.0000	0	0	140.0000	0	0.4000	1.0000	0	7.0000	1.0000
56.0000	1.0000	3.0000	130.0000	256.0000	1.0000	2.0000	142.0000	1.0000	0.6000	2.0000	1.0000	6.0000	2.0000
59.0000	1.0000	4.0000	110.0000	239.0000	0	2.0000	142.0000	1.0000	1.2000	2.0000	1.0000	7.0000	2.0000
60.0000	1.0000	4.0000	140.0000	293.0000	0	2.0000	170.0000	0	1.2000	2.0000	2.0000	7.0000	2.0000
63.0000	0	4.0000	150.0000	407.0000	0	2.0000	154.0000	0	4.0000	2.0000	3.0000	7.0000	2.0000
59.0000	1.0000	4.0000	135.0000	234.0000	0	0	161.0000	0	0.5000	2.0000	0	7.0000	1.0000
53.0000	1.0000	4.0000	142.0000	226.0000	0	2.0000	111.0000	1.0000	0	1.0000	0	7.0000	1.0000
44.0000	1.0000	3.0000	140.0000	235.0000	0	2.0000	180.0000	0	0	1.0000	0	3.0000	1.0000
61.0000	1.0000	1.0000	134.0000	234.0000	0	0	145.0000	0	2.6000	2.0000	2.0000	3.0000	2.0000
57.0000	0	4.0000	128.0000	303.0000	0	2.0000	159.0000	0	0	1.0000	1.0000	3.0000	1.0000
71.0000	0	4.0000	112.0000	149.0000	0	0	125.0000	0	1.6000	2.0000	0	3.0000	1.0000
46.0000	1.0000	4.0000	140.0000	311.0000	0	0	120.0000	1.0000	1.8000	2.0000	2.0000	7.0000	2.0000
53.0000	1.0000	4.0000	140.0000	203.0000	1.0000	2.0000	155.0000	1.0000	3.1000	3.0000	0	7.0000	2.0000
64.0000	1.0000	1.0000	110.0000	211.0000	0	2.0000	144.0000	1.0000	1.8000	2.0000	0	3.0000	1.0000
40.0000	1.0000	1.0000	140.0000	199.0000	0	0	178.0000	1.0000	1.4000	1.0000	0	7.0000	1.0000

Figure 8.11 Training dataset with target.

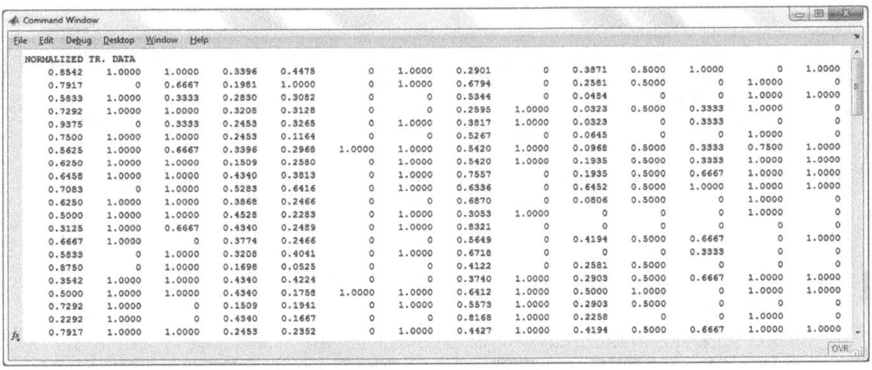

0.8542	1.0000	1.0000	0.3396	0.4475	0	1.0000	0.2901	0	0.3871	0.5000	1.0000	0	1.0000
0.7917	0	0.6667	0.1981	1.0000	0	1.0000	0.6794	0	0.2581	0.5000	0	1.0000	0
0.5833	1.0000	0.3333	0.2830	0.3082	0	0	0.5344	0	0.0484	0	0	1.0000	1.0000
0.7292	1.0000	1.0000	0.3208	0.3128	0	0	0.2595	1.0000	0.0323	0.5000	0.3333	1.0000	0
0.9375	0	0.3333	0.2453	0.3265	0	1.0000	0.3817	1.0000	0.0323	0	0.3333	0	0
0.7500	1.0000	1.0000	0.2453	0.1164	0	0	0.5267	0	0.0645	0	0	1.0000	0
0.5625	1.0000	0.6667	0.3396	0.2968	1.0000	1.0000	0.5420	1.0000	0.0968	0.5000	0.3333	0.7500	1.0000
0.6250	1.0000	1.0000	0.1509	0.2580	0	1.0000	0.5420	1.0000	0.1935	0.5000	0.3333	1.0000	1.0000
0.6458	1.0000	1.0000	0.4340	0.3813	0	1.0000	0.7557	0	0.1935	0.5000	0.6667	1.0000	1.0000
0.7083	0	1.0000	0.5283	0.6416	0	1.0000	0.6336	0	0.6452	0.5000	1.0000	1.0000	1.0000
0.6250	1.0000	1.0000	0.3868	0.2466	0	0	0.6870	0	0.0806	0.5000	0	1.0000	0
0.5000	1.0000	1.0000	0.4528	0.2283	0	1.0000	0.3053	1.0000	0	0	0	1.0000	0
0.3125	1.0000	0.6667	0.4340	0.2489	0	1.0000	0.8321	0	0	0	0	0	0
0.6667	1.0000	0	0.3774	0.2466	0	0	0.5649	0	0.4194	0.5000	0.6667	0	1.0000
0.5833	0	1.0000	0.3208	0.4041	0	1.0000	0.6718	0	0	0	0.3333	0	0
0.8750	0	1.0000	0.1698	0.0525	0	0	0.4122	0	0.2581	0.5000	0	0	0
0.3542	1.0000	1.0000	0.4340	0.4224	0	0	0.3740	1.0000	0.2903	0.5000	0.6667	1.0000	1.0000
0.5000	1.0000	1.0000	0.4340	0.1758	1.0000	1.0000	0.6412	1.0000	0.5000	1.0000	0	1.0000	1.0000
0.7292	1.0000	0	0.1509	0.1941	0	1.0000	0.5573	1.0000	0.2903	0.5000	0	0	0
0.2292	1.0000	0	0.4340	0.1667	0	0	0.8168	1.0000	0.2258	0	0	1.0000	0
0.7917	1.0000	1.0000	0.2453	0.2352	0	1.0000	0.4427	1.0000	0.4194	0.5000	0.6667	1.0000	1.0000

Figure 8.12 Normalized training dataset with target.

of major vessels colored through fluoroscopy (value zero to three), thal (three = normal; six = fixed defect; seven = reversible defect), trest blood pressure (in mmHg upon hospital admission), serum cholesterol (mg/dl), Thalach – risky heart rate achieved, and old peak – ST depression prompted by exercise.

Figures 8.13 and 8.14 show the test data and normalized test data with the target from the implementation of the CVD process in ANN-SHO.

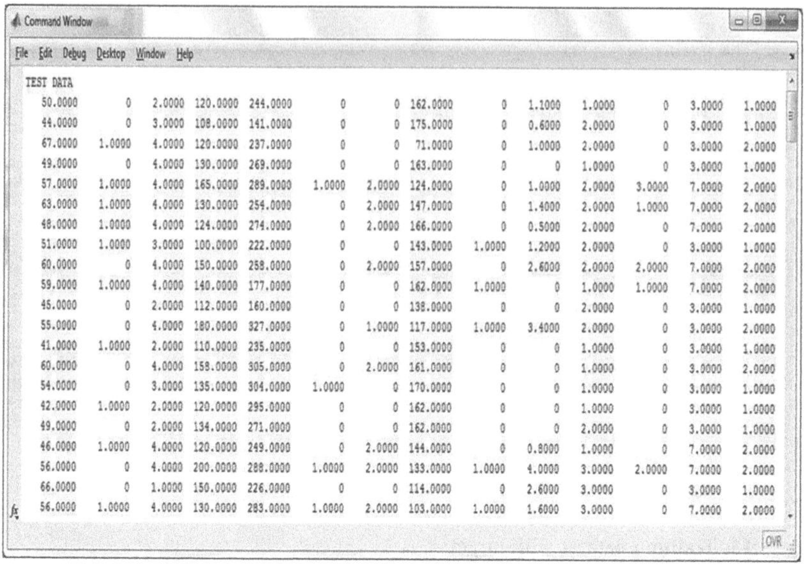

Command Window

File Edit Debug Desktop Window Help

TEST DATA

50.0000	0	2.0000	120.0000	244.0000	0	0	162.0000	0	1.1000	1.0000	0	3.0000	1.0000
44.0000	0	3.0000	108.0000	141.0000	0	0	175.0000	0	0.6000	2.0000	0	3.0000	1.0000
67.0000	1.0000	4.0000	120.0000	237.0000	0	0	71.0000	0	1.0000	2.0000	0	3.0000	2.0000
49.0000	0	4.0000	130.0000	269.0000	0	0	163.0000	0	0	1.0000	0	3.0000	1.0000
57.0000	1.0000	4.0000	165.0000	289.0000	1.0000	2.0000	124.0000	0	1.0000	2.0000	3.0000	7.0000	2.0000
63.0000	1.0000	4.0000	130.0000	254.0000	0	2.0000	147.0000	0	1.4000	2.0000	1.0000	7.0000	2.0000
48.0000	1.0000	4.0000	124.0000	274.0000	0	2.0000	166.0000	0	0.5000	2.0000	0	7.0000	2.0000
51.0000	1.0000	3.0000	100.0000	222.0000	0	0	143.0000	1.0000	1.2000	2.0000	0	3.0000	1.0000
60.0000	0	4.0000	150.0000	258.0000	0	2.0000	157.0000	0	2.6000	2.0000	2.0000	7.0000	2.0000
59.0000	1.0000	4.0000	140.0000	177.0000	0	0	162.0000	1.0000	0	1.0000	1.0000	7.0000	2.0000
45.0000	0	2.0000	112.0000	160.0000	0	0	138.0000	0	0	2.0000	0	3.0000	1.0000
55.0000	0	4.0000	180.0000	327.0000	0	1.0000	117.0000	1.0000	3.4000	2.0000	0	3.0000	2.0000
41.0000	1.0000	2.0000	110.0000	235.0000	0	0	153.0000	0	0	1.0000	0	3.0000	1.0000
60.0000	0	4.0000	158.0000	305.0000	0	2.0000	161.0000	0	0	1.0000	0	3.0000	2.0000
54.0000	0	3.0000	135.0000	304.0000	1.0000	0	170.0000	0	0	1.0000	0	3.0000	1.0000
42.0000	1.0000	2.0000	120.0000	295.0000	0	0	162.0000	0	0	1.0000	0	3.0000	1.0000
49.0000	0	2.0000	134.0000	271.0000	0	0	162.0000	0	0	2.0000	0	3.0000	1.0000
46.0000	1.0000	4.0000	120.0000	249.0000	0	2.0000	144.0000	0	0.8000	1.0000	0	7.0000	2.0000
56.0000	0	4.0000	200.0000	288.0000	1.0000	2.0000	133.0000	1.0000	4.0000	3.0000	2.0000	7.0000	2.0000
66.0000	0	1.0000	150.0000	226.0000	0	0	114.0000	0	2.6000	3.0000	0	3.0000	1.0000
56.0000	1.0000	4.0000	130.0000	283.0000	1.0000	2.0000	103.0000	1.0000	1.6000	3.0000	0	7.0000	2.0000

OVR

Figure 8.13 Test dataset with target.

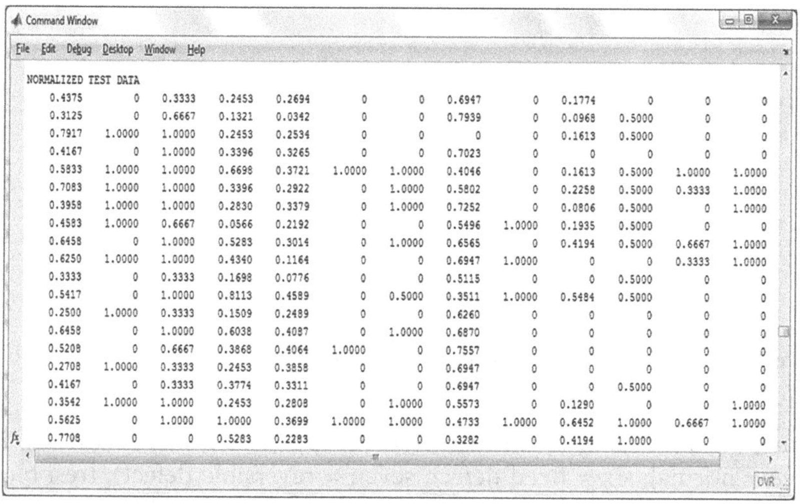

Command Window

File Edit Debug Desktop Window Help

NORMALIZED TEST DATA

0.4375	0	0.3333	0.2453	0.2694	0	0	0.6947	0	0.1774	0	0	0
0.3125	0	0.6667	0.1321	0.0342	0	0	0.7939	0	0.0968	0.5000	0	0
0.7917	1.0000	1.0000	0.2453	0.2534	0	0	0	0	0.1613	0.5000	0	0
0.4167	0	1.0000	0.3396	0.3265	0	0	0.7023	0	0	0	0	0
0.5833	1.0000	1.0000	0.6698	0.3721	1.0000	1.0000	0.4046	0	0.1613	0.5000	1.0000	1.0000
0.7083	1.0000	1.0000	0.3396	0.2922	0	1.0000	0.5802	0	0.2258	0.5000	0.3333	1.0000
0.3958	1.0000	1.0000	0.2830	0.3379	0	1.0000	0.7252	0	0.0806	0.5000	0	1.0000
0.4583	1.0000	0.6667	0.0566	0.2192	0	0	0.5496	1.0000	0.1935	0.5000	0	0
0.6458	0	1.0000	0.5283	0.3014	0	1.0000	0.6565	0	0.4194	0.5000	0.6667	1.0000
0.6250	1.0000	1.0000	0.4340	0.1164	0	0	0.6947	1.0000	0	0	0.3333	1.0000
0.3333	0	0.3333	0.1698	0.0776	0	0	0.5115	0	0	0.5000	0	0
0.5417	0	1.0000	0.8113	0.4589	0	0.5000	0.3511	1.0000	0.5484	0.5000	0	0
0.2500	1.0000	0.3333	0.1509	0.2489	0	0	0.6260	0	0	0	0	0
0.6458	0	1.0000	0.6038	0.4087	0	1.0000	0.6870	0	0	0	0	0
0.5208	0	0.6667	0.3868	0.4064	1.0000	0	0.7557	0	0	0	0	0
0.2708	1.0000	0.3333	0.2453	0.3858	0	0	0.6947	0	0	0	0	0
0.4167	0	0.3333	0.3774	0.3311	0	0	0.6947	0	0	0.5000	0	0
0.3542	1.0000	1.0000	0.2453	0.2808	0	1.0000	0.5573	0	0.1290	0	0	1.0000
0.5625	0	1.0000	1.0000	0.3699	1.0000	1.0000	0.4733	1.0000	0.6452	1.0000	0.6667	1.0000
0.7708	0	0	0.5283	0.2283	0	0	0.3282	0	0.4194	1.0000	0	0

OVR

Figure 8.14 Normalized test dataset with target.

Figure 8.15 illustrates the ANN hidden layer random weights and output layer trained weights. The hidden layer's arbitrary weights are in between -20 and +20. Then, the trained hidden neural network weight is achieved.

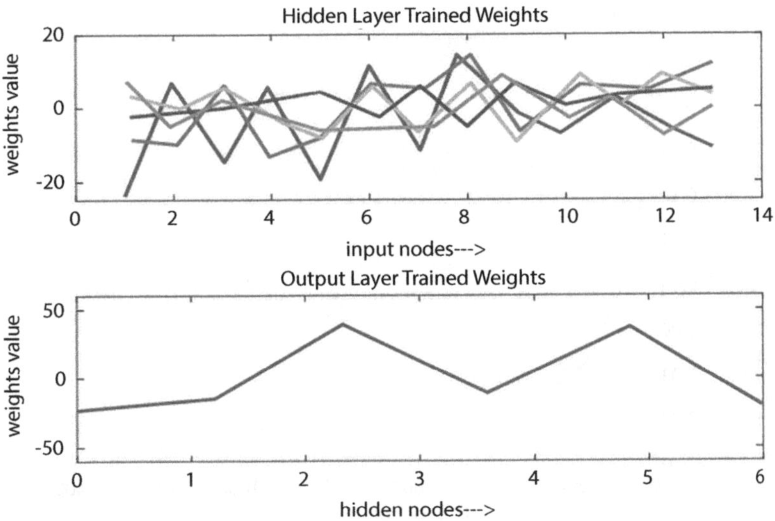

Figure 8.15 Hidden and output layer trained weights.

The suggested ANN-SHO technique is used to train the neural network's weights. The objective is calculated based on the MSE error. For the minimized MSE, the optimal weight values are achieved. Figure 8.16 illustrates the error estimation of SHO algorithm in the CVD process. Figure 8.17

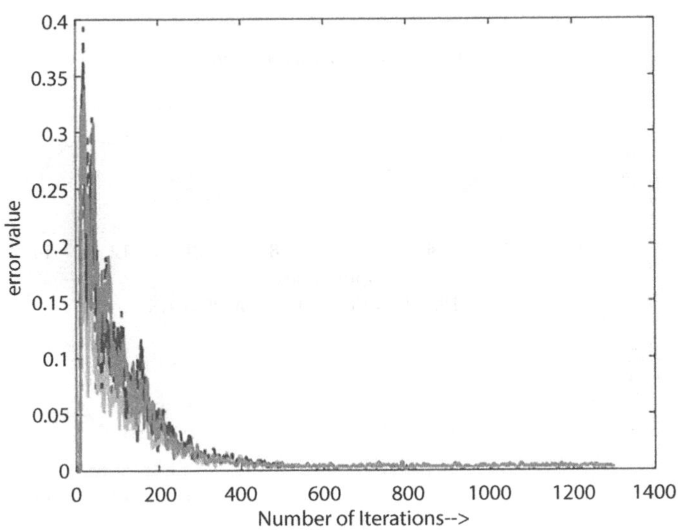

Figure 8.16 Error plot for the ANN-SHO.

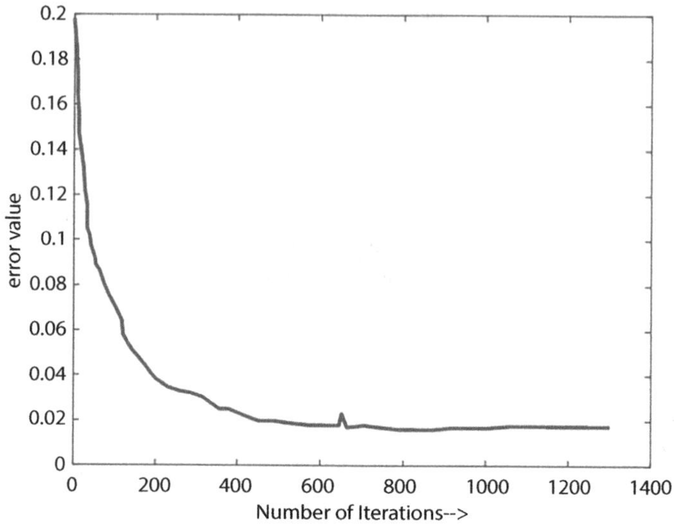

Figure 8.17 Convergence plot for the proposed ANN-SHO.

displays the convergence of the proposed ANN-SHO algorithm. The minimum optimal MSE is achieved as 0.237.

Figure 8.18 illustrates the hidden and output layer trained weights with respect to the hidden nodes and their corresponding weights. The different

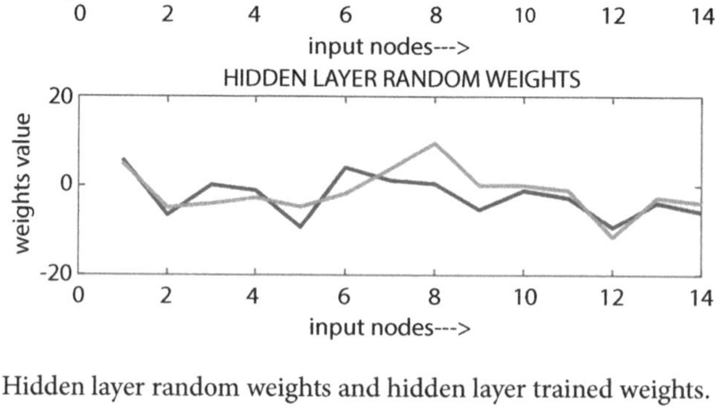

Figure 8.18 Hidden layer random weights and hidden layer trained weights.

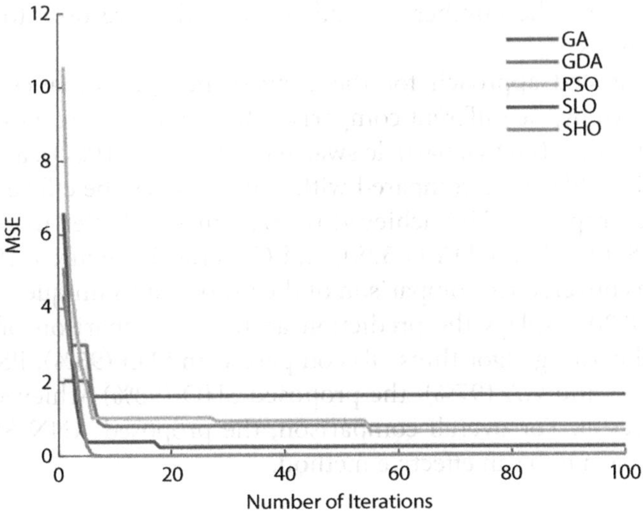

Figure 8.19 Convergence comparison of the proposed method.

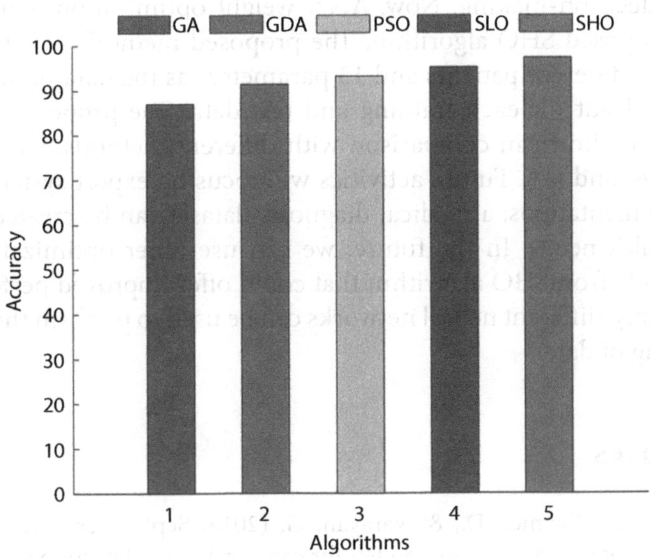

Figure 8.20 Prediction accuracy.

colors represent the number of hidden nodes that are used to train the neural network.

The suggested approach for the convergence performance effectiveness is shown by the different comparison techniques. Here, GA, gradient decent algorithm (GDA), particle swarm optimization (PSO), and sea lion optimization (SLO) are compared with SHO to show the efficiency of the work. The proposed SHO achieved 0.237, which is better than the SLO (0.835), PSO (1.023), GDA (1.328), and GA (1.862). Figure 8.19 demonstrates the convergence comparison of the proposed technique.

Figure 8.20 displays the prediction accuracy comparison of the proposed and existing algorithms. To compare with SLO (95%), PSO (92%), GDA (90%), and GA (97%), the proposed SHO (98%) achieved a better accuracy result. For overall comparison, the proposed ANN-SHO technique is shown to be an effective method.

8.6 Conclusion

The ANN structure and SHO algorithm are discussed in this paper. There is no universally accepted method for predicting heart diseases. It will be necessary to do a literature review of the numerous reset mechanisms that have been discovered to date. At the very least, these studies aid in the confirmation of the condition. Patients will not have to deal with uncertainty or poor decision-making. Now, ANN weight optimization is performed by the proposed SHO algorithm. The proposed method's effectiveness is verified by different patients and 13 parameters as the dataset. It can thus be checked out for each training and test data. The proposed method's efficiency is shown in comparison with different methods, namely, SLO, PSO, GDA, and GA. Future activities will focus on expert system designs and implementations; a medical diagnosis dataset can be created to meet the module's needs. In the future, we can use other optimization algorithms aside fromSHO algorithm that could offer improved performance. Additionally, different neural networks can be used to perform the training and testing of data.

References

1. Rajan, D., Beymer, D., & Narayan, G. (2018, September). Generalization studies of neural network models for cardiac disease detection using limited

channel ECG. In *2018 Computing in Cardiology Conference (CinC)*, 45, 1-4. IEEE.

2. Gopika, P., Krishnendu, C. S., Chandana, M. H., Ananthakrishnan, S., Sowmya, V., Gopalakrishnan, E. A., &Soman, K. P. (2020). Single-layer convolution neural network for cardiac disease classification using electrocardiogram signals. In *Deep Learning for Data Analytics*,21-35. Academic Press.
3. Suhail, M. M. (2021). Asymmetric Back Propagation Neural Network-Based Automatic Cardiac Disease Detection Using Electrocardiogram Signal. *Turk. J. Comput. Math. Educ. (TURCOMAT)*, 12(10), 2621-2631.
4. Jalali, S. M. J., Karimi, M., Khosravi, A., &Nahavandi, S. (2019). An efficient neuroevolution approach for heart disease detection. In *2019 IEEE International Conference on Systems, Man and Cybernetics (SMC)*, 3771-3776. IEEE.
5. Nashif, S., Raihan, M. R., Islam, M. R., & Imam, M. H. (2018). Heart disease detection by using machine learning algorithms and a real-time cardiovascular health monitoring system. *World J. Eng. Technol.*, 6(4), 854-873.
6. Chavda, P., Bhavsar, H., Pithadia, Y., &Kotecha, R. (2019). Early detection of cardiac disease using machine learning. In *2nd International Conference on Advances in Science & Technology (ICAST).*
7. Lin, Y. J., Chuang, C. W., Yen, C. Y., Huang, S. H., Huang, P. W., Chen, J. Y., & Lee, S. Y. (2019, March). Artificial intelligence of things wearable system for cardiac disease detection. In *2019 IEEE International Conference on Artificial Intelligence Circuits and Systems (AICAS)*, 67-70. IEEE.
8. El-Kased, R. F. (2018). Immuno-analytical approach and its application for cardiac disease marker detection. *J. Immunoassay Immunochem.*, 39(5), 538-550.
9. Curigliano, G., Lenihan, D., Fradley, M., Ganatra, S., Barac, A., Blaes, A., ...& ESMO Guidelines Committee. (2020). Management of cardiac disease in cancer patients throughout oncological treatment: ESMO consensus recommendations. *Ann. Oncol.*, 31(2), 171-190.
10. Khan, A. H., Hussain, M., & Malik, M. K. (2021). Cardiac disorder classification by electrocardiogram sensing using deep neural network. *Complexity*, 2021.
11. Arabasadi, Z., Alizadehsani, R., Roshanzamir, M., Moosaei, H., &Yarifard, A. A. (2017). Computer aided decision making for heart disease detection using hybrid neural network-Genetic algorithm. *Comput. Methods Programs Biomed.*, 141, 19-26.
12. Shwetha N., Gangadhar N., Neelagar M.B., Shilpa K.C. (2022) ANN-based Hybridization Approach for Detection of Cardiac Disease. In: Shakya S., Balas V.E., Kamolphiwong S., Du KL. (eds) *Sentimental Analysis and Deep Learning. Advances in Intelligent Systems and Computing*, vol 1408. Springer, Singapore. https://doi.org/10.1007/978-981-16-5157-1_20.
13. N. Shwetha and M. Priyatham, "Performance Analysis of Self Adaptive Equalizers using EPLMS Algorithm," *2020 Fourth International Conference*

on I-SMAC (IoT in Social, Mobile, Analytics and Cloud) (I-SMAC), Palladam, India, 2020, pp. 872-876, doi: 10.1109/I-SMAC49090.2020.9243512.
14. Shwetha N., Priyatham M. (2021) Convergence Analysis of Self-Adaptive Equalizers Using Evolutionary Programming (EP) and Least Mean Square (LMS). In: Bindhu V., Tavares J.M.R.S., Boulogeorgos AA.A., Vuppalapati C. (eds) *International Conference on Communication, Computing and Electronics Systems.* Lecture Notes in Electrical Engineering, vol 733. Springer, Singapore. https://doi.org/10.1007/978-981-33-4909-4_48.
15. Shwetha N., Priyatham M. (2021) Performance Analysis of Self Adaptive Equalizers Using Nature Inspired Algorithm. In: Smys S., Balas V.E., Kamel K.A., Lafata P. (eds) *Inventive Computation and Information Technologies.* Lecture Notes in Networks and Systems, vol 173. Springer, Singapore. https://doi.org/10.1007/978-981-33-4305-4_37.
16. Shwetha N, Manoj Priyatham and Gangadhar N, Adaptive Filter Equalizer Optimization using Hybrid Approach, *Int. J. Adv. Res. Eng. Technol.,* 12(1), 2021, pp. 473-483.

9

The Role of Artificial Intelligence, Machine Learning, and Deep Learning to Combat the Socio-Economic Impact of the Global COVID-19 Pandemic

Biswa Ranjan Senapati[1]*, Sipra Swain[1] and Pabitra Mohan Khilar[2]

[1]*Department of Computer Science and Engineering, Siksha 'O' Anusandhan (Deemed to be University), Bhubaneswar, Odisha, India*

[2]*Department of Computer Science and Engineering, National Institute of Technology, Rourkela, Odisha, India*

Abstract

The COVID-19 pandemic has created an emergency across the globe. Within a short span, it has changed the habits and lifestyles of people. It massively affected every aspect of an individual's life and propensities. Its socio-economic impact is seen in every country in the world. In this chapter, the objective is to study the overall impact of the coronavirus in our society and the role of various technologies, technological devices, and technological applications in the battle against the coronavirus infection. Artificial intelligence, computer vision, 3D printing, vehicular ad hoc network (VANET), blockchain, telehealth technology, drones, robots, open source technology, mobile apps, and video conferencing are discussed in the technologies, technological devices, and technological applications. The overall impact includes both positive and negative effects on various sectors like health, transportation, Internet data consumption, economy, sports, entertainment, agriculture, and environment. The severe impact of COVID-19 and the role of technology, particularly artificial intelligence, machine learning, and deep learning, in combating the virus are specified here, and various statistical data for this survey are also presented.

**Corresponding author*: biswa.rnjn@gmail.com

Chandra Singh, Rathishchandra R. Gatti, K.V.S.S.S.S. Sairam, Manjunatha Badiger, Naveen Kumar S. and Varun Saxena (eds.) Modeling and Optimization of Signals Using Machine Learning Techniques, (173–208) © 2024 Scrivener Publishing LLC

Keywords: Artificial intelligence, coronavirus, COVID-19, deep learning, drone, machine learning

9.1 Introduction

COVID-19 turned into a global health threat within a very short time. A wave of COVID-19 infections struck almost all countries including the USA, Russia, Spain, the UK, Iran, and India. Until now, two waves of the COVID-19 infection have been observed in the whole world. A month-wise comparison of the two waves of the COVID-19 infection is presented in Table 9.1.

Even in developing countries like India, positive cases of the coronavirus are rising very rapidly. In India, the first active case was detected on January 30, 2020. At present (up to June 2021), the total positive cases in India are more than 39,777,853.

If we look at the cumulative data up to June 2021, the number of positive cases for the whole world is approximately 18.2 Cr, and the total number of death cases is approximately 39.5 L [2]. The effect of the coronavirus is so dangerous that the World Health Organization declared this outbreak as a global pandemic. This pandemic not only affected the health of the people, but it also changed the habits and lifestyles of the people. The spread of the coronavirus forced governments across many countries to take various safety and protective measures for the well-being of the people. Quarantine as the restriction for people who are in contact with infected people, remote working for public offices and software companies, and the closing of schools and universities are some of the outcomes of COVID-19

Table 9.1 Comparative month-wise coronavirus statistics for the whole world [1].

Month	Cumulative positive cases in 2020	Cumulative deaths in 2020	Cumulative positive cases in 2021	Cumulative deaths in 2021
January	9,847	213	103,333,479	2, 225, 773
February	85,403	2,924	114,350,611	2, 528, 993
March	856,955	42,089	129,317,945	2, 807, 094
April	3,303,296	230,804	151,517,785	3, 171, 896
May	6,226,409	370,854	170,865,698	3, 544, 118

in the whole world. Along with the health crisis, other sectors like education, research, transportation, and agriculture were severely affected all over the world. This creates panic among the common people and becomes a challenge for the government to maintain the safety of the people.

As per our knowledge, most of ongoing research on this COVID-19 pandemic emphasizes a particular field. There is no research study about the overall impact of this outbreak in various sectors. Apart from this, maintaining social distancing and thinking of the development of the country is only possible with the help of technology. This paper discusses the overall impact of COVID-19 (positive and negative) and the role of various technologies used to overcome this pandemic.

The significant contributions of this chapter are mentioned as follows:

1. Discussions on the socio-economic impact of the coronavirus across the globe
2. A statistical survey of both the bad and the good impacts of COVID-19
3. A review of different technologies to reduce the impacts of COVID-19
4. The role of artificial intelligence (AI), deep learning, and machine learning in the fight against COVID-19.

The rest of the chapter is organized as follows. Section 9.2 discusses the coronavirus, its origin, the mode of spreading, and the steps taken by the government to contain the transmission of the coronavirus. Section 9.3 presents the socio-economic impacts on various sectors because of COVID-19. Section 9.4 focuses on the benefits due to the impacts of COVID-19. Section 9.5 details the use of different technologies to combat this global pandemic. Section 9.6 specifies particularly the role of AI, machine learning, and deep learning against COVID-19. Section 9.7 summarizes the different related works. Finally, section 9.8 concludes this review paper.

9.2 Discussions on the Coronavirus

This section briefly discusses the coronavirus, the origin of the coronavirus, COVID-19, the mode of spreading of the coronavirus, and symptoms of COVID-19.

9.2.1 Coronavirus

Coronaviruses are a large family of viruses that can affect the health of animals and human beings. *Corona* in Latin means crown. The coronavirus is

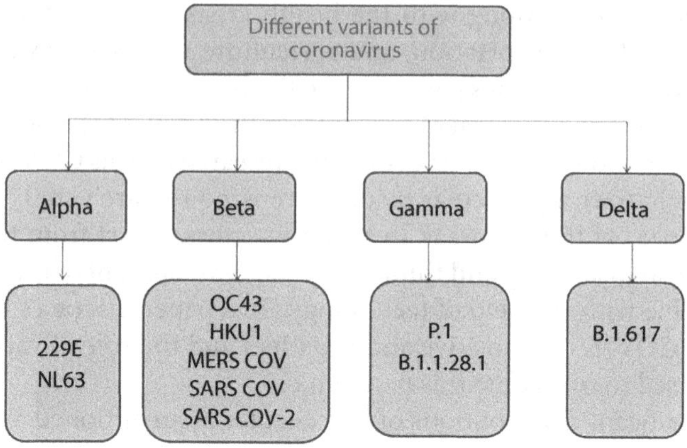

Figure 9.1 Different variants of the coronavirus.

named as such because its projection on the surface appears like a crown. Researchers first identified the coronavirus in 1937, which causes a type of bronchitis in birds. In human beings, coronaviruses were found in the noses of people suffering from a common cold in 1960 [3]. Different respiratory infections occur in human beings, like the common cold, Middle East respiratory syndrome (MERS), and severe acute respiratory syndrome (SARS), due to coronaviruses. The different variants of the coronavirus that can affect human health are presented in Figure 9.1 [4].

Gamma P.1 is another variant of the coronavirus detected in January 2021, which is found to be more contagious than the earlier strain.

The Delta (B.1.617.2) strain has mostly affected young people, and its effect was mostly found in mid-April in almost 43 countries like the US, the UK, Australia, Singapore, and India.

9.2.2 COVID-19

The infectious disease caused by the most recently discovered coronavirus is known as COVID-19. The name of the novel coronavirus responsible for COVID-19 is SARS-COV-2. This virus was first identified in Wuhan, China. At present, the greatest number of deaths occurred in the USA. People from around 215 countries are affected by COVID-19.

9.2.3 Origin of COVID-19 and Its Symptoms

The outbreak of SARS-COV-2 started at the Hunan seafood market of Wuhan City in China. At the Hunan market, live animals are sold like

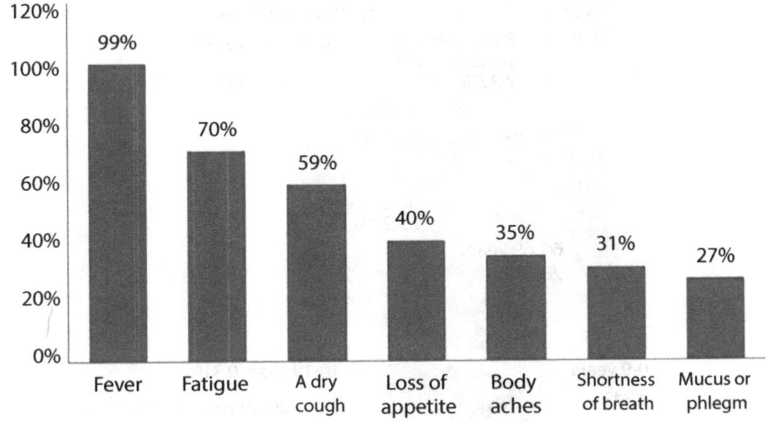

Figure 9.2 Symptoms-wise rate of infection.

frogs, bats, snakes, and rabbits [5, 6]. On January 12, 2020, the National Health Commission of China declared the COVID-19 as viral pneumonia. By the sequence-based analysis of isolates from the patient, the virus was declared a novel coronavirus. The most common symptoms of COVID-19 are fever, cough, and tiredness. Some mild symptoms of COVID-19 are nasal congestion, headache, sore throat, loss of taste or smell, diarrhea, rash on-screen, discoloration of fingers or toes, etc. The result of this survey is shown in Figure 9.2 [7].

9.2.4 Mode of Spreading

The total number of deaths in the whole world due to COVID-19 is around 364,041. However, the number of deaths varies for different age groups. Figure 9.3 shows the number of deaths in various age groups of people. There is a larger number of deaths for the older group as compared to the younger group of people. Looking at the mortality, the analysis of the different modes of spreading is essential. COVID-19 is mainly spread by the droplets produced as a result of the coughing and sneezing of the infected person suffering from COVID-19.

The spreading may be due to direct contact or indirect contact. In direct contact, when the normal person comes nearer to the infected person, then the virus is transmitted. In indirect contact, by touching the droplets present on surfaces or clothes with the hand, and then touching the mouth and nose with the hand, the virus is transmitted to a person [8].

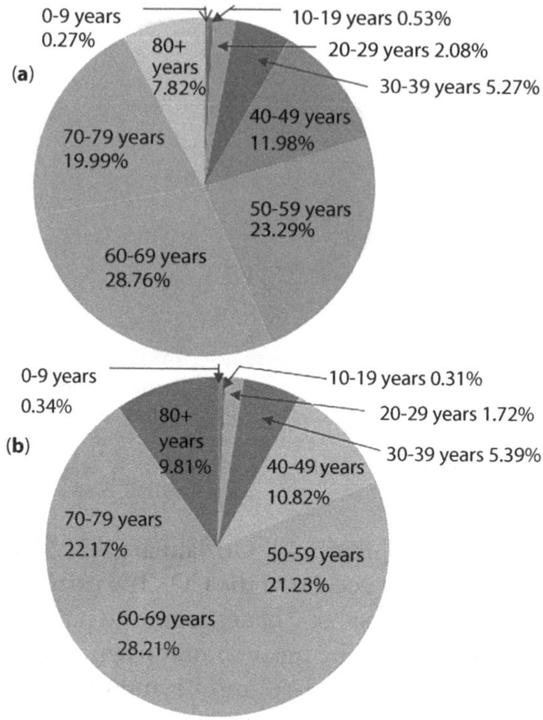

Figure 9.3 The comparison of mortality rates in the first and second waves. (a) Percentage of death in the first wave for different age groups. (b) Percentage of death in the second wave for different age groups.

9.2.5 Steps Taken by the Government to Prevent the Spread of COVID-19

The only way to restrain the COVID-19 transmission is social distancing. In India, according to the guidelines of the Health Ministry, states and union territories have divided their corresponding area into red, orange, and green zones. Then, district administrations and local urban bodies have demarcated containment and buffer zones inside the red and orange zones by considering the local transmission history [9, 10]. Brief descriptions of the different zones are as follows.

- **Red Zone:** These are the areas with the highest caseloads.
- **Orange Zone:** These are the regions that have reported a restricted number of cases before and no rise in total cases as of late.

- **Green Zone:** These are the regions with no positive cases until now or zero active cases in the most recent 21 days.
- **Containment Zone:** These zones are decided by considering the number of positive cases, contact tracing the history, and the population density of a particular area as the parameters. The area around a 3-km radius from the epicenter, which satisfies the above three parameters, is declared as the containment zone. Outward movement from the containment zone is prohibited, except for some necessary utilities. All streets connecting with the control zone will be watched by the police.
- **Buffer Zone:** These are the adjoining blocks of the affected district or rural districts of the affected city.

Dividing the region into various zones helps to maintain social distancing that helps to reduce the spreading of COVID-19.

Looking at the death rate and the increase in active cases, researchers from different countries carry out their research to create vaccines. Many countries have found different vaccines. A list of vaccines authorized for emergency use by the different countries is presented in Table 9.2.

The spread of the coronavirus has not only affected the health of the people but also has some impact on various fields.

The overall impact due to COVID-19 is classified into two categories. These are good impacts and bad impacts. The good impact is further classified into health benefits and other benefits. The bad impact is broadly classified into social impact and economic impact. In the social impact,

Table 9.2 Coronavirus vaccines for emergency use in different countries.

Country	Vaccine
United Kingdom, Sweden	Oxford–AstraZeneca COVID-19 vaccine
Germany, United States	Pfizer–BioNTech COVID-19 vaccine
United States, Netherlands	Johnson and Johnson COVID-19 vaccine
China	BBIBP-CorV
United States	Moderna COVID-19 vaccine
Russia	Sputnik V COVID-19 vaccine
India	Covaxin, Covishield

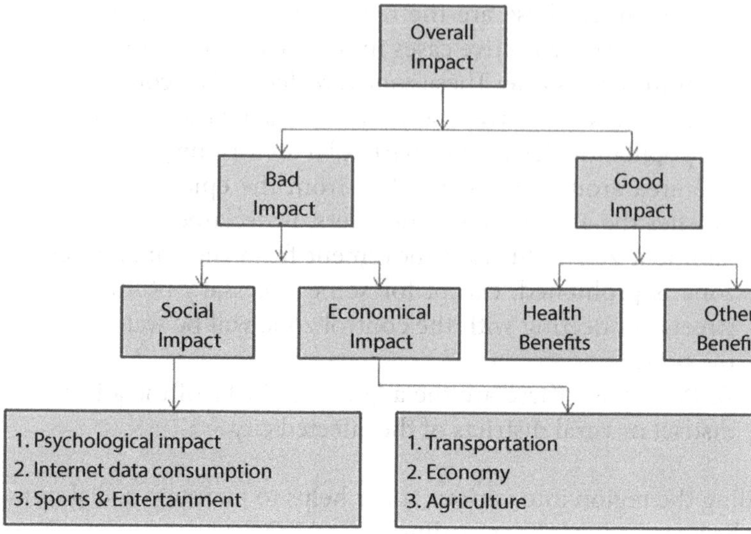

Figure 9.4 Classification of the overall impact due to COVID-19.

psychological and mental impact, Internet data consumption, sports, and entertainment are discussed. Economy, transportation, and agriculture are discussed in the economic impact. The schematic representation of the overall impact is illustrated in Figure 9.4.

The bad impact of COVID-19 on various sectors is discussed in the next section.

9.3 Bad Impacts of the Coronavirus

COVID-19 has had an adverse impact on almost all sectors. The loss in various sectors can be easily identified from the world stock market. The return of various sectors of the world stock market is shown in Figure 9.5 [11]. In almost all sectors, the stock market return is negative. The different social and economic impacts due to COVID-19 are discussed next.

9.3.1 Social Impact

The outbreak of COVID-19 has not only affected global health but also has a social impact worldwide. The different social impacts of COVID-19 are discussed in the next subsection.

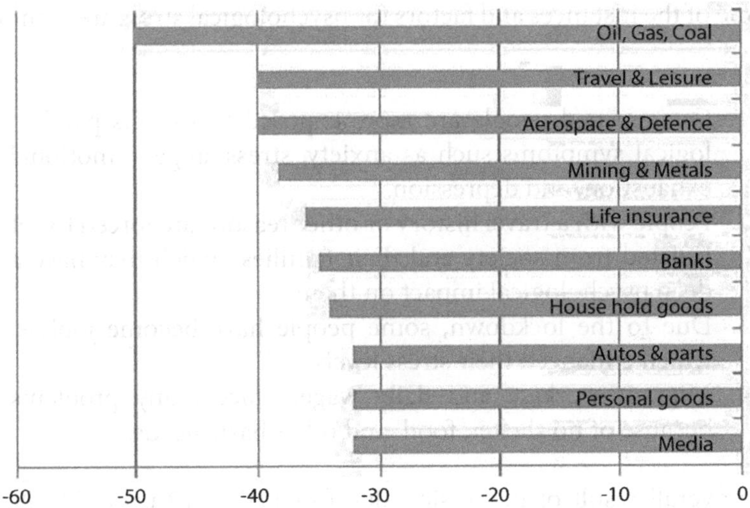

Figure 9.5 Returns of various sectors on the world stock market.

9.3.1.1 *Mental Health and Psychological Impacts Due to COVID-19*

To control the further spread of the coronavirus, most governments declared lockdowns and shutdowns in the countries. This also affects the health of the people psychologically and mentally. A study on the psychological and mental effects was conducted in China consisting of 194 cities [12]. In this study, around 1200 people participated. Using the impact of event scale-revised (IES–R), the psychological impact was accessed. Using the depression, anxiety, and stress scale (DASS-21), the mental health status was accessed. The overall result of the psychological impact of using IES-R is represented in Table 9.3.

From the results, it is clear that due to COVID-19, the majority of people are psychologically affected.

Table 9.3 Result of the psychological impact.

State of psychological impact	Percentage
Minimal psychological impact	24.5%
Mild psychological impact	21.7%
Moderate or severe psychological impact	53.8%

Some of the instances and factors for psychological stress are mentioned here:

1. Quarantined people are more exposed to various psychological symptoms such as anxiety, stress, anger, emotional exhaustion, and depression.
2. People with a travel history or other reasons are forced to get isolated from society and their families, which may have a deep psychological impact on them.
3. Due to the lockdown, some people have become jobless, which enhanced their stress levels.
4. Migrant workers and daily wagers face many problems because of no shelter, food, and other basic needs.

The overall result of depression due to COVID-19 using DASS-21 is shown in Table 9.4.

Some of the reasons for depression are mentioned here:

1. At present, nobody can predict the end period of this pandemic, and this has had a serious impact on the mental health of health workers.
2. Many health workers and front-line COVID-19 fighters stress over spreading the infection to different patients, friends, and family as they are more exposed to the virus.
3. In this emergency, many medical professionals do not have time to consult a therapist about their anxiety, insomnia, and depression.

Table 9.4 Result of the mental health impact.

State of depression	Percentage
Minimal depression	69.7%
Mild depression	13.8%
Moderate depression	12.2%
Severe depression	4.3%

4. Stigma and discrimination can affect the mental health of healthcare workers, sanitary workers, police, administrations, etc. who are more exposed to the coronavirus infection.
5. Certain groups of people like adults and people having health issues are getting much more worried and concerned about the spread of COVID-19.

As long as the spreading of the coronavirus continues, the psychological and mental health effects on people will not be stabilized.

9.3.1.2 Impact on Internet Data Consumption Due to COVID-19

So far, online usage has been raised significantly due to the lockdown in so many countries because of COVID-19. Internet traffic is more than 25% to 30% during this pandemic period. From the beginning of this pandemic until the present, traffic through the WebEx video conferencing service has increased by 21 times in China [13]. The number of users using WebEx video conferencing has almost doubled in countries like Japan, South Korea, and Singapore. In India, Internet consumption increased by 13% [14]. The daily Internet data consumption in India is around 308 petabytes (PB). Cloud-flare tracks the internet traffic growth in various cities of the world from the first week of January to the last week of March, which is represented in Table 9.5.

Vodafone, which operates in around 64 countries, predicted around a 50% rise in data traffic. Ericsson predicted a 20% to 70% rise in voice traffic throughout this period. Various factors for the increase in Internet data are as follows:

1. Remote work for many employees due to lockdowns in many countries.
2. E-learning due to the closing of educational institutions during this pandemic.
3. Almost all age groups are using the Internet for online games and videos for recreational purposes while staying at home.
4. Searching for COVID-19 data regularly through the Internet has also increased Internet data.

The types of usage of the Internet and the increase in its use in percentage during COVID-19 in comparison to the pre-COVID-19 period are shown in Table 9.6 [14].

Table 9.5 Internet traffic growth in various places.

Country or city	Internet traffic growth
Berlin	11%
London	22%
UK	17%
New York	40%
San Francisco	48%

Table 9.6 Percentage increase in various usages of the Internet.

Types of usage	Percentage increase
Gaming	115%
VPN	49%
Video	36 %
Downloads	39%
Web	27%
Social	-12%
Voice minutes of use	25%

The demand for the increased usage of the Internet during the COVID-19 pandemic is a challenge to service providers. Apart from this network, security will be a challenging task.

9.3.1.3 *Impact on Sports and Entertainment Due to COVID-19*

The sports and entertainment fields are also affected during this pandemic period. Many international, national, and regional sports events like marathons, football, ball tournaments, athletics championships, basketball, handball, ice hockey, rugby, cricket, and weightlifting are either canceled or postponed.

Euro 2020 and the Tokyo Olympic Games were postponed to 2021. Football leagues, NBA leagues, and Formula 1 races were suspended until further notice. The most popular cricket league, the Indian Premier League (IPL), was suspended until further notice. Many international crickets and domestic crickets were canceled due to COVID-19, which incurred a huge loss to the players, organizers, and sporting service industries associated with travel, catering, media broadcasting, etc. Movies, a popular mode of entertainment, were also affected. Making and releasing new movies were not possible during this pandemic. Canada's Cineplex Inc. closed 165 movie theaters. The only positive point for the entertainment industry is the significant hike in video streaming companies because of the lockdown.

9.3.2 Economic Impact Due to COVID-19

COVID-19 has an adverse effect on many sectors like transportation, tourism, agriculture, and global trade, which affects the economy of many countries. The different sectors that have affected the economy are discussed in the next section.

9.3.2.1 Impact on Transportation Due to COVID-19

One of the primary victims of COVID-19 is the transportation sector. The transportation mode is classified into three categories: land (rail and road), air, and water. A huge loss has occurred in these three modes of transportation during the COVID-19 pandemic. Canada is operating at 79% of the normal commercial transportation activities, and Maritimes is operating at 70% of the normal transportation activities. The US is operating at 83% of the normal transportation activities [15]. Due to the lockdown, around 200 trains were canceled in India, incurring a huge loss [16]. Within cities, the different ways of transportation like metros, city buses, auto/e-rickshaws, and taxis were also stopped due to the lockdown in many countries. The cancellation of international and domestic flights during the lockdown resulted in a loss of USD 3 to 3.6 billion for the Indian aviation industry [17]. The marine transportation system was also affected due to COVID-19. Some of the effects of COVID-19 on the marine transportation system are the delay in the arrival of shipment goods, the accumulation of goods in the port beyond the accumulation limit, the requirement of extra time for loading and unloading goods, etc. [18]. Also, the tourism sector was affected due to the cancellation of tickets for various modes of transportation. Although all the transportation sectors will be functional within some time, the fear of spreading the coronavirus will still be there.

9.3.2.2 Impact on the Economy Due to COVID-19

The economy of a country is dependent on various other sectors like tourism, transportation, education, and utilization of human resources. However, all these sectors were severely affected due to COVID-19 in almost all countries. In the US, job loss is high. Airbus suspended production in Spain and France. Many small industries were closed due to the lockdown. Many employees and workers were affected financially. Due to the lockdown in many countries, the production and consumption of various goods were decreased. There are sharp falls in the global financial market. According to the estimation of the Organisation for Economic Co-operation and Development (OECD), the global economic growth is tumbling to 2.4% for the entire year when contrasted with the 2.9% in the year 2019. The economic impact in terms of the percentage of Gross Domestic Product (GDP) during this pandemic period has decreased for almost all countries [11]. The prediction of the economic impact is a challenging task, and the economic loss will continue as long as the severity of COVID-19 has not decreased.

9 3.2.3 Impact on Agriculture Due to COVID-19

The impact on agriculture due to COVID-19 is complex. The problem in agriculture due to the lockdown is the non-availability of laborers and transportation facilities to carry out agricultural work and to sell the products respectively. Also, due to the fear of spreading the coronavirus, consumers do not go to the market frequently to purchase agricultural products like vegetables, fish, fruits, dairy products, etc. which become rotten or wasted in the marketplace. Thus, due to COVID-19, the food supply chain that connects the farmer and the consumer through various processes like manufacturing, packaging, distribution, and storage is affected.

The lockdown in various countries during this pandemic also had some benefits to the people and the environment, which are discussed in the next section.

9.4 Benefits Due to the Impact of COVID-19

A detailed gain in several fields is described in this section. Most of the benefits acquired due to the lockdown, shutdown, social distancing, and restricted movement were directly or indirectly connected to our health.

9.4.1 Health Benefits

This subsection briefly discusses cleaner air, a reduction in smoking and drinking alcohol, and some other benefits.

9.4.1.1 Cleaner Air

These days, polluted air is one of the major causes of premature death, as well as several health issues, such as respiratory infections, heart disease, chronic obstructive pulmonary disease, stroke, and lung cancer. Now, the worldwide lockdown, restricted movement, and prohibition in all commercial and industrial activities make the atmosphere purer and give a chance for people to breathe easily. A statistical analysis of some major air pollutants, in those cities during the period from January 1 to April 9, 2020, is mentioned here [19]:

1. PM2.5 and PM10: The maximum decline in the monthly concentration of PM2.5 and PM10 is shown in Figures 9.6 and 9.7, respectively.
2. NO_2 and O_3: The maximum decline in the monthly concentration of NO_2 and O_3 is shown in Figures 9.8 and 9.9, respectively

In India, to slow down the spreading of the COVID-19 outbreak, the government declared a nationwide lockdown on March 25, 2020, for the first time. It restricted the movement of 1.3 billion people and banned all commercial and industrial activities. Research says that from March 16

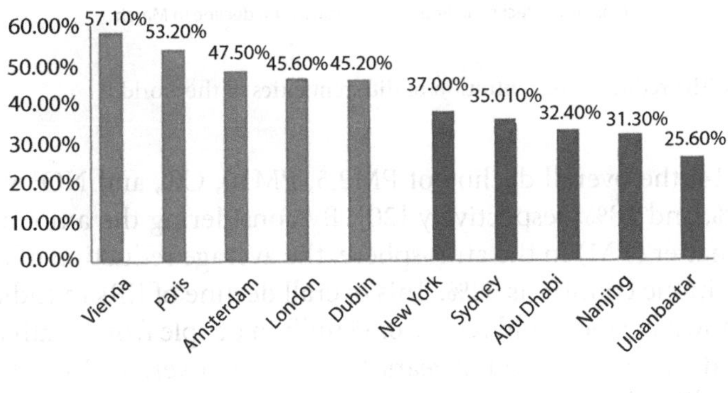

Figure 9.6 The reduced concentration of PM2.5 in different cities of the world.

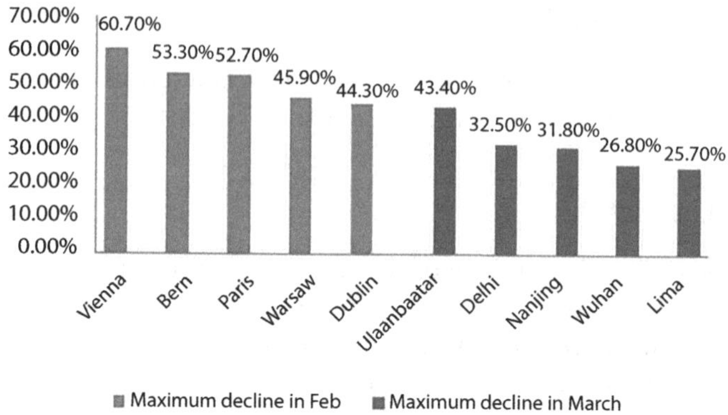

Figure 9.7 The reduced concentration of PM10 in different cities of the world.

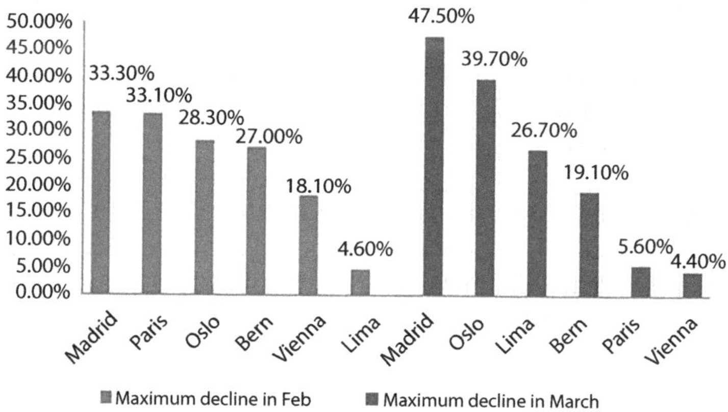

Figure 9.8 The reduced concentration of different cities of the world.

to April 14, the overall decline of PM2.5, PM10, CO, and NO^2 were 43%, 31%, 10%, and 18%, respectively [20]. By considering the amount of particulate matter (PM) in the atmosphere, the average reduction of excessive risk (ER) in the country is 52%. This overall decline of ERs in India during the lockdown period could rescue 0.65 million people from death in a year. Compared with the previous 3 years, there was an overall 30% reduction in the air quality index (AQI).

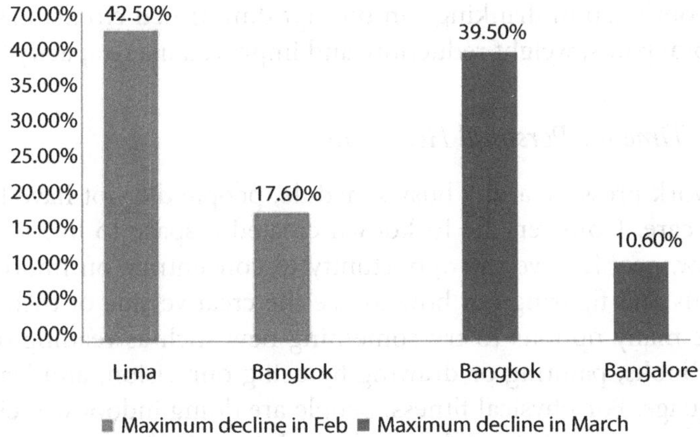

Figure 9.9 The reduced concentration of O_3 and NO_2 in different cities of the world.

9.4.1.2 Limited Smoking

By knowing the direct effect of the coronavirus on our respiratory systems, around 300,000 people have been convinced to stop smoking and have pushed another 550,000 people against smoking. The lungs start to fix themselves very quickly after they quit smoking and the function increases up to 10% within nine months, according to the views of the National Health Service. Following a year, the danger of coronary illnesses will be reduced to half, and in the following 10 years, the possibility of having lung diseases will be decreased to half for those who stopped smoking [21].

9.4.1.3 Drinking Alcohol is Down for a Few

Many people's drinking habits were disturbed due to the cancellation of parties, clubs, pub trips and the shutting down of alcohol shops in different cities, etc. According to research, the following conclusions are drawn:

1. In every five people, only one is drinking more frequently.
2. One in three people has reduced his/her habit of drinking since the lockdown.
3. Out of 20 people, one person has become abstinent.

A 1-month gap in drinking can prompt diminished circulatory strain, less cerebral pains, weight reduction, and improved liver capacity [21].

9.4.1.4 *Time for Personal Healthcare*

Due to work pressure and a busy schedule, people did not have time for personal care. However, the lockdown created a space to live life differently. Now, people have the opportunity to concentrate on improving as individuals and figuring out how to use the creative side of their minds. There are many options to try something new such as reading different kinds of books, painting or drawing by using our vision, and learning a new language. For physical fitness, people are doing indoor exercises and consuming homemade food during the lockdown [21].

9.4.2 Other Benefits Due to the Lockdown

- **Reduced water pollution:** Shutting down almost all industrial activities during the lockdown caused a significant improvement in the water quality of rivers. Out of 36 monitoring points in Ganga, 27 monitoring points were recorded as favorable for bathing and breeding of wildlife and fishes [22].
- **Natural resources conservation:** The consumption of natural resources like petroleum, diesel, and gasoline was reduced because of restricted transportation, industry shutdown, and a halt in cargo movement.

The effect of COVID-19 has both good and bad impacts on people. However, the bad impact is more visible compared to the good impact. Despite the global pandemic, the development of the country must go on, and the effect of COVID-19 must decrease. For this, the role of technology is very important. Different technologies used for the benefit of the people during this pandemic are discussed in the next section.

9.5 Role of Technology to Combat the Global Pandemic COVID-19

In this vital situation, technology makes a way to connect people virtually and becomes a solution to various problems. Different technologies,

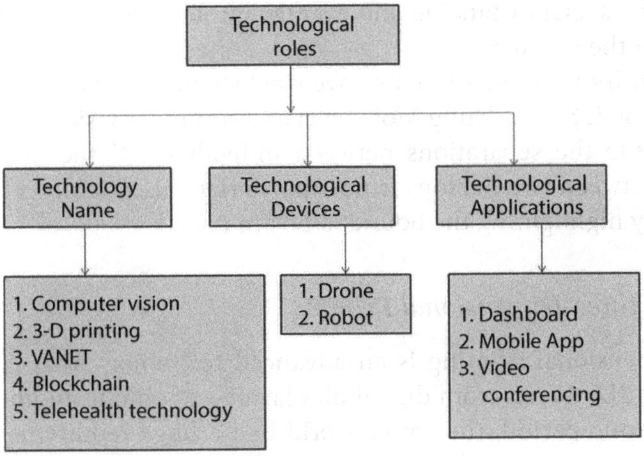

Figure 9.10 Technological roles during COVID-19.

technological devices, and technological applications are used for the upliftment of different sectors and the fight against COVID-19. The overall technological role is mentioned in Figure 9.10.

The overall role of technology is briefly discussed in the next subsection.

9.5.1 Use of Different Technologies

In this subsection, various technologies like computer vision, 3D printing, vehicular ad hoc networks (VANETs), blockchain, and telehealth technology are discussed.

9.5.1.1 Computer Vision

For safety against COVID-19, computer vision also plays an important role in checking social distancing, the need for sanitization, etc. [23].

- Numina.co has built up a device that estimates what, where, when, and how objects move in roads and open spaces. New York uses this tool for monitoring social distancing.
- Voxel51, a start-up of computer vision, utilizes its cutting-edge computer vision models and the collected live videos from some of the busy streets in the world for calculating the physical distancing index (PDI). This index computes

the social distancing and average human activity over time in the world.
- In India, some students have used single-shot object detection (SSD) utilizing MobileNet and computer vision to compute the separations between individuals. If the distance between two persons is not appropriate, then it alerts them by highlighting the bounded box in red.

9.5.1.2 Three-Dimensional Printing

Three-dimensional printing is an advanced technology that enables the making of 3D objects from digital files layered upon one another. During this pandemic period, the whole world had a huge requirement for different protective types of equipment, masks, ventilators, and other medical supplies. Three-dimensional printing is used to reduce the production pressure of the above requirements. It helps to create the prototypes of the final product by using metal and plastic. The production of different medical supplies by some industries using 3D technologies is discussed here. Airbus, Apple, Nike, and Adidas have produced face shields using this technique. In the US, HP and Helpful Engineering have manufactured nasopharyngeal test swabs and ventilators, respectively. In Italy, Isinnova helped to produce valves for reanimation devices. In India, Boson Machines and Divide by Zero Technologies created face shields, face masks, and face mask pressure relievers. Ethereal Machines has also produced ventilator splitters in India [24].

9.5.1.3 Vehicular Ad Hoc Network (VANET)

The VANET is an emerging research technology that is widely used for various applications because of the presence of a communication unit (on-board unit) in the vehicle, availability of various sensors for different applications, availability of 5G technology for faster data transmission, availability of different optimization approaches to optimize vehicular network parameters, faulty communication unit identification, etc. [25, 26]. Different applications of VANET include environmental parameters monitoring [27], automatic toll tax collection [28], automatic parking service [29], route diversion during congestion [30], and automatic fire monitoring [31]. During this global pandemic, using vehicular technology deviation of the route from the containment zone, the identification of the location of the people suffering from COVID-19 by ambulance can be easily achieved through vehicular technology.

9.5.1.4 *Blockchain*

For the management of the COVID-19 pandemic, blockchain technology is utilized in various ways. Concerning the pandemic, the blockchain acts as an important technology for COVID-19 by tracking the spread of the pandemic. The IBM blockchain is taking part in MiPasa, a project supporting an open data hub intended to rapidly give the early detection of the coronavirus and infection hotspots. Blockchain technology helps to raise the end-to-end visibility in various supply chains such as the food supply chain and the pharmaceutical supply chain. It provides a trusted platform to drastically improve processes and share information in near real-time between businesses, governments, health organizations, doctors, researchers, and the public [32].

9.5.1.5 *Telehealth Technology*

To limit the rate of infection, particularly inside emergency clinics, telehealth can be a cost-effective technology that intends to contain the coronavirus from spreading. It permits patients to be observed and diagnosed remotely by doctors using an audiovisual, real-time, and two-way interactive communication system. The growth of the telehealth market in the US increased year-over-year by 64.3%. China used a mobile telehealth system (MTS) in the hospital to take care of the patients treated in isolated wards. In India, digital platforms for health consultancy like Practo and Foreign Outpatient Department (OPD) have been seeing a surge since the start of March 2020 [33].

9.5.2 Technological Devices

The two most popular technological devices widely used during this pandemic are drones and robots. These are briefly discussed in the next subsection.

9.5.2.1 *Drones*

Drones or Unmanned Aerial Vehicles (UAVs) play a vital role in managing the situation in various ways such as monitoring the movement of people, surveillance, sanitizing an area, shooting videos, and capturing images of a place. The list of some uses of UAVs across the world, including in India, is mentioned here [34, 35].

- **Surveillance:** In India, Gujarat police used drones to keep track of the people's movements across states. Delhi police used drones to monitor social distancing in Azadpur Mandi, Asia's largest fruit and vegetable market. Now, approximately 450–500 drones have been working across India in different states.

- **Broadcast:** To broadcast the COVID-19 guidelines, drones were used in China and many European countries exclusively in rural areas. In the starting phase of the lockdown, at Srinagar's busy Lal Chowk, India, UAVs were used to make the announcements regarding the COVID-19 guidelines on public movement.
- **Spraying disinfectant:** To prevent the further spreading of the coronavirus, various countries like China, Indonesia, Colombia, the Philippines, and India have used drones to disinfect the cities across each country.
- **Medicine and essential goods deliveries:** In Wuhan, China, drones were used for supplying medicines, samples for tests, and other medical accessories. Apart from this, drones have been used to deliver groceries in China, the US, and Australia.
- **Temperature screening:** During lockdown, drones were equipped with infrared cameras for checking the temperature of people at home in China.
- **Other uses:** Due to shutdowns in some parts, media organizations have been using drones to do live video updates and capture images to cover the current news.

9.5.2.2 Robots

- In Kerala, India, Srishti Robotics, an industrial robot programming training institute, and a group of manufacturers from Vimal Jyothi Engineering College, Kannur, have developed a robot named Nightingale-19 for the benefit of patients and health workers against COVID-19. This facilitates a video system for interaction. Through this video system, patients can interact with their relatives. Also, the video system helps medical professionals check the health condition of patients while maintaining social distancing. The robot has a bearing capacity of 25 kg in weight (approximately six patients' food and medicine) and can operate a maximum of 1 km in distance from the controller.

- Robots are used to perform the COVID-19 test quickly. Spain has deployed several robots to automate the testing process. They can give results of up to 80,000 tests in one day [36].
- Robots are widely used for disinfecting in countries including India.
- Other uses of robots are to remind people about social distancing at some entry points of public places, provide a medium to teach students during the closure of schools, help in preparing masks (in Spain), etc. [36].

9.5.3 Technological Applications

Various technological applications like dashboards, mobile apps, and video conferencing during the pandemic period are discussed in the next subsection.

9.5.3.1 Open-Source Technology

- In Kerala, the Corona Safe Network is an open-source network to help the public. This network is planned by a multidisciplinary group of trendsetters and volunteers who are working to support the government during this crisis. The

Table 9.7 A list of dashboards available in India.

Name	Description
Bing COVID	Microsoft's Bing search engine has developed a site for COVID-19 information. Along with the coronavirus cases, it provides a map that highlights the coronavirus hotspots of the world.
COVID-19 India	This dashboard is only created to show the infection statistics such as total active cases, deaths, and recoveries in India.
Ministry of Health Webpage	This page provides other information like the helpline toll-free number and the helpline email IDs, along with the coronavirus cases update.

content of this project is available in seven languages. The main motive of this network is to provide various digital options for helping people and to provide COVID-19 statistics for public awareness.

As an open-source technology, various dashboards are used for tracking the daily and cumulative records of worldwide COVID-19 cases. Some of the highlighted dashboards of India are described in Table 9.7 [37].

Some of the top-listed dashboards of the overall world are mentioned in Table 9.8 [38].

Table 9.8 A list of worldwide dashboards.

Name	**Description**
Johns Hopkins University dashboard	It collects data from organizations like the World Health Organization (WHO) and the Centers for Disease Control (CDC) in the US, China, and Europe to provide detailed information about the number of confirmed, suspected, recovered, and death cases.
Microsoft Bing COVID-19 dashboard	It pulls information from the WHO, the CDC, and the European Centre for Disease Prevention and Control (ECDC). Here, the public can easily access the news of the latest infection, as well as COVID-19 cases.
Centers for Disease Control dashboard	It likewise gives a thorough outline of the spread of coronavirus cases, particularly in the US. Here, the only drawback is that its content is not updated frequently.
WHO dashboard	It provides worldwide coronavirus data in a minimalistic form, as well as timetable diagrams for demonstrating the cases of COVID-19.
The New York Times dashboard	It has a coronavirus site with many maps and charts that show the most recent and freely accessible information on the spread of COVID-19.

9.5.3.2 *Mobile Apps*

Throughout the world, along with India, many mobile apps are designed for contact tracing and tracking corona-infected people. The following are some apps used in different countries [39]:

- A start-up in the UK has developed an app for the public, named the C19 COVID Symptom Tracker, to self-report their symptoms. It assists to distinguish the zones which are more exposed to the coronavirus.
- South Korea launched an application called Corona 100m, which can map the position of a corona-infected person and warn the people who are within 100 m closure to the patient.
- In China, Alibaba and Tencent designed an application that acts as an e-pass to decide people's entry into various public places like malls and subway stations. This app is based on a big data-driven algorithm, and it considers the health condition and travel history of an individual as some parameters to make decisions.

India has also developed different contact tracing and patient monitoring mobile apps, which are described next [40].

- **Aarogya Setu:** The National Informatics Centre under the Ministry of Electronics and Information Technology (MeitY) launched this app on April 2, 2020 to bring control over the spread of the coronavirus. It utilizes an advanced mobile phone's Bluetooth and GPS systems to warn people at the time they come to a closure of 6 ft in distance from a corona-infected person. The alerts are created by scanning the location-specific records of the government. Now, the installation of Aarogya Setu has become mandatory for people in various sectors, such as for the people in the containment zone, airports, railways, employees in the private sector, health workers, and frontline COVID-19 fighters.
- **Quarantine Monitor:** This app was developed by the Tamil Nadu eGovernance Agency to keep an eye on the individuals in quarantine centers according to the official database of the state.

- **MahaKavach:** In Maharashtra, this app is used to inform the local health authorities in case any suspected person in quarantine violates the rule.
- **COVA Punjab:** During the lockdown, the Punjab government launched a multipurpose app named COVA Punjab. In public places, this app can provide information to a person about the presence of COVID-19-affected persons. The government also distributes basic goods to people in need using this app.
- **Test Yourself:** Both Goa and Puducherry governments are using this app for public healthcare. It sends the client's self-assessment tests at whatever point they make appointments at medical service units.
- **GoK Direct - Kerala:** This app helps to spread trustworthy and fully proven information about COVID-19.

9.5.3.3 Video Conferencing

Because of the worldwide lockdown, video conferencing is the only way to establish communication. Also, many national or international meetings on coronavirus are held through video conferencing. Some of the most usable video conferencing tools are Zoom, Google Meet, Microsoft Teams, etc. [41].

Although all technology plays a crucial role during this pandemic, the role of AI, machine learning, and deep learning is very significant and vast. The role of the abovementioned technologies is discussed in the next section.

9.6 The Role of Artificial Intelligence, Machine Learning, and Deep Learning in COVID-19

This section mentions the role of AI, machine learning, and deep learning to tackle various situations during the COVID-19 pandemic. Many researchers are working with different methods of AI approaches to overcome the global pandemic situation. Various methods used for AI are presented in Figure 9.11. Different approaches based on the various methods to fight the coronavirus battle are mentioned in the next subsection.

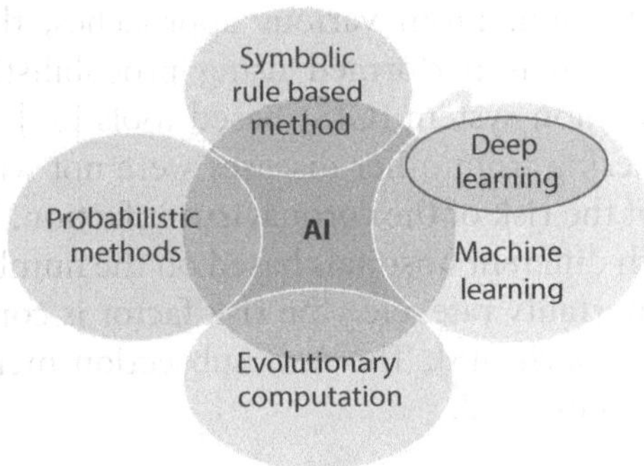

Figure 9.11 Various methods of AI.

9.6.1 Symbolic Rule-Based Method

The symbolic rule-based AI is also called classical AI, rule-based AI, or good old-fashioned AI. The benefit of symbolic-based AI is the easy visualization of its logic. Some examples of symbolic rule-based AI are expert systems, rule engines, and knowledge graphs. Predicting the spread of the effect of the coronavirus is a challenging task. A combination of the mixture of expert systems is designed to predict the COVID-19 patients correctly [42]. Also, during the pandemic period, the normal lives of people have been affected severely. Many people have lost their jobs, many are suffering from financial constraints. During this situation, an analysis of the sentiments of people is essential. The sentiments of the people are analyzed through various social media with the help of automata theory and natural language processing [43]. The next subsection discusses various probabilistic methods of AI to battle against COVID-19.

9.6.2 Probabilistic Method

For the given set of input data and the model for computation, the probabilistic approach can predict the value of unknown quantities and can also perform the classification of the predicted quantities. Basically, this approach helps in the determination of the forecasting of uncertain data, learning from the limited data or noisy data, learning from complex models, etc.. One of the popular examples of the probabilistic approach is the Naïve Bayesian model. The probabilistic method also plays a vital role during COVID-19, which is discussed here. Different countries adopt different approaches to tackle the COVID-19 situation. Even in a country, different states are approaching different methods to cope with the COVID-19 situation. From various approaches, the determination of a suitable approach is performed using probabilistic modeling and Geographic information system (GIS)-based tools [44]. During the pandemic, people suffering from other diseases were not willing to go to the hospital because of the risk of the coronavirus infection. To determine the risk associated with different hospitals based on the number of COVID-19 patients and the mortality rate, etc., the risk factor is computed using the Naïve Bayesian approach [45]. The next subsection mentions the evolutionary computation method.

9.6.3 Evolutionary Computation Method

An evolutionary approach is a metaheuristic approach. In each iteration, evolutionary computation generates a solution set that is called the population. Many researchers are working on the evolutionary computation method to battle the COVID-19 situation. A multi-objective genetic algorithm is proposed to estimate the early transmission of COVID-19 for epidemic prevention [46]. To predict the COVID-19 outbreak, a multi-gene genetic programming (MGGP) approach is proposed [47]. A prediction model using genetic programming is proposed to analyze the possible impact of COVID-19 in India and predict future behavior [48, 49]. The next subsection discusses various machine learning methods of AI to battle COVID-19.

9.6.4 Machine Learning Approach

Machine learning is the subset of AI that acts as a promising tool in the pandemic to predict the possible active case and mortality rate and speed up the preparation of vaccines by predicting the effect of the vaccine, etc.. The overview of machine learning is presented in Figure 9.12. The role of machine learning in the COVID-19 pandemic is discussed as follows.

The only way to control the spread of the coronavirus and its effect is the use of the vaccine. The machine learning approach using the reverse vaccinology approach is used by many researchers in different countries to speed up the vaccine development process. The detection of COVID-19 at the early stages is essential to take safety measures to cure the disease. Normally, for the detection of COVID-19, an RT-PCR test is used. The result of this test takes 4–5 hours or more. The false-negative rate for the test is higher. One prediction model is proposed using machine learning for the early detection of COVID-19. This approach shows a higher specificity of 0.95 and an improved sensitivity rate of 0.82. A machine learning approach is proposed to analyze the X-ray image of the chest. Based on

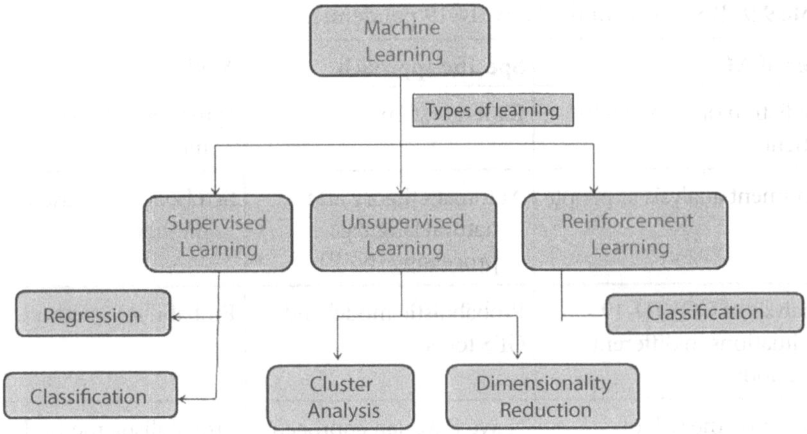

Figure 9.12 Overview of machine learning.

the image analysis, whether the person is infected by COVID-19 or not is detected [50]. To predict the future count of COVID-19 cases and to control the situation before going out of control, one prediction model is proposed using machine learning. The machine learning approach used for this purpose is data visualization, polynomial regression, and support vector regression [51]. The machine learning-based framework plays a vital role in determining the role of existing drugs to work effectively against SARS-CoV-2 in clinical trials.

9.6.5 Deep Learning Approach

Manual COVID-19 case detection by radiologists using chest computed tomography (CT) examination is time-consuming and resource-consuming. Also, there is a fear of the spreading of the coronavirus among the radiologists and paramedical staff. Deep learning helps in the quick detection of COVID-19 cases by processing and analyzing medical image data. The COVID-19 detection neural network (COVNet), a three-dimensional deep learning method, is used for the detection of COVID-19 using volumetric CT chest images. The area under the receiver operating the characteristics curve has a value of 0.96, which indicates the capability of the model for the detection of COVID-19 [52]. The protein structure during COVID-19 was determined by a deep learning approach called AlphaFold. This approach was developed by the team of Google DeepMind [53].

The overall summary of the role of AI in the fight against COVID-19 is presented in Table 9.9.

Table 9.9 Role of AI in the COVID-19 pandemic.

Role of AI	Specific approach	Methods of AI
Prediction of COVID-19 patient	Expert system	Symbolic rule-based method
Sentiment analysis of people	Automata theory and natural language processing (NLP)	Symbolic rule-based method
Analysis of COVID-19 situations in different regions	Probabilistic model and GPS tools	Probabilistic method
Predicting the risk of hospitalization during the COVID-19 pandemic	Naïve Bayesian approach	Probabilistic method
Estimation of the early transmission of COVID-19	Multi-objective genetic algorithm	Evolutionary computation method
Sentiment analysis of the people during the COVID-19 period	Automata theory and NLP	Evolutionary computation method
Prediction of the COVID-19 outbreak	Multi-gene genetic programming	Evolutionary computation method
Prediction model to analyze the possible impact of COVID-19 in India	Genetic programming	Evolutionary computation method
Speed up of the vaccine development process	Reverse vaccinology	Machine learning-based approach
Detection of COVID-19 at the early stages	Image analysis	Machine learning-based approach
Prediction of the future count of COVID-19 cases	Data visualization, polynomial regression, and support vector regression	Machine learning-based approach
Quick detection of COVID-19 cases	Three-dimensional neural network	Deep learning-based approach
Determination of the protein structure	AlphaFold	Deep learning-based approach

9.7 Related Studies

Within a short span, many research articles on COVID-19 were published, and these articles helped to carry out further research until the impact of COVID-19 has not decreased. In the beginning, the coronavirus spread in China [54], then it spread to other countries in the world [55, 56]. The symptoms of COVID-19 and the necessary preventive measures were discussed [57]. Lau *et al.* discussed the beneficial effect of the lockdown in Wuhan, China [58]. The various negative impacts of the coronavirus on different sectors were also studied by researchers like the impact on the economy [59], the impact on the network [60], the impact on the global supply chain [61], and the impact on the global trade [62]. The importance of digital technology, AI, and blockchain during this pandemic was also discussed [63]. This paper provided a brief study on the positive and negative impacts of COVID-19 on the different sectors, and the role of various technologies used to overcome the harmful effects of COVID-19.

9.8 Conclusion

The worldwide spread of COVID-19 occurred so quickly that it is reducing the overall growth of many countries. The whole world has been suffering from a new type of crisis in which no one was prepared. Almost everything was changed in various respects. These changes brought both bad and good impacts on society. Almost all sectors like health, education, economy, transportation, entertainment, tourism, sports, and a lot more were severely affected in all countries; however, nature has begun to rebuild itself due to the lockdown in many countries. In this emergency, the world has used technology as a weapon to fight COVID-19. In this paper, a brief discussion on the overall impact, i.e., the positive and the negative, was given, and the role of various technologies like AI, computer vision, machine learning, robotics, and blockchain were mentioned to bring control over the coronavirus infection. In the future, anti-viral drugs with the help of different technologies for COVID-19 should be a challenge for the whole world. After we overcome this global pandemic, which will surely happen, the whole world must take the necessary steps to face similar challenges in the future.

References

1. https://covid19.who.int/ (2020).
2. https://www.worldometers.info/coronavirus/.
3. https://www.medicalnewstoday.com/articles/256521/covid-19.
4. https://www.cdc.gov/coronavirus.
5. Wang, C., Horby, P.W., Hayden, F.G., Gao, G.F.: A novel coronavirus outbreak of global health concern. *Lancet,* 395(10223), 470–473 (2020).
6. Shereen, M.A., Khan, S., Kazmi, A., Bashir, N., Siddique, R.: Covid-19 infection: Origin, transmission, and characteristics of human coronaviruses. *J. Adv. Res.,* (2020).
7. https://www.webmd.com/lung/covid-19-symptoms.
8. https://www.mohfw.gov.in/pdf/PreventionandManagementofCOVID19.
9. https://www.google.com/amp/s/english.jagran.com/lite/india/coronavirus-pandemic-what-is-covid-19-containment-zones-how-is-it-created-and-what-restrictions-are-imposed-10010483.
10. https://www.google.com/amp/s/www.timesnownews.com/amp/india/article/corona-virus-zones-and-their-meanings-covid-19-containment-plan-what-are-red-orange-green-zones.
11. Fernandes, N.: Economic effects of coronavirus outbreak (covid-19) on the world economy. Available at SSRN 3557504 (2020).
12. Wang, C., Pan, R., Wan, X., Tan, Y., Xu, L., Ho, C.S., Ho, R.C.: Immediate psychological responses and associated factors during the initial stage of the 2019 coronavirus disease (covid-19) epidemic among the general population in china. *Int. J. Environ. Res. Public Health,* 17(5), 1729 (2020).
13. https://economictimes.indiatimes.com/tech/interne/view-invest-in-fibre-network-to-support-data consumption-upswing-during-covid pandemic.
14. https://www.hindustantimes.com/india-news/india-s-internet-consumption-up-during -covid-19-lockdown-shows-data.
15. https://www.geotab.com/blog/impact-of-covid-19/.
16. https://www.indiatoday.in/business/story/coronavirus-in-india-railways-cancels-nearly-200-trains-staring-at-rs-450-crore-loss.
17. https://www.urbantransportnews.com.
18. https://www.locktoninternational.com/apac/marine-transport-and-cargo-disruption-due-covid-19-breakout.
19. Shrestha, A.M., Shrestha, U.B., Sharma, R., Bhattarai, S., Tran, H.N.T., Rupakheti, M.: Lockdown caused by covid-19 pandemic reduces air pollution in cities worldwide (2020).
20. Sharma, S., Zhang, M., Gao, J., Zhang, H., Kota, S.H., et al.: Effect of restricted emissions during covid-19 on air quality in india. *Sci. Total Environ.,* 728, 138878 (2020).
21. https://thebodycamp.com/the-top-10-benefits-of-the-world-reset-or-lockdown/.

22. https://www.google.com/amp/s/www.downtoearth.org.in/blog/amp/covid-19-lockdown-a-ventilator-for-rivers-70771.
23. https://analyticsindiamag.com/covid-19-computer-vision/.
24. https://www.google.com/amp/s/www.dqindia.com/3d-printing-helping-fightcovid-19/amp/.
25. Senapati, B.R., Khilar, P.M.: Optimization of performance parameter for vehicular adhoc network (vanet) using swarm intelligence. In: *Nature Inspired Computing for Data Science*, pp. 83–107. Springer (2020).
26. Senapati, B.R., Mohapatra, S., Khilar, P.M.: Fault detection for vanet using vehicular cloud. In: *Intelligent and Cloud Computing*, pp. 87–95. Springer (2021)
27. Senapati, B.R., Swain, R.R., Khilar, P.M.: Environmental monitoring under uncertainty using smart vehicular ad hoc network. In: *Smart intelligent computing and applications*, pp. 229–238. Springer (2020).
28. Senapati, B.R., Khilar, P.M., Sabat, N.K.: An automated toll gate system using vanet. In: *2019 IEEE 1st International Conference on Energy, Systems and Information Processing (ICESIP)*. pp. 1–5. Springer (2019).
29. Senapati, B.R., Khilar, P.M.: Automatic parking service through vanet: A convenience application. In: *Progress in Computing, Analytics and Networking*, pp. 151–159. Springer (2020).
30. Jayapal, C., Roy, S.S.: Road traffic congestion management using vanet. In: *2016 International Conference on Advances in Human Machine Interaction (HMI)*. pp. 1–7 (2016).
31. Senapati, B.R., Khilar, P.M., Swain, R.R.: Fire controlling under uncertainty in urban region using smart vehicular ad hoc network. *Wirel. Pers. Commun.* 116(3), 2049–2069 (2021).
32. https://www.ibm.com/blogs/blockchain/2020/04/blockchains-role-in-covid-19 -response-and-recovery/.
33. https://www.google.com/amp/s/health.economictimes.indiatimes.com/amp/news/health-it/covid-19-lockdown-2-0-telemedicine-in-india-to-see-continued -grwoth/75172147.
34. https://www.google.com/amp/s/www.ndtv.com/india-news/covid-19-heres-how-drones-helping-india-in-fight-against-coronavirus-2211888/.
35. Swain, S., Mohan Khilar, P., & Ranjan Senapati, B. An effective data routing for dynamic area coverage using multidrone network. *Trans. Emerging Telecommun. Technol.*, e4532.
36. https://www.eurobotics.net/eurobotics/newsroom/press/robots-against-covid-19.html.
37. https://www.indiatvnews.com/technology/news-covid-19-all-coronavirus-dashboards-to-track-positive-cases-death-toll-india-603191/.
38. https://www.zdnet.com/article/how-to-track-the-coronavirus-dashboard-delivers-real-time-view-of-the-deadly-virus/.
39. https://m.economictimes.com/tech/software/how-countries-are-using-technology -to-fight-coronavirus/amp-articleshow/74867177.cms/

40. https://yourstory.com/2020/04/coronavirus-contact-tracing-patient-monitoring -apps-india.
41. http://www.google.com/amp/s/www.forbes.com/sites/enriquedans/2020/03/31/lets-take-advantage-of-coronavirus-to-get-comfortable-using-video-call-technology/amp/.
42. Castillo Ossa, L.F., Chamoso, P., Arango-L´opez, J., Pinto-Santos, F., Isaza, G.A., SantaCruz-Gonz´alez, C., Ceballos-Marquez, A., Herna´ndez, G., Corchado, J.M.: A hybrid model for covid-19 monitoring and prediction. *Electronics,* 10(7), 799 (2021).
43. Khan, R., Shrivastava, P., Kapoor, A., Tiwari, A., Mittal, A.: Social media analysis with ai: sentiment analysis techniques for the analysis of twitter covid-19 data. *J. Crit. Rev.,* 7(9), 2761–2774 (2020).
44. Bherwani, H., Anjum, S., Kumar, S., Gautam, S., Gupta, A., Kumbhare, H., Anshul, A., Kumar, R.: Understanding covid-19 transmission through bayesian probabilistic modeling and gis-based voronoi approach: a policy perspective. *Environ. Dev. Sustainability,* 23(4), 5846–5864 (2021).
45. Karaismailoglu, E., Karaismailoglu, S.: Two novel nomograms for predicting the risk of hospitalization or mortality due to covid-19 by the na¨ıve bayesian classifier method. *J. Med. Virol.,* 93(5), 3194–3201 (2021).
46. Yousefpour, A., Jahanshahi, H., Bekiros, S.: Optimal policies for control of the novel coronavirus disease (covid-19) outbreak. *Chaos, Solitons Fractals,* 136, 109883 (2020).
47. Niazkar, M., Niazkar, H.R.: Covid-19 outbreak: Application of multi-gene genetic programming to country-based prediction models. *Electron. J. Gen. Med.,* 17(5) (2020).
48. Salgotra, R., Gandomi, M., Gandomi, A.H.: Time series analysis and forecast of the covid-19 pandemic in india using genetic programming. *Chaos, Solitons Fractals,* 138, 109945 (2020).
49. Singh, D., Kumar, V., Kaur, M., et al.: Classification of covid-19 patients from chest ct images using multi-objective differential evolution–based convolutional neural networks. *Eur. J. Clin. Microbiol. Infect. Dis.,* 39(7), 1379– 1389 (2020).
50. Elaziz, M.A., Hosny, K.M., Salah, A., Darwish, M.M., Lu, S., Sahlol, A.T.: New machine learning method for image-based diagnosis of covid-19. *PLoS One,* 15(6), e0235187 (2020).
51. Tamhane, R., Mulge, S.: Prediction of covid-19 outbreak using machine learning. *Int. Res. J. Eng. Technol. (IRJET),* 7(5) (2020).
52. Li, L., Qin, L., Xu, Z., Yin, Y., Wang, X., Kong, B., Bai, J., Lu, Y., Fang, Z., Song, Q., *et al.*: Artificial intelligence distinguishes covid-19 from community acquired pneumonia on chest ct. *Radiology,* (2020).
53. Alimadadi, A., Aryal, S., Manandhar, I., Munroe, P.B., Joe, B., Cheng, X.: Artificial intelligence and machine learning to fight covid-19 (2020).

54. Novel, C.P.E.R.E., *et al.*: The epidemiological characteristics of an outbreak of 2019 novel coronavirus diseases (covid-19) in china. *Zhonghua liu xing bing xue za zhi= Zhonghua liuxingbingxue zazhi,* 41(2), 145 (2020).
55. Anderson, R.M., Heesterbeek, H., Klinkenberg, D., Hollingsworth, T.D.: How will country-based mitigation measures influence the course of the covid-19 epidemic? *Lancet,* 395(10228), 931–934 (2020).
56. Fanelli, D., Piazza, F.: Analysis and forecast of covid-19 spreading in china, italy and france. *Chaos, Solitons Fractals,* 134, 109761 (2020).
57. Adhikari, S.P., Meng, S., Wu, Y.J., Mao, Y.P., Ye, R.X., Wang, Q.Z., Sun, C., Sylvia, S., Rozelle, S., Raat, H., *et al.*: Epidemiology, causes, clinical manifestation and diagnosis, prevention and control of coronavirus disease (covid-19) during the early outbreak period: a scoping review. *Infect. Dis. Poverty,* 9(1), 1–12 (2020).
58. Lau, H., Khosrawipour, V., Kocbach, P., Mikolajczyk, A., Schubert, J., Bania, J., Khosrawipour, T.: The positive impact of lockdown in wuhan on containing the covid-19 outbreak in china. *J. Travel Med.,* 27(3), taaa037 (2020).
59. Atkeson, A.: What will be the economic impact of covid-19 in the us? rough estimates of disease scenarios. Tech. Rep., National Bureau of Economic Research (2020).
60. Favale, T., Soro, F., Trevisan, M., Drago, I., Mellia, M.: Campus traffic and e-learning during covid-19 pandemic. *Comput. Networks,* p. 107290 (2020).
61. Ivanov, D.: Predicting the impacts of epidemic outbreaks on global supply chains: A simulation-based analysis on the coronavirus outbreak (covid-19/sars-cov-2) case. *Transp. Res. Part E Logist. Transp. Rev.,* 136, 101922 (2020).
62. Baldwin, R., Tomiura, E.: Thinking ahead about the trade impact of covid-19. *Economics in the Time of COVID-19,* 59 (2020).
63. Ting, D.S.W., Carin, L., Dzau, V., Wong, T.Y.: Digital technology and covid-19. *Nat. Med.,* 26(4), 459–461 (2020).

10

A Review on Smart Bin Management Systems

Bhoomika Patel H. C.*, Soundarya B. C. and Pooja M. R.

Vidyavardhaka College of Engineering, Mysuru, Karnataka, India

Abstract

In urban areas, smart dustbins are becoming increasingly necessary. The goal of the study is to create an intelligent alert system for waste bins. The present waste management systems are intensive in resource, not efficient, and outdated. Using the Internet of Things, we can manage this waste by placing sensors in the middle of the waste materials to monitor the levels of waste as they fill gradually and notify the city's waste assortment authority when the waste bins are ready to be emptied and start to fill with unwanted items. Waste management systems, as we all know, are crucial to maintaining a clean and hygienic environment. An overview of the different strategies employed in smart bin management systems is given in this chapter.

Keywords: Waste management, IOT, bin, manure, vehicle, smart technology

10.1 Introduction

The dustbin, also known as a container, is used to collect waste and is meant to be kept there until it is disposed of. Failure to do so could result in serious problems with solid waste contamination, including in private homes, business foundations, streets, and agriculture. Additionally, plastic waste is a more toxic type of waste that takes thousands of years to decompose in soil. Waste management can also be solid waste including household hazardous waste (HHW), development and destruction garbage, modern/

**Corresponding author*: bhoomijanu9@gmail.com

Chandra Singh, Rathishchandra R. Gatti, K.V.S.S.S.S. Sairam, Manjunatha Badiger, Naveen Kumar S. and Varun Saxena (eds.) Modeling and Optimization of Signals Using Machine Learning Techniques, (209–218) © 2024 Scrivener Publishing LLC

Figure 10.1 Different kinds of waste.

business waste, dangerous waste lights, controlled clinical waste, utilized electronic gear, utilized oil and waste tires. Other than sensors, neural networking also will be done to manage waste management solutions.

Garbage is gathered in dustbins (Figure 10.1), also known as containers, which are meant to hold waste until it is disposed of. Failure to do so could result in a serious problem with solid waste contamination—affecting things like private homes and business foundations—such as trash from development and destruction, garbage from streets, scrap from vehicles, and agricultural waste. Additionally, plastic waste is a more toxic type of waste that takes thousands of years to decompose in soil.

10.1.1 Internet of Things (IoT)

Internet of Things gadgets are utilized to get the information, investigate the distinct information, and form experiences into how the city, utilizing less resources, can both accumulate waste and complete reusing as effectively as could be expected.

Ultrasonic sensors (UI): As referenced, assortment services ordinarily null garbage cans on a normal pivot, if they are full, which waste fuel, time, and cash. A squander bin with ultrasonic sensors is skilled to recognize the fill level. A smart waste management system is depicted in Figure 10.2. A full dustbin has an impact on the garbage collection authority, which uses that data to help plan collection routes that take just the full bins into consideration. To monitor various boundaries, such as area, temperature, and robbery, different sensors combine movement sensors, GPS sensors,

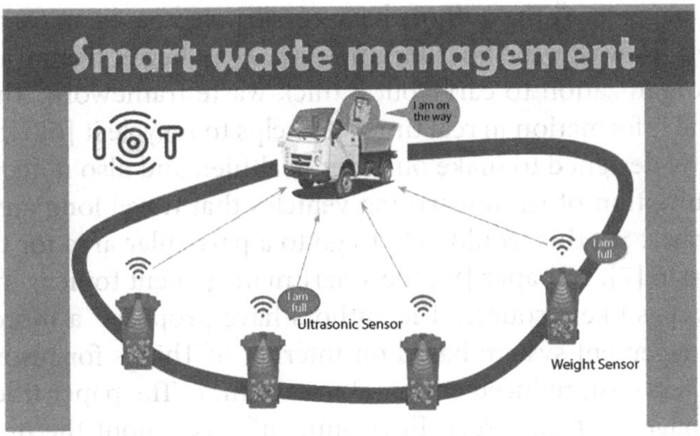

Figure 10.2 Smart waste management system.

light sensors, and vibration sensors. There are benefits of smart waste management with IoT. Urbiotica is a company that designs smart cities. It has installed smart management sensors on trash cans in places like Israel, France, Barcelona, and others.

10.2 Related Work

A smart garbage container based on a hardware kit that rides on the garbage container top and includes sensor nodes, a router, and an Android application is intended for waste management. When the container is filled, the sensor computes and transfers all waste conditions, sending all excess waste data to the base station [1]. The primary cause of society's problems is the vast amounts of waste products. It must be appropriately fixed because it causes issues for the health and development system [2]. Neural networks played a significant role in the garbage system and its correction in the paper [3]. The system is not reliable for us. In order to remedy the issue, numerous other challenges must be discussed, including the need to clean and classify the waste in landfills. Next, sort through and clean the waste from places like courts, hospitals, and schools. Through the use of sensors and the containers mentioned, the authors have provided a smart trash management system. Low-power and long-range data transfer are represented by LoRo communication [4]. This technique assists in locating and classifying waste according to paper, plastic, and metal. A multi-layer waste hierarchy is explained by writers Belal Chowdhury and Morshed U. Chowdhury [5]. In the paper titled "Design a Smart Waste Bin for Smart

Waste Management", the authors have constructed bases on technological interaction for waste, the board framework. Here began from waste bin by utilizing organization to carry out a thick waste framework. This system collects the information in real time and helps to recycle it [6]. This system was smartly designed to make our lives healthier, and also it decreases the fuel consumption of waste-carrying vehicles that travel long areas to collect the waste. So, this would help to go to a particular area for the collection of waste [7]. In paper [8], the smart management topic is introduced for landscapes like resource. The authors have proposed a more efficient waste management system based on Internet of Things for resource optimization, less cost, reduced cost, and to save time. The paper titled "Smart WasteManagement for Green Environment" talks about the problems of overflow of waste containers in public places which creates health issues to people and lodging a complaint on the trust drivers who is responsible for collecting waste in public places. The main goal of this system is to create a smart management system using bootstrap platform [9]. The authors S.S. Navghane1, M.S. Killedar, and Dr. V.M. Rohokale worked on developing a smart trash, the board framework utilizing IR sensor, microcontroller, Wi-Fi module, etc. This system will ensure to clean all the containers as maximum. If the container does not clean correctly, it lists Wi-Fi router, mobile, garbage dustbin, level detector, weight sensor, micro controller [10]. This management also helps to record the false reports to decrease corruption, and it will decrease the number of trips of the trucks that collect in a waste management system. The paper is established to decrease garbage and to overcome this problem by using a NODE MCU which is very small in size and does not use WIFI connectivity [11]. In paper [12], they used ultrasonic range sensor in a smart management system to know how much waste had been collected, and this data is sent to waste management centers. To overcome this operation, they are using ATmega328P microcontroller. The authors have introduced a new answer for waste assortment utilizing Arduino Uno with Arduino Ethernet safeguard innovation and ultrasonic sensor framework to overcome the problem of overflowing garbage and to communicate this SMS to a particularly authorized person [13]. The smart bin concept is depicted in Figure 10.3.

In paper [14], the authors have used raspberry_pi to provide continuous Internet connection to the system for updating the data in a cloud database using Android application. The paper used ultrasonic sensor to detect a row of waste and using Arduino node MCU. It will guide the data to the server and will check the fulfillment of the collected garbage level. This system sees the effectiveness for reducing in a waste management system [15]. The system shows a reduction of power consumption and

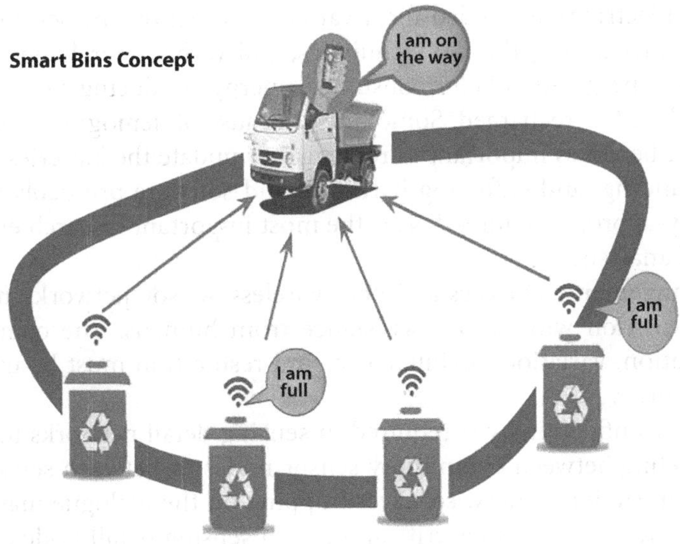

Figure 10.3 Full message intimation.

maximum operational time [16]. It checks the garbage flow in the outside environment. The authors used IoT (Internet of Things) to overcome the problem of garbage system using ZigBee network, and MQTT (Message Queue Telemetry Transport) protocol was used to check whether the container is filled or not [17]. The machine learning algorithm is used to make changes in existing method of waste management. The machine learning will improve the classification and accuracy from 86.8% and 47.9% to 99.1% and 98.2%. Comparing an engineering model, it increased the forecast for time of recycle container [18]. Similar techniques have been used in security systems as well [19]. In paper [20], the authors have used IOT-based case studies to demonstrate various applications of the same. Lastly, in contrast with the existing pattern of a physically designed model, the best performing arrangement likewise worked on the nature of figures for exhausting the season of reusing holders.

10.3 Challenges, Solution, and Issues

A. Challenges

The demanding situations which can be faced when operating with wireless sensor networks (WSN) are listed below:

Energy: Electricity is needed for a variety of functions by sensors. Data series, information gathering techniques, and verbal interchange of facts all provide energy. After being consumed, energy-producing batteries need to be replaced or recharged. Sometimes, because of demographic circumstances, it becomes important to replenish or update the batteries. Laying out, expanding, and enforcing hardware and software protocols that are electricity-appropriate for WSNs is the most important research endeavor for WSN analysts.

Self-management: Once established, wireless sensor networks must be able to function without any assistance from humans. The community configuration, variation, maintenance, and restoration must be under its own authority.

Security: Confidentiality is required in sensing detail networks to secure data traveling between community sensor nodes or between sensors and the bottom station; otherwise, eavesdropping on the dialogue may occur. In sensor networks, it is far critical for each sensing detail node and also the base station to be able to validate that the information obtained was truly sent by a trustworthy sender and not by a human who deceived legitimate nodes into accepting phony understanding. A misleading information will change the way a network is expected to operate. The integrity of the records must be preserved. Data does not change, and precise comprehension must be achieved at the consumer stop.

Quality of service (QoS): The level of carrier provided to its customers by using sensor networks is referred to as quality of service. Because WSNs are used in several real-time and critical applications, it is much critical for the network to provide appropriate QoS.

Fault tolerance: Even if one node breaks while the community is operating, the sensor community must remain functioning. In the event of a difficulty, the network should be able to respond by modifying its assets. In that instance, a well-green routing algorithmic program is used as alternative in the community's final shape.

Limited memory and storage space: A sensor is a small device with just a reduced quantity of memory and storage space for the code. To create an efficient security system, the code size of the safety set of rules must be limited.

B. Solutions

Some of the solutions to overcome the problems faced by WSN are as follows:

- **Data freshness:** There is a desire to ensure the freshness of each message even though comprehension of confidentiality and data integrity is guaranteed. Informally, expertise freshness assures that no outdated messages have been repeated and that the facts are current. To resolve this issue with the counter that is connected to the back, may be changed into the packet to ensure experience freshness.
- **Secure localization:** A device's application community frequently agrees on a device's capacity to discover every sensor in the community automatically. A sensor network created to find defects can employ accurate location information to pinpoint where a fault is located.
- **Privacy:** The sensor networks have similar pressing privacy problems as other older networks. The sensing detail networks that were first deployed for a good reason may also afterwards be employed in immediate ways. It is vitally important to pay attention to the existence of sensor nodes and knowledge gathering.
- **Secure routing:** In device networks, routing and record forwarding play a key role as a medium for facultative communication.
- **Data availability:** A node's capacity to utilize resources and the community's availability for message communication are determined by availability. However, the entire community involved in detecting detail may be in danger if the bottom station or cluster leader becomes unavailable. As a result, availability is of utmost significance for sustaining the operating network of the companion diploma.

C. Issues

Some of the issues as per the survey are as follows:

- It needs a variety of preservation costs.
- The elderly in nearby neighborhoods face a precarious position due to excessive garbage disposal in public spaces.
- It implements several technologies, including Java, relational databases, and others.
- The bins are cleanest when they are fully filled.
- The boxes fill up rapidly on special days, so we have to extend the gathering times.

- It is used to explore, collect, and send information online.
- Although not net, it is not practicable. High gas prices when the amount of the gap is covered by the bin collector.
- Different types of stakeholders are included, and their interests are given the utmost priority.

10.4 Advantages

The proposed strategy has a lot of advantages and is strong enough to be implemented in all areas of a developing state. The benefits stem from how simple and useful it is to use. This will provide a surface for better functioning technology in addition to enhancing the streets in which we live.

- Effective and efficient operation.
- Purer environments
- Concerns with better health
- Odor- and pollution-free surroundings
- Smart cities
- Technology development
- Tourist attraction.

Conclusion

Smart waste management (SWM) introduces an intelligent approach to control waste while discussing the waste management procedure for various types of trash. The entire Internet of Things dustbin is made to automate and effectively monitor the operation. The device gauges the amount of rubbish within the trash can, automatically locks it, and notifies the local authorities of its position so that corrective action can be done to ensure timely garbage collection and prevent filth in the neighborhood. In contrast to conventional bins, smart bins can optimize garbage collection routes, monitor fill levels in real time, and encourage waste segregation, all of which contribute to increased efficiency, less environmental impact, and a cleaner environment.

References

1. Murugaanandam, S., V. Ganapathy, and R. Balaji. "Efficient IOT based smart bin for clean environment." *2018 International Conference on Communication and Signal Processing (ICCSP)*. IEEE, 2018.
2. Soh, Zainal Hisham Che, *et al.* "Smart waste collection monitoring and alert system via IoT." *2019 IEEE 9th symposium on computer applications & industrial electronics (ISCAIE)*. IEEE, 2019.
3. Yang, Zhihu, and Dan Li. "WasNet: A neural network-based garbage collection management system." *IEEE Access* 8 (2020): 103984-103993.
4. Chowdhury, Belal, and Morshed U. Chowdhury. "RFID-based real-time smart waste management system." *2007 Australasian telecommunication networks and applications conference*. IEEE, 2007.
5. Wijaya, Aksan Surya, Zahir Zainuddin, and Muhammad Niswar. "Design a smart waste bin for smart waste management." *2017 5th International Conference on Instrumentation, Control, and Automation (ICA)*. IEEE, 2017.
6. Balamurugan, S., *et al.* "Design of smart waste management system." *2017 International conference on Microelectronic Devices, Circuits and Systems (ICMDCS)*. IEEE, 2017.
7. Chaudhari, Sangita S., and Varsha Y. Bhole. "Solid waste collection as a service using IoT-solution for smart cities." *2018 International Conference on Smart City and Emerging Technology (ICSCET)*. IEEE, 2018.
8. Fei, Teh Pan, *et al.* "SWM: Smart waste management for green environment." *2017 6th ICT International Student Project Conference (ICT-ISPC)*. IEEE, 2017.
9. Navghane, S. S., M. S. Killedar, and V. M. Rohokale. "IoT based smart garbage and waste collection bin." *Int. J. Adv. Res. Electron. Commun. Eng. (IJARECE)* 5.5 (2016): 1576-1578.
10. Sonawane, N. D., *et al.* "Smart Waste Management Using IOT Powered Dustbin." (2019).
11. Neema, Sakshi, and Kaushal Gor. "Smart Waste Management Using IoT." (2022).
12. Chugh, Harshita. "Dushyant singh2, Shahensha shaik3, Ashwani Singla," IOT Based Smart Bin""." *Int. Res. J. Eng. Technol. (IRJET)* e-ISSN: 2395-0056.
13. Garach, Pooja V., and Rikin Thakkar. "Design and implementation of smart waste management system using FOG computing." *Int. Res. J. Eng. Technol. (IRJET)* 5 (2018): 3888-3891.
14. Folianto, Fachmin, Yong Sheng Low, and Wai Leong Yeow. "Smartbin: Smart waste management system." *2015 IEEE Tenth International Conference on Intelligent Sensors, Sensor Networks and Information Processing (ISSNIP)*. IEEE, 2015.

15. Catarinucci, Luca, *et al.* "IoT-aware waste management system based on cloud services and ultra-low-power RFID sensor-tags." *IEEE Sens. J.* 20.24 (2020): 14873-14881.
16. SKarthikeyan, S., *et al.* "IoT enabled waste management system using ZigBee network." *2017 2nd IEEE International Conference on Recent Trends in Electronics, Information & Communication Technology (RTEICT)*. IEEE, 2017.
17. Lu, Jia-Wei, *et al.* "Smart and green urban solid waste collection systems: advances, challenges, and perspectives." *IEEE Sens. J.* 11.4 (2015): 2804-2817.
18. Rutqvist, David, Denis Kleyko, and Fredrik Blomstedt. "An automated machine learning approach for smart waste management systems." *IEEE Trans. Ind. Inform.* 16.1 (2019): 384-392.
19. Das, S., Pooja, M.R., Anusha, K.S. (2023). Implementation of Women's Self-Security System Using IoT-Based Device. In: Awasthi, S., Sanyal, G., Travieso-Gonzalez, C.M., Kumar Srivastava, P., Singh, D.K., Kant, R. (eds) Sustainable Computing. Springer, Cham.
20. DAS, SONIA, MR POOJA, and KS ANUSHA. "IOT Based Case Study: Applications and Challenges." *2021 5th International Conference on Computing Methodologies and Communication (ICCMC)*. IEEE, 2021.

11

Unlocking Machine Learning: 10 Innovative Avenues to Grasp Complex Concepts

K. Vidhyalakshmi[1]* and S. Thanga Ramya[2]

[1]*Veltech Multi Tech Dr. Rangarajan Dr. Sakunthala Engineering College, Avadi, Tamilnadu, India*
[2]*RMK Engineering College, Chennai, Tamilnadu, India*

Abstract

The mathematical statement that expresses facts in the context of a problem is a model often known as a machine learning (ML) algorithm. Let us have a look at 10 alternative approaches for demystifying ML and offering a knowledge route used by individuals unfamiliar with fundamental ideas, each with clear descriptions, images, and examples. The objective is to move from facts and information to knowledge. For illustration, if an online business needs to forecast profit sales in the coming part, it can use ML techniques that forecast sale relation with prior sales with related data. For example, a windmill maker may utilize video surveillance to monitor critical equipment and send data to algorithms that might spot faults. The 10 methods listed below provide an overview — as well as a foundation upon which we can construct as your ML acquaintance and experience grows:

1. Regression
2. The classification system
3. The clustering
4. Adaptive learning
5. The ensemble methods
6. The reinforcement
7. The dimensionality reduction
8. Neural networks and deep learning
9. Natural language processing (NLP)
10. The word embeddings

**Corresponding author*: vidhyamecse@gmail.com

Chandra Singh, Rathishchandra R. Gatti, K.V.S.S.S.S. Sairam, Manjunatha Badiger, Naveen Kumar S. and Varun Saxena (eds.) Modeling and Optimization of Signals Using Machine Learning Techniques, (219–244) © 2024 Scrivener Publishing LLC

Let us first learn the difference between supervised and unsupervised ML techniques. We employ the supervised (with label) ML algorithm when we need to predict, and otherwise explain, a part of data. We do this by predicting a result based on a new input by combining data from past inputs and outcomes. For example, supervised ML algorithms might aid a service provider in forecasting in the coming month, a new client count who will use the service. Machine learning without supervision, on the other hand, looks for ways to link and organize data without depending on a goal variable to anticipate the result. In other words, it evaluates data for quality and uses the findings to group similar items into two clusters. You might, for example, utilize unsupervised learning techniques to assist a store in grouping items by means of like characteristics without needing to specify which traits to use ahead of time.

Keywords: Machine learning, NLP, reinforcement learning, deep learning and neural network, ensemble method, clustering, regression, dimension of reduction

11.1 Regression

The regression type is included in the supervised ML category. A regression problem is used in machine learning (ML) to predict (forecast) the importance of a numeric variable based on its experimental values. A number,

Table 11.1 Costs of used cars: Regression example data.

Cost (Rate)	**Old**	**Distance**	**Mass (Weight)**
(US$)	(years)	(km)	(pounds)
13000	22	46786	1265
13500	23	72837	1165
13900	24	41611	1165
14900	26	47000	1165
13650	30	36500	1270
12850	32	60000	1270
16800	27	93612	1245
18700	30	74889	1245
21400	27	18700	1165
12850	23	71238	1105

such as a decimal or a floating-point number, can be used as the output value. Quantities like the amount and sizes are frequently utilized. The input variables could be distinct or real-valued [1].

11.1 **Example:** The following data on car costs is shown in Table 11.1 as an example of a regression problem.

Estimate the cost of a 25-year-old car with a given distance of 53240 kilometers and a mass (weight) of 1200 pounds. The aim to forecast the value of the numeric variable "cost" is a case of a regression problem.

11.1.1 General Approach

Let a represent the input variables and c represent the outcome variable. The common approach to solve regression in ML is to consider a representation that is a mathematical correlation [4] between a and c including a certain parameter in the form of:

$$c = f(a, \theta)$$

The regression is denoted by f (a, θ). The ML technique optimizes the variable in θ set in a way that has minimized error in approximation; that is, the value of the dependent (c's) variable evaluation is as near to the actual values provided in the training set as possible.

Example:

The model may be valid if the key in variables are "Old (Aged)," "Distance," and "Mass," and the production variable is "cost."

$$\mathrm{C} = \mathrm{f}\,(\mathrm{a}, \theta)$$

$$\text{Cost}=a_0+a_1 \times (\text{Old})+ a_2 \times (\text{Distance}) + a_3\times(\text{Mass})$$

Where $x=$ (Old, Distance, Mass) indicates the set of input variables and $\theta= (a_0, a_1, a_2, a_3)$ indicates the set of model parameters.

11.1.2 Different Regression Models [11]

To produce predictions, you can use a variety of regression approaches. The quantity (number), self-determining (independent) variables, dependent variables, and the form of the line used for regression are the four main differences between these methodologies. The following are a few of them.

Regression with a linear model: It only has one continuous type of independent parameter. x, the dependent variable parameter, and the independent variable parameter share a relationship *c*.

$$c = b + mx$$

Multivariate (several) linear regression: It has numerous independent (autonomous) variables, such as x_1, x_2, x_n, and the relationship with the independent variables and the dependent variable is supposed to be

$$C = a_0 + a_1 x_1 + \ldots + a_n x_n$$

Polynomial regression: The postulated model has only one continuous independent variable, x.

$$Y = a_0 + a_1 x + \ldots + a_n x^n$$

Logistic regression: The dependent variable in logistic regression is binary, which means it can only have two values: 0 and 1. Certain probability distributions are assumed in the model.

11.2 Classification

11.2.1 Definition

Classification is the task of locating which category a new statement belongs to by using a training set of information made up of observations (or instances) whose category membership has been identified by way of ML [2].

11.2.2 Example

Consider the following data in the table for classification.

The training set of the data is shown in Table 11.2. "Score1" and "Score2" are two qualities. "Output" is the name of the class label. "Success" and "Unsuccess" are the two possible options for the class label. Circles

Table 11.2 A classification problem (example data).

Score1	30	23	10	30	16	32	33	21
Score2	42	27	46	55	18	52	41	42
Output	Success	Unsuccess	Unsuccess	Success	Unsuccess	Success	Success	Success

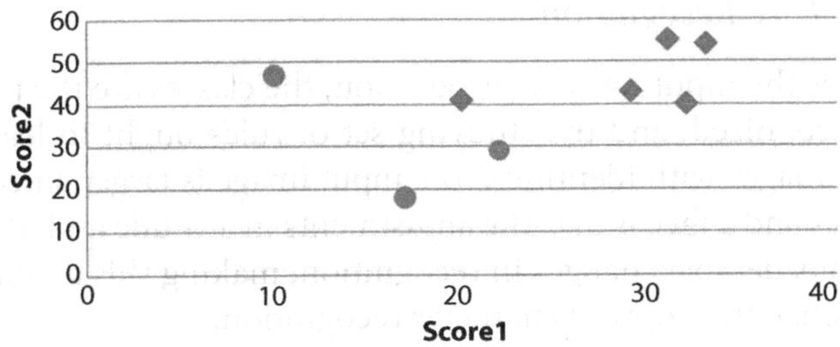

Figure 11.1 Representation of data in graphical form.

represent data in the "Unsuccess" class, whereas diamonds represent data in the "Success" class. The square dot's class label must be determined [9].

The data can be separated into two groups or classes: those whose class label is "Success" and those whose class label is "Unsuccess." Let us pretend we do not know anything on the subject of the data that is not contained in the table. The following is how the crisis (issue) can presently be stated: What should the "final result" for new data, such as "Score1 = 25" and "Score2 = 36" be? In other words, should the new observation be attributed to one of the two groups or classes? Figure 11.1 represents a situation in graphical form.

To solve this issue, we must determine the rule, formula, or technique that was utilized in conveying the values to the class label "Output" using only the data provided. The categorization problem is the problem of locating this rule, formula, or procedure. Still, the broad structure of the rule, function, or technique will not be identified in most cases. As a result, multiple distinct rules, etc., may be required.

11.2.3 Day-to-Day Example

11.2.3.1 Optical Character Recognition (OCR)

The issue of categorization is the OCR issue, which refers to the difficulty of identifying codes of characters from photographs. This is an example of a case in which the number of classes equals the number of characters we wish to recognize. When the characters are handwritten, the situation becomes much more intriguing. People write in a variety of styles. Characters may be written in a variety of ways; characters can be written large or little, slanted or straight, and using a pen or pencil, and there are many different visuals that correlate to the same letter.

11.2.3.2 Face Recognition

The image is the input for face recognition, the classes are the people who must be recognized, and the studying set of rules ought to learn how to relate face images with identities. The input image is larger since there are more classes and a face is 3D. The adjustments in attitude and illumination produce considerable changes in recognition, making this a more complicated challenge than optical character recognition.

11.2.3.3 Recognition of Speech

In these classes are words that can be pronounced out loud, and the input is acoustic.

11.2.3.4 Medical Findings [8]

The relevant information about the patient that we have is the input in medical analysis, and the diseases are the classifications. The age, gender, previous medical history, and current symptoms of the patient are all inputs. Some tests on the patient may not have been completed, leaving these fields empty.

11.2.3.5 Extraction of Acquaintance

Classification algorithms may also be used to extract information. The rule is easy reproduction that describes factual information; by viewing it, the process that underpins the data can be figured out.

11.2.3.6 Compression

Compression can be accomplished using classification rules. We achieve a simpler explanation by fitting a rule to the data, which a smaller amount of memory is stored and a smaller amount of work is put toward the process.

11.2.3.7 Additional Examples

Some more examples based on classification problems are discussed as follows:

a) In a hospital emergency room, 17 factors, such as blood pressure, age, and so on, are measured in newly admitted patients. The decision on whether to admit the patient to the intensive care unit must be made. Only patients who are likely to live for at least a month are given priority due to the high expense of the ICU. Others are referred to as "high-risk patients," while others are referred to as "low-risk individuals." The problem is determining how to construct a rule that will designate a person as "low-risk" or "high-risk."
b) The credit card employer gets masses of thousands of packages for brand-new credit scorecards. Several attributes, such as annual salary, age, and so on, are included in the applications. The challenge is to come up with a rule to categorize the applicants into those who are creditworthy, those who are not, and those who need more investigation.
c) Digital photographs created with special instruments have been used by astronomers to catalog faraway objects in the sky. The objects should be labeled as stars, galaxies, nebulas, and so forth. The data is quite noisy and dim. The challenge is to come up with a rule that will allow a faraway object to be appropriately named.

11.2.4 Discriminant

In a classification, the discriminant is an imperative or function for assigning labels (output) to new interpretations. Consider the information in Table 11.3 and the difficulties in categorizing it.

The following guidelines can be applied to the new data classification:

if Score1 + Score2 ≥ 60,
Then "Success or completed"
Else "Unsuccess or unsuccessful"

Alternatively, subsequent rules may be analyzed with uncertain values for S, S1 and S2, and then their values may be estimated using some method.

If Score1 $\geq$ 21 AND Score2 $\geq$ 41 THEN "Success or Success" ELSE "Unsuccess or not success"

If Score1 + Score2 $\geq M$, Then "Success" Else "Unsuccess".

If Score1$\geq m_1$ & Score2 $\geq m_2$ Then "Success" Else "Unsuccess"

Take a look at a financial organization that loans money to its consumers. Prior to lending money, the corporation wants an evaluation report of the risks related to accompanying the loan. Suppose that the firm finds a risk based on two parameters: yearly income and customer investment's money.

Let c1 represent the annual revenue and c2 represent the yearly savings for a party. Utilizing the previous information, create a rule with appropriate parameters for 1 and 2.

IF $c_1 > \theta_1$ AND $c_2 > \theta_2$ Then "low-risk" Else "high-risk". This rule is a discriminant example based on the previous information, and a rule of the consequent can be made:

If $c_2 = 0{,}2c_1 > 0$ Then "low-risk" Else "high-risk". This rule can be considered as the discriminant in this example. f(c1, c2) =c2-0,2c1 is a function. It is also possible to think of it as a discriminant.

11.2.5 Algorithms

For classifications, there are numerous ML methods. The algorithms listed here are some of the most well-known:

- Random forest algorithm,
- Support vector machine algorithm,
- Logistic regression,
- K-Nearest Neighbour algorithm,
- Decision tree algorithm, and
- Naïve Bayes algorithm.

Remarks

- A classification challenge necessitates the classification of examples in a combination of two or more classes.
- Input variables could be discrete.
- An issue with having two classes is a binary classification problem.
- A multi-class classification issue is having more than two classifications.
- An instance assigned with multiple (more) classes is defined as a multi-label classification problem.

11.3 Clustering

Clustering is combining a set of similar items in the same group (called a cluster) that are (in some sense) more comparable to different ones in different groups (clusters). Clustering is a primary goal of investigative data mining. It is utilized in industries as diverse as pattern recognition, information retrieval, image analysis, bioinformatics, data compression, computer graphics, and ML. It may be done in a variety of ways, each with its different idea of what defines a cluster and how to identify them quickly. Clusters are commonly thought of as Clusters[3] having the fewest distances between individuals, dense areas of the data space, and so on [3].

11.3.1 Data Examples Using Natural Clusters

In the number of applications, samples will automatically be divided into various clusters or groups.

1. Take, for example, optical character recognition: The digit 7 is written in two ways: in the United States, it is written as '7,' but in Europe, it is written with a horizontal bar in the middle (something like 7). When a sample includes representatives from both continents, it is divided into two groups or clusters; one for the 7 in the US and the other for the European 7.
2. There is no one universal prototype for voice recognition because the same word might be spoken differently depending on pronunciation, accent, gender, age, and other factors. All the numerous ways to say a word should be represented in a wide sample of utterances.

11.4 Clustering (k-means)

11.4.1 Outline

The most fundamental unsupervised learning algorithm for dealing with the clustering problem is the k-means clustering algorithm. Assume that you need to partition a specified information set into a specific number of clusters, notify k. To begin, choose sites of k at random as the "centers" of the clusters. For each one, every specified data point is then assigned to the nearest center. We now replace the center with the average of the data points associated with each of the centers. The technique is repeated until the centers converge on some fixed positions. The clusters in the dataset are formed by the data points closest to the centers. The related center represents each cluster.

11.4.2 Example

We show the process with only two variables so that points in a coordinate plane can geometrically represent the data points and cluster centers. The distance (separation) between the two locations (x1, x2) and (y1, y2) will be computed using the basic analytical geometry distance formula.

11.4.2.1 Problem

Divide the following information into two clusters and calculate the data points for each cluster using the k-means clustering algorithm. The data and the scatter plot are given in the Table 11.3 and Figure 11.2 respectively.

11.4.2.2 Solution

1. Because there are two clusters required in the problem, we use k2.
2. We choose two sites at random as the first cluster centers. Let us look at v1 = (2, 3) and v2 = (1, 2).
3. The distance between cluster centers and the specified data points is calculated.

$$\sqrt{(x_1 - y_1)^2 + (x_2 - y_2)^2}$$

The dataset in the table was clustered with the k-means procedure with k = 2. The following clusters and cluster centers are produced by 1:

Table 11.3 Example of k-means algorithm data.

x_1	2	2	1	3	4	5
x_2	3	1	1	2	3	5

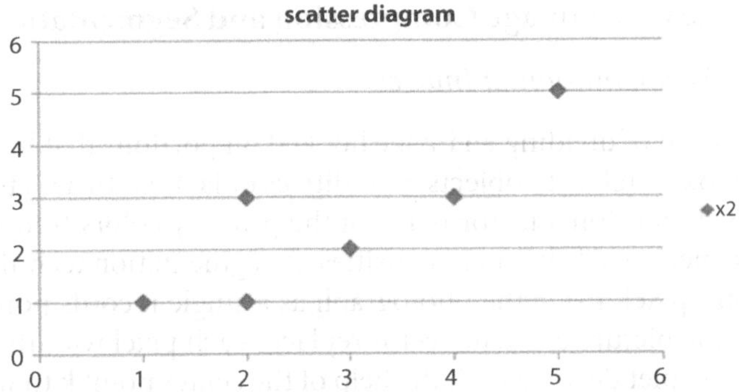

Figure 11.2 Data scatter plot.

V1 = x1; x2; x3; x4 represents Cluster 1: x1; x2 x3; x4 (2.00; 1.75)

Cluster 2: {x5; x6} represented by v2 = (2.00; 4.75)

11.4.3 Some Methods for Initialization

The approaches for selecting the initial vi's are as follows.

- For the first vi's, randomly select k data points.
- To get the k initial vi's, add small random vectors to the average of all data.
- Calculate the major component's range. Divide it into equal intervals of k. The k group of data can be divided, and the average of these groups can be used as the beginning centers.

11.4.4 Disadvantages

Although the k-means technique is quick, accurate, and easy to understand, it has a number of disadvantages prior to the determination of the number of cluster centers required by the learning process. The initial vi's determine the ultimate cluster centers.

We get different outcomes when we express data in different ways. The results will differ depending on whether you use cartesian or polar coordinates.

Using Euclidean distance measures, the underlying causes can be given disproportionate weighting. The learning process provides the squared error function's local optima. Choosing the initial cluster centers at random may not yield the best results. The algorithm does not work with categorical data.

11.4.5 Use Case: Image Compression and Segmentation

11.4.5.1 Segmentation of Images

It is the method of dividing an image hooked on portions that have a similar visual look or relate to objects or additives of factors. In a picture, every pixel is a 3-dimensional factor made of the primary colors (red, blue, and green) in experienced channel intensities. A segmentation technique takes care of every pixel inside the photograph as a single records point. In any value of k, the picture element vector replaces each pixel with the primary color depth triplet decided with the help of the center point k to which that pixel turned into is assigned. For a value of k, the technique represents the photograph use of a palette of the simplest k shades. It ought to be noted that using k-means for photograph segmentation is a rudimentary technique. The assignment of image segmentation is large.

11.4.5.2 Compression of Data

Data can also be compressed using the clustering process. Its types are lossless and lossy data compressions. Lossless data compression aims to be able to reconstruct original data from the compressed data without any loss or error, while lossy data compression allows some reconstruction errors in exchange for higher compression levels than lossless data compression. The k-means techniques can be used to solve the problem of lossy data compression in the following way. The only way to identify the clusters to which all the N information (data) points is assigned is stored for each of the N data points. Further, the point of the k cluster centers is saved; μ_k uses significantly less data if k is substantially less than N. The nearest center μ_k is then used to estimate each data point.

New instances can be compressed in a similar way by locating the nearest μ_k first and then saving the label k rather than the original. The vectors Îijk are known as code-book vectors, and this structure is often referred to as vector quantization. Corresponding label or information can be fetched by using codebook.

11.5 Reduction of Dimensionality

The quantity of inputs decides the difficulty of a classifier or regressor. It specifies a classifier or regressor's complexity in terms of time and space, as well as the training instances required [6].

11.5.1 Introduction

Datasets containing a large number of variables are common in many learning tasks. When the number of variables surpasses the number of observations, variable overflow develops. Analogous scenarios have evolved in mass spectrometry, image processing, internet search engines, time series analysis, computerized text examination, and other scientific disciplines. There are certain difficulties when dealing with such high-dimensional data. Before ML methods can be successfully implemented, the number of input variables is usually minimized. Reduction of dimensionality is the process of lowering the number of variables under concern by acquiring a lesser collection of primary information in statistics and ML. There are two possibilities when it comes to reducing dimensionality.

11.5.1.1 Feature Selection

When picking features, we look for the k features out of a total of n that provide the greatest data (information), and ignore the remaining n k dimensions.

11.5.1.2 Feature Extraction

In the extraction of features, we wish to find a fresh set of k features in a combination of the original n features. Whether or not the output data is employed, approaches can be supervised or unsupervised. Unsupervised and supervised linear projection approaches include principal component analysis (PCA) and linear discriminant analysis (LDA), respectively.

11.5.1.3 Error Measures

We need a measure of the model's inaccuracy in both techniques.

The mean squared error (MSE) or the root mean squared error (RMSE) for regression issues is seen as a measure of error. The MSE is calculated by dividing the total square of the differences between the predicted and actual target variables by the number of data points. The rate of misclassification, when dealing with classification issues is a measure of error. This is how it is spelled out:

$$\text{Rate of misclassification} = \frac{\text{misclassified examples in (number)}}{\text{Examples in total (number)}}$$

11.5.2 Benefits of Reducing Dimensionality

We are interested in lowering dimensionality for a variety of reasons. Because the quantity of input dimensions(d) and the size of the data sample (N), influence the complexity of most learning algorithms, we are interested in reducing the problem's dimensionality for less memory and processing. The complexity of the inference process during testing is minimized by lowering d. We save the expense of extracting an input when it is determined to be unneeded.

- On little datasets, simpler models are more reliable. Simpler models have less variance; therefore, they are less affected by sample features like noise and outliers.
- We have a greater understanding of the process that underpins the data when we can describe it with fewer features, which helps us to extract knowledge.
- Data can be displayed and examined graphically for structure and outliers when it can be mentioned in a few dimensions without compromising information.

11.5.3 Subset Selection

The method of selecting a subset of relevant characteristics (variables, predictors) for use in model development is identified as subset selection in ML. This technique is referred to as feature selection, variable selection, and attribute selection.

There are four reasons why feature selection techniques are used:

- o Models can be simplified to make them easier to interpret for academics and users.
- o Less time is spent on training.
- o The dimensionality curse is avoided.
- o There is better generalization due to less overfitting.

11.5.3.1 Selecting Forward

We begin with no variables and then gradually add them, starting with the one that reduces the error the most at each stage and continues until no more additions reduce the error (or decreases it only slightly).

11.5.3.2 Remarks

We halt this step if the accumulation of any features does not lower E (error). If the error reduction is too minimal, we may choose to stop the operation sooner if a user-defined threshold based on the application limits exists. Because we must train and test the system at a time complexity of O (n2) in order to lower the dimensions from n to k, this approach may be costly.

11.5.3.3 Selection in Reverse

We begin with the set that contains all characteristics in sequential backward selection and delete the one feature that generates the least error at each stage.

11.6 The Ensemble Method

"A group of things or people acting or taken collectively as a whole, notably a group of musicians who frequently play together," the word ensemble literally implies. Ensemble method learning in ML has the following two steps:

1. Using the data provided, create various models for tackling a specific problem.
2. Combine the models built to get better outcomes.

The various models can be selected in a variety of ways [10]:

- o Different algorithms, such as the k-NN algorithm, the decision tree algorithm, and the Naïve Bayes algorithm, can be used to generate the models.
- o Models can be formed in the same way that artificial neural network (ANN) algorithms can by assigning multiple initial values to the algorithm's parameters.

Ensemble learning approaches generate models that can be integrated in a variety of ways.

- In classification tasks, simple majority voting is used. Each representation makes a forecast (vote) for every test case, and the final output prediction gets greater than 50% of the votes.

- In classification issues, we calculate the predictions of the improved models in many periods using weighted majority voting. It is up to us to devise an appropriate set of weights.
- The simple averaging method gives average forecasts for each instance of the test dataset in prediction problems.
- Weighted averaging is used in prediction issues. The forecast of each representation is multiplied by the weight, and the average is derived using this method.

Ensemble approaches apply the same concept of integrating numerous predictive models (supervised ML) to obtain better predictions than each single model delivered by its individual. For instance, the random forest method is an ensemble technique that merges many decision trees that have been learned with a variety of data sets. A random forest's forecasts are of higher quality than those of other approaches. Consider ensemble methods that are the same as a path for lowering the volatility and bias of a single ML model. This is significant because a model may be correct in some situations and incorrect in others. With a different model, relative accuracy may be reversed. By merging the two models, the quality of forecasts is improved.

Ensemble approaches are used by the vast majority of elite Kaggle competition winners. The most well-known ensemble methods are random forest, LightGBM, and XGBoost.

11.6.1 Random Forest

An ensemble learning technique in which many decision trees are built and mixed to obtain a more precise forecast is a random forest, as depicted in Figure 11.3.

11.6.2 Algorithm

The random forest algorithm is described in detail. The random forest algorithm generates a large number of classification trees. Every tree is constructed as per the following techniques: if the training set has N occurrences, select N examples at random commencing the original data but with replacements. This trial will be used for training sets. Here, M is the input parameter at each node, m is selected at random from the M, and the preeminent split on m is used for the split node. Throughout the

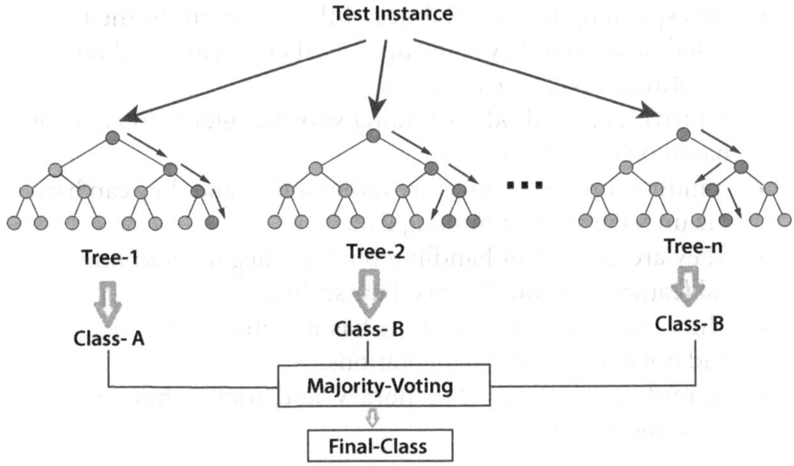

Figure 11.3 Example of a random forest by way of majority voting.

expansion of the forest's many trees, the m remains constant. Every tree is given every opportunity to reach its full latent. Based on a vector input to classify a new object, the vector input is placed among the forest's trees. Each tree assigns a categorization, and we refer to this as the tree's "vote." The classification is decided by the forest [9].

11.6.3 Benefits and Drawbacks

11.6.3.1 Benefits

A few of the most significant benefits of a random forest are listed below:

- It works swiftly on huge databases and can handle hundreds of input variables without eliminating them.
- It calculates the relevance of numerous categorization variables.
- It provides a method for anticipating missing data that works well and retains accuracy even when a considerable amount of data is missing.
- The output forests will be retained for other data in the future, and it works well with huge databases.
- Prototypes are created to demonstrate the relationship between the variables and classifications.

- By expanding the possibilities of the precedent to the unlabeled data, data views, unsupervised clustering, and outlier identification can be achieved.
- It provides a method for finding variable interactions that is based on experimentation.
- Random forest run times are rather quick, and they can handle unbalanced and missing data.
- They are capable of handling binary, category, and numerical features without the need for scaling.
- The random forest algorithm has a plethora of great, free, and open-source implementations.
- Almost every major ML library and toolkit have robust implementations.

11.6.3.2 Drawbacks

- Random forest techniques are unable to foresee beyond the collection of the training information in regression and over-fit information sets.
- Random forest models have the potential to be quite huge.
- Random forest is opaque and difficult to decipher in black boxes. They can take up hundreds of megabytes of memory, and the process is lengthy.

11.6.4 Deep Learning and Neural Networks

Compared to linear logistic regressions, which are identified as linear structures, the goal of neural networks (NNs) [5] is to detain non-linear patterns [13] in information by increasing the layers of the parameters to the structure. The basic NN includes three inputs, a single hidden layer with five parameters, and an output layer, as shown in the diagram [12].

Indeed, the arrangement of an NN is diversely adequate that we can construct our renowned linear and logistic regression models using them. A type of ML that utilizes NNs through multiple hidden layers to handle a variety of designs is deep learning. An NN with a hidden layer is depicted in Figure 11.4.

11.6.4.1 Definition

Deep learning is the process of training a deep neural network (DNN), which is an NN with many hidden layers.

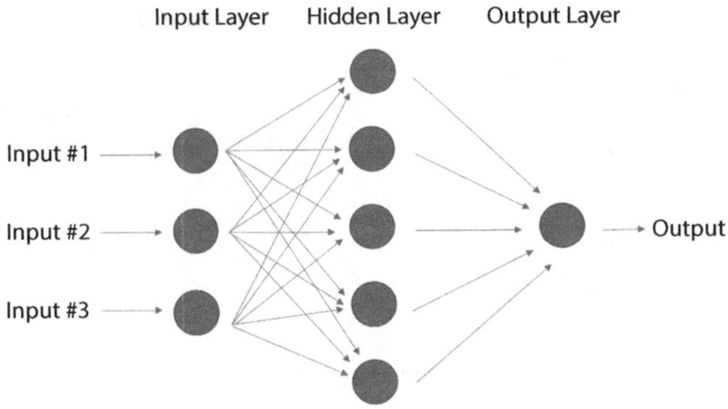

Figure 11.4 A neural network (NN) with a hidden layer (source: Wikipedia).

11.6.4.2 Remarks

The phrase "deep" is a technological term in the context of "deep learning." The number of layers of a NN is referred to as this. A shallow network contains only one hidden layer, whereas a deep network has many levels. Basic qualities mix the other, to generate improved feature from one layer. A DNN can learn data features in a so-called feature hierarchy using multiple hidden layers. In a network with numerous layers, features are processed through additional mathematical functions than in networks with less layers, making them more computationally expensive for training. The computational intensity of deep learning is one of its distinguishing qualities. A shallow neural network is shown in Figure 11.5, and in Figure 11.6, a DNN with three hidden layers is shown.

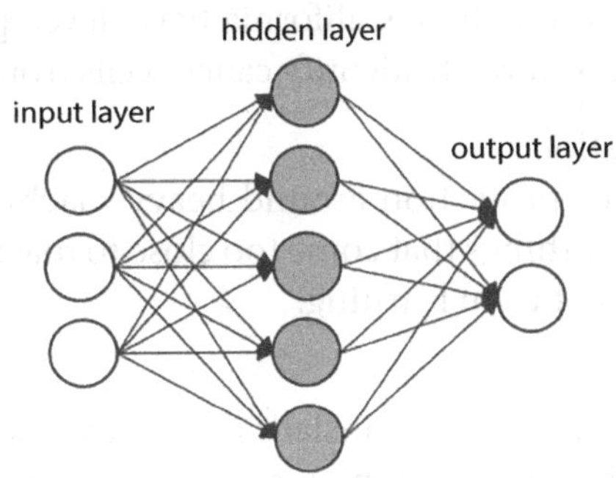

Figure 11.5 Shallow NN.

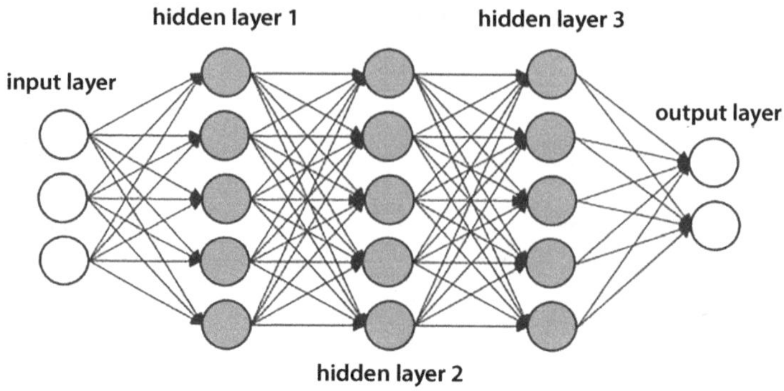

Figure 11.6 DNN model with three hidden layers.

11.6.5 Applications

Deep learning algorithms are employed in a variety of industries, including automated driving, medical equipment, Defense and aerospace, Industrial automation, Electronics [7].

1. Autonomous driving
Automotive experts employ a method of deep learning to spot components, such as stop signs and traffic lights, automatically. Deep learning also allows automobiles to recognize pedestrians and helps to reduce number of collisions.

2. Defense and aerospace
Deep learning is used to recognize items observed by satellites and to determine the safe and dangerous regions for troops.

3. Medicinal field
Scholars employ deep learning to identify cancer cells automatically. Researchers in the University of California have developed a deep learning algorithm that could accurately identify cancer cells from a larger data set.

4. Industrial automation
To increase employee protection around heavy machinery, automatically recognizing people or things that come too close to machines is now doable using the techniques of deep learning.

5. Electronics:
Automatic hearing and voice translation employ deep learning. Voice-activated home assistants that remember your preferences, for example, are powered by deep-learning applications.

11.6.6 Artificial Neural Network

An ANN uses a model based on our understanding of how the human brain responds to stimulation from sensory inputs to define the link between a set of input signals and an output signal. To solve learning problems, an ANN employs a network of artificial neurons or nodes, similar to how the brain constructs a massive parallel processor using a network of interconnected cells called neurons.

11.6.6.1 Biological Motivation

Let us have a look at how a biological neuron works. The operation of a biological neuron is depicted schematically in Figure 11.7. Incoming signals are received by a metabolic process in the cell via the cell's dendrites. You can use this strategy to weigh an impulse based on its relevance or frequency. When the input impulses reach a certain threshold, the cell fires and the output signal is conveyed down the axon through an electrochemical process. The electric signal is converted to a chemical signal at the axon's terminals, which is then conveyed to surrounding neurons via a tiny gap known as a synapse. Neuronal networks form the basis of biological learning systems. The human brain is made up of a network of around 1011 neurons, each of which is connected to an average of 104 additional neurons.

Although neuron switching speeds are substantially slower than computer switching speeds as shown in Figure 11.8, we are capable of making complicated decisions in a timely manner. As a result, the ability of the biological brain model that carries out vast parallel operations that spread

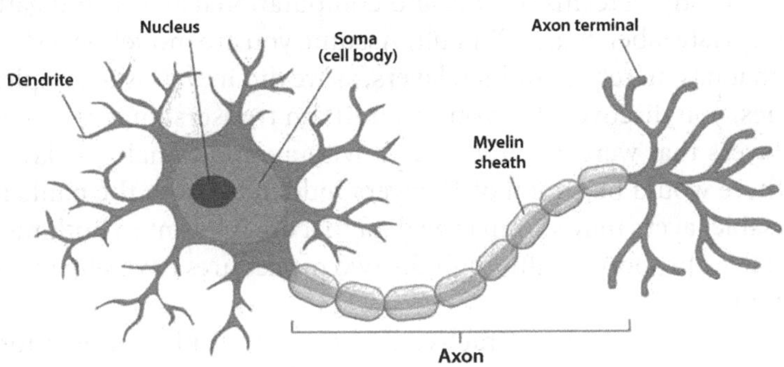

Figure 11.7 A neuron (in Wikipedia, The Free Encyclopedia).

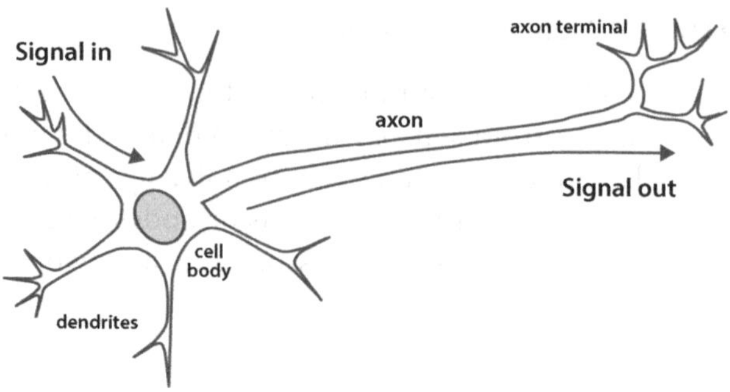

Figure 11.8 Flow of signals in a biological neuron.

across many neurons is assumed to be the cause of their information-processing capacities. The goal of developing ANN systems is to implement this form of highly parallel computations.

11.7 Transfer of Learning

Assume you are a data scientist in the retail industry. You have spent months teaching a top model how to identify shirts, tees, and polos in photos. Your most recent objective is to create a model that categorizes clothing photos into four groups: jeans, casual, cargo, and pants. Are you able to apply what you have learned from the first model to the second? Yes, you can do so using the previous model as a guide.

Transfer learning has the advantage of requiring lower information for training the NN, which is the main reason that deep learning algorithm training is costly in terms of time and computational assets, and gathering the appropriate label data is difficult. Assume you are modeling a shirt with an NN that has merely 20 hidden layers, as we did in the last example. After a few tries, you discover that you can train on trousers images by combining 18 layers that were already created with an additional new layer. As a result, there would be a total of 19 layers hidden beneath the pants model. The reusable layers may summarize data that are relevant to both methods, such as fabric properties, although the two procedures have separate inputs and outputs.

Transfer learning is an attractive and more well-liked model for standard deep-learning applications as image and text classification.

11.8 Learning Through Reinforcement

The process of receiving an agent to act in the environment in order to take advantage of its rewards is called reinforcement learning [14].

In contrast to most ML techniques, a learner (the computer) is not taught what actions to take; instead, it must figure out which activities provide the best reward by doing them. Actions can have an impact on not just the current reward but also future situations, and, as a result, all following rewards. Consider how we can teach a pet animal a trick. We cannot say how to do it. If it obeys our directions, it will be rewarded or penalized. It must figure out what it did to deserve the reward or punishment. A similar strategy might be used to train computers to perform certain tasks, including chess backgammon playing, work scheduling, and operating robot limbs. It differs from supervised learning (SL) in that it is not monitored. An example provided by a certified expert and used for learning is referred to as supervised learning. Consider a mouse attempting to locate cheese chunks that have been hidden in a maze. The longer the mouse is exposed to the maze, it improves at identifying the location of the cheese. The mouse may appear to be wandering aimlessly at first, but with practice, it learns which behaviors bring it closer to the cheese. The mouse's training method is comparable to how we use reinforcement learning (RL) to educate a system or a game. In general, RL is an ML method that assists an agent in remembering past occurrences. RL maximizes an overall reward by documenting behaviors and utilizing a brute forcing technique in a closed environment. The mouse is the agent in this scenario, and the maze represents the environment. There are four options for movement: forward, backward, left, and right. The prize is the cheese. Because it does not require prior information, you can utilize RL when you have small to no prior data on the subject (unlike traditional ML methods). As you go from the data in an RL framework, you learn. RL is particularly good at games that need "perfect information," such as Go and chess. The agent and the environment provide instant feedback in games, which allow the model to learn quickly. The disadvantage of RL is that it can take a long time to train if the problem is complex. In 2016, AlphaGo, an RL-type algorithm, destroyed the top human Go player, just like IBM's Deep Blue did in 1997 with the best human chess player. The current RL pioneers are the DeepMind teams in the United Kingdom. The OpenAI Five team became the first AI to defeat a world champion team in the video game Dota 2, which they chose since no RL algorithms could win it.

11.9 Processing of Natural Languages

Human language is used to represent a large portion of the world's facts and information. Imagine skimming through tens of hundreds of books, blogs, and articles in less than a minute. Despite their incapacity to comprehend human text, computers may be taught to execute specific tasks. For example, we can instruct our phones to autocomplete text messages or restore misspelled words. It is feasible to educate a machine to converse with a person in a simple manner. Rather than being an ML approach, NLP [15] is a technique for processing text for use in ML. Examine many text documents in various forms (word, internet blogs, etc.). Most of these textual data will have mistakes, missing characters, and other words that must be removed. The most extensively used text processing application is the Natural Language ToolKit (NLTK) developed by Stanford University scholars. Calculating the occurrences of each word in each text file is the easiest way for turning texts to numerical forms. Consider an integer matrix in which column representing a text file and row represent a word. A term frequency matrix (TFM) is a matrix form of word frequencies. We may construct a new common matrix representation of a text document by dividing each entry on the matrix by the weight of how important each word is within the corpus of texts. The process of ML term frequency inverse document frequency (TFIDF) is a commonly used metric.

11.10 Word Embeddings

The TFM and TFIDF are numerical forms of text files that simply consider the frequency and weighted frequencies. Word embeddings, on the other hand, can track the word context within a file. By quantifying the similarity of words based on their context, word embeddings allow arithmetic functions in words. The Word2Vec maps words in corpus to numerical vectors using an NN. These vectors can then be utilized to locate synonyms, perform word arithmetic, and represent the text file (by calculating the average of all word vectors in a file). Consider the case where we estimate word embeddings using a suitably large corpus of text. Assume the corpus consists of the words "queen," "man," "king," and "woman." Assume the vector ('word') is a numerical form vector that gives the word 'word.' Use arithmetic functions with vectors to estimate the vector ('woman') [10]:

$$v(\text{'king'}) + v(\text{'woman'}) - (\text{'man'}) \sim vector(\text{'queen'})$$

Word representations can be used to find commonalities in two words by calculating the cosine relationship in them in vector form. The cosine similarity of two vectors determines the angle between them. To compute word embeddings, we employ ML approaches; however, this is usually a requirement before adding an ML technique on top. Let us pretend having access to tens of thousands of users' tweets. Assume we already have knowledge on which of these Twitter users just bought a house. To analyze the possibility of a new Twitter user acquiring a home, we may utilize Word2Vec in conjunction with a logistic regression model. Either design your own word embedding or use a pre-made one. You can use pre-trained word vectors in more than 150 different languages using FastText.

11.11 Conclusion

This chapter examines 10 key ML methodologies, from the most fundamental to the most cutting-edge. By reading this in depth and fully comprehending their principles, a solid foundation can be laid for future research into more sophisticated algorithms and procedures. You can read and comprehend the remaining chapters of the book if you use this chapter as a guide. For a better understanding of each method, we have provided definitions, example approach types, and real-time instances of strength and weakness, and we hope that you will find this information beneficial while applying or confronting the method in real-time.

References

1. Shetty, Shruthi H., *et al.* "Supervised Machine Learning: Algorithms and Applications." *Fundamentals and Methods of Machine and Deep Learning: Algorithms, Tools and Applications* (2022): 1-16.
2. Sarker, I.H. Machine Learning: Algorithms, Real-World Applications and Research Directions. *SN Comput. Sci.*, 2, 160 (2021). https://doi.org/10.1007/s42979-021-00592-
3. Ardabili SF, Mosavi A, Ghamisi P, Ferdinand F, Varkonyi-Koczy AR, Reuter U, Rabczuk T, Atkinson PM. Covid-19 outbreak prediction with machine learning. *Algorithms*, 2020;13(10):249.
4. He, K., Ren, S., Sun, J., & Zhang, X. (2016). *CoRR*, abs/1512.03385.
5. A, Harisha *et al.* 'A Performance Evaluation of Convolution Neural Networks for Kinship Discernment: An Application in Digital Forensics'. 1 Jan. 2022: 379 – 386.

6. Witten, Ian H., Eibe Frank, Mark A. Hall [2016] *Data Mining: Practical Machine Learning Tools and Techniques*, 4th edition, 467.
7. Nasri, Abdelghaffar *et al.*, "A Smart Gas Sensor Using Machine Learning Algorithms: Sensor Types Based on IED Configurations, Fabrication Techniques, Algorithmic Approaches, Challenges, Progress, and Limitations: A Review", *IEEE Sens. J.*, vol.23, no.11, pp.11336-11355, 2023.
8. Abraham, Shiji, *et al.* "Diagnosing Patient Health Conditions and Improving the Patient Experience: An Application of AI and ML." *Healthcare and Knowledge Management for Society 5.0*, CRC Press, 2021. 21-35.
9. S. Angra and S. Ahuja, "Machine learning and its applications: A review," *2017 International Conference on Big Data Analytics and Computational Intelligence (ICBDAC)*, Chirala, Andhra Pradesh, India, 2017, pp. 57-60, doi: 10.1109/ICBDACI.2017.8070809.
10. Hinton, G.E., Krizhevsky, A., Srivastava, N., Sutskever, I., & Salakhutdinov, R. (2014). *J.Mach. Learn. Res.*, 15, 1929-1958
11. Hastie, T., Tibshirani, R., & Friedman, J. (2009). "The Elements of Statistical Learning: Data Mining, Inference, and Prediction." Springer.
12. Bengio, Y., Goodfellow, I., Courville, A., & Bengio, Y. (2016). "Deep Learning." MIT Press.
13. Bishop, C. M. (2006). "Pattern Recognition and Machine Learning." Springer.
14. Sutton, R. S., & Barto, A. G. (2018). "Reinforcement Learning: An Introduction." MIT Press.
15. Manning, C. D., Raghavan, P., & Schütze, H. (2008). "Introduction to Information Retrieval." Cambridge University Press.

12

Recognition Attendance System Ensuring COVID-19 Security

Praveen Kumar M.*, **Ramya Poojary, Saksha S. Bhandary and Sushmitha M. Kulal**

Sahyadri College of Engineering & Management, Mangaluru, Karnataka, India

Abstract

Attendance is an essential requirement for every organization, but maintaining the attendance data daily can be a tedious and time-consuming task. There are many automated methods available for recording attendance, such as biometrics, voice recognition, eye detection, and radio frequency identification (RFID). This project provides a smart and efficient method for recording attendance with the help of facial recognition. Knowing that the primary identification of any human is his/her face, face recognition is considered an accurate system, which overcomes uncertainties like high cost, proxy attendance, and time consumption. The coronavirus disease 2019 (COVID-19) pandemic brought to light all the weaknesses and inadequacies in our current systems. Fixing these inadequacies leads to a future made up of safer systems that are less likely to fail in such emergencies. The proposed system utilizes a face recognizer library for carrying out facial recognition and storing attendance within a database. The data saved in the database can be accessed and downloaded as a PDF for the user or the admin, so that the attendance report generation becomes easier and efficient. Overall, the project is a face-detecting, temperature-sensing, report-generating, and all-in-one attendance system that is low-cost and easily integrated into the current system to re-innovate the existing project methodology and make it available at a low cost.

Keywords: Security, COVID-19, voice recognition, recat

**Corresponding author*: praveenkm17@gmail.com

Chandra Singh, Rathishchandra R. Gatti, K.V.S.S.S.S. Sairam, Manjunatha Badiger, Naveen Kumar S. and Varun Saxena (eds.) Modeling and Optimization of Signals Using Machine Learning Techniques, (245–270) © 2024 Scrivener Publishing LLC

12.1 Introduction

Ever since COVID-19 made its unexpected albeit precarious arrival into society, daily lives have changed massively. To address the problem of using biometric fingerprint detection; to take accurate temperature readings using the touch-safe product; and send alerts to personnel with high-temperature readings, which may indicate COVID-19 symptoms; and to make a cost-effective, COVID-19-shielded device that can be implemented in any situation, which is low maintenance and low cost, we decided to come up with a solution called "Covid Secure Recat," which works on image processing to identify staff. It uses a machine learning (ML)-trained face detection algorithm to use datasets of the user's face, which we can easily extract from a 10-second video to accurately detect and recognize the user and mark their attendance. It also has an built-in temperature sensor that provides accurate measurements to send out alerts to the staff contact list in case of high-temperature readings for certain personnel. Testing the accuracy when compared with the database for the face detection, recognition, and temperature-sensing of the product in real-time on live faces and storing the attendance and temperature values in the device. All these actions are integrated into a Raspberry Pi module with camera and sensor peripherals. Developing the product framework on top of the face recognition model containing temperature sensing, database managing, and website updating features. The extracted data are backed up to a server connected to the Pi, and from there, extracted to a web interface where the data will be aptly represented and accessed by the designated users and the management for any changes and web-generated auto reports. The online one-step portal website is updated with the day's attendance and temperature records and creates monthly downloadable reports for the admin. The COVID-19 pandemic brought to light all the weaknesses and inadequacies in our current systems. Fixing these inadequacies leads to a future made up of safer systems that are less likely to fail in such emergencies. A smart attendance management system is created with the intention to solve the problems of existing manual systems and contacting the system while recording attendance during COVID-19. By integrating face recognition into the process, we aim to solve these issues and try to implement a possible practical approach to the problem.

12.2 Literature Survey

Kaur, Amandeep, and Ashish Jasuja [1] stated that the medical world of science is an unfolding area accelerating with new technologies as the

Internet of Things (IoT) turned into reality. The demand of people coming up with more supervision at considerably remote health monitoring and low clinical costs could be one of the practical solutions that can fulfill the demand. A system was made to keep track of the body temperature and pulse rate of the human body with resolute sensors along with Raspberry Pi and other devices. Overall, the system can support remote health monitoring and probably also become a part of wearable technology.

Sollu, Muhammad Bachtiar, Tan Suryani, and Benyamin Bontong [2] observed that currently, the monitoring system development has an eye on wireless sensor networks to develop a new system based on this sensor technology which is easy to monitor. Wireless-based technology, which is used for monitoring systems, is widely applied in health monitoring, military operations, and natural disasters. Body temperature and heartbeat monitoring systems are some of the health service applications used for the examination of vital signs, which are fundamental for performing early treatment safely upon knowing the physical condition of the human body. However, some of the problems present in the health services make the medical staffs takes a long time to examine patients; the retrieval of patient data is conventional; and cable media are used for equipment. Thus, the usage of the body temperature and heart rate monitoring systems built out of Raspberry Pi provided test results showing that the success rate obtained by the heartbeat detection by the system was up to 97.78% and the body temperature was up to 99.73% [3, 4].

Nandhini R., S. P. Chokkalingam, and Duraimurugan N. [5] observed that, in the changing world with high definition (HD) video monitor and other technologies, the conversion of video frames into images where the face of the person could be easily recognized during attendance and the database can also be easily and automatically reflected.

Van Natta, Meredith *et al.* [6] observed that as the COVID-19 pandemic sweeped the globe entirely and profoundly affects society, businesses, and governments, people are resorting to breakthrough biometric technologies to prevent transmission and protect economic prospects. Remote fever diagnosis, thermal facial recognition, and immunity certificates created via cell phones are now not only possible but are being used.

Mundial, Imran Qayyum *et al.* [7] noted that during the COVID-19 outbreak, face masks are an essential component of our daily lives. The face mask is thought to safeguard and prevent us from contracting the coronavirus in our daily lives. As a result, to train machines on state-of-the-art facial recognition feature vector, a dataset of faces, which are covered with masks, was collected, reflecting a method that can increase the capabilities of existing facial recognition technology using masked faces. The proposed method has a recognition accuracy of up to 97%.

Okerea for, Kenneth *et al.* [8] studied the usage of scanners by biometric machines to verify a person's identity by analyzing the patterns of their physiological or behavioral traits. Few biometric machines are contactless, requiring no direct physical contact with the scanner for the user's biometric pattern to be properly read and quantified, whereas others, such as fingerprint verification used in biometrics, the person is required to have physical contact with the biometric machine for a particular period for the user's biometric pattern to be properly read and quantified. As a result, the risk of infection by hazardous microbial pathogens has increased. The inoculation of dangerous pathogens into the respiratory system is also increased by physical contact, resulting in the onset of numerous infectious disorders. Presenting the risk of infectious disease transmission via touch-based biometric devices, as well as the essential control measures, must be done to prevent the spread of certain infectious diseases, including COVID-19 [9].

Costanzo, Sandra, and Alexandra Flores [10] inferred that non-contact temperature monitoring with integrated sensors is adopted as a solution, which is based on the integration of a capacitive humidity sensor and an infrared thermometer. This device can provide a quick and accurate tool for remotely monitoring both body and ambient temperature during pandemics like COVID-19 while avoiding direct contact with others. The results of relevant simulations on variations in body temperature, relative humidity, and ambient temperature are presented.

12.3 Software Requirements

12.3.1 Operating System - Windows 7 and Above

An operating system (OS) is a system that manages and maintains the system's functionality by hardware guidance. The OS must be Windows 7 and above, and it should be updated to the latest version.

12.3.2 IDE-Visual Studio Code

Microsoft's Visual Studio Code (Figure 12.1) is a free-source coding editor for Windows, macOS, and Linux. Some of its key features are debugging, intelligent code completion, syntax highlighting, snippets, embedded Git, and code refactoring. This IDE (integrated development environment) is

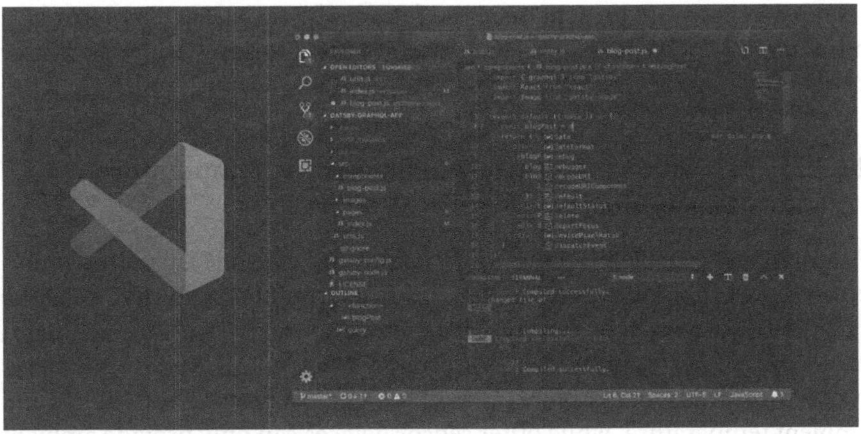

Figure 12.1 Visual Studio Code as the code editor software.

being used to create a website where admin and users may download attendance reports.

12.3.3 Programming Languages: Python, HTML, CSS, JS, and PHP

Python programming language is used to develop machine learning algorithms for facial recognition and other machine learning concepts of the project. The HTML, CSS, JS, and PHP programming languages are used to develop a website where HTML, CSS, and JS are used to build the front-end of the website and PHP is used for the backend of the website.

12.4 Hardware Requirements

12.4.1 Three Processors and Above

A hard disk or hard drive is an electromechanical storage device for storing and retrieving digital data with the help of magnetic storage and rapidly rotating platters which are coated with magnetic material. The minimum capacity of a hard disk should be 5 GB since a lot of data have to be stored.

12.4.2 RAM - 2GB (Minimum Capacity)

The RAM is one of the most fundamental components of super-fast computing and is the space for temporary data storage, which a computer requires to access at present or in the future. The RAM capacity must be a minimum of 2 GB to ensure efficient and faster working of the system or computing device.

12.4.3 MLX90614 IR (Infrared) Sensor for Temperature Measurement

A temperature sensor [Figure 12.2] is a device that measures the ambient temperature and turns the data into electronic data so that it can be recorded, monitored, or communicate temperature changes. The temperature is recorded, monitored, and displayed using an MLX90614 temperature sensor. This sensor differs from others as it can provide both object and ambient temperature, while other sensor devices can only provide ambient temperature. We also employed the IR-based MLX90614 to ensure that no physical contact is required, and we can monitor the item temperature rather than the ambient temperature [11].

12.4.4 Pi Camera

The Pi camera module, which supports Raspberry Pi (Figure 12.3), is a mobile/portable lightweight camera that communicates using the MIPI camera serial interface protocol. It is commonly utilized in machine learning applications, surveillance projects, and image processing [12].

Figure 12.2 MLX90614 IR temperature sensor connected to the Raspberry Pi.

Figure 12.3 Pi camera with Raspberry Pi.

12.4.5 Raspberry Pi

The Raspberry Pi was produced as a line of single-board computers by the Raspberry Pi Foundation. The Raspberry Pi is a low-cost computer with a set of general-purpose input output (GPIO) ports that can be used to control physical computing components and understand the IoT. The Raspberry Pi 4 is a follow-up to prior models, particularly in terms of speed and performance. It is a much more energy-efficient and cost-effective machine [13,14]. Raspberry Pi 4 model is depicted in Figure 12.4.

12.4.6 OLED Display

The organic light emitting diode (OLED) (Figure 12.5), also referred to as the organic electroluminescent (organic EL) diode, is a LED in which the

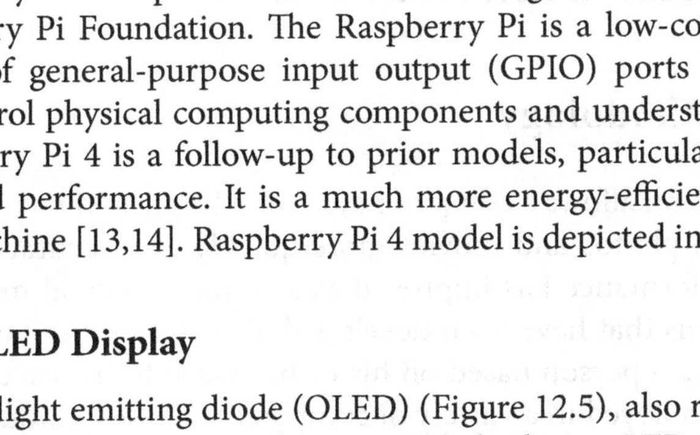

Figure 12.4 Raspberry Pi 4 model B.

Figure 12.5 OLED display (where a message and a temperature are shown).

layer of emissive electroluminescent is an organic compound film, which, in response to electric current, emits light.

12.5 Methodology

Face recognition allows a computer system to locate and recognize human faces in both photos and movies more quickly and accurately. Face recognition performance has improved as a result of several methodologies and algorithms that have been developed. Face recognition is the process of recognizing a person based on his or her facial traits. Face characteristics can be employed in a variety of computer vision techniques, including face detection, and a variety of video surveillance applications, including expression detection. Face recognition has recently attracted researchers' attention because of its prospective capabilities to detect and recognize faces, which might be applied to a wide range of practical real-time applications such as criminal detection, system security, and identity verification, among others. There are two stages to the facial recognition method [15, 16].

Face detection is when a picture is taken as input and searched for any face at all, after which image processing is applied to clean up the facial image so that the captured face can be recognized easily. Face recognition is the technique of detecting facial features and processing the face before comparing it to a database of known faces to determine the person's identity [17]. Configuration of facial recognition and storing data in the database is depicted in Figure 12.6.

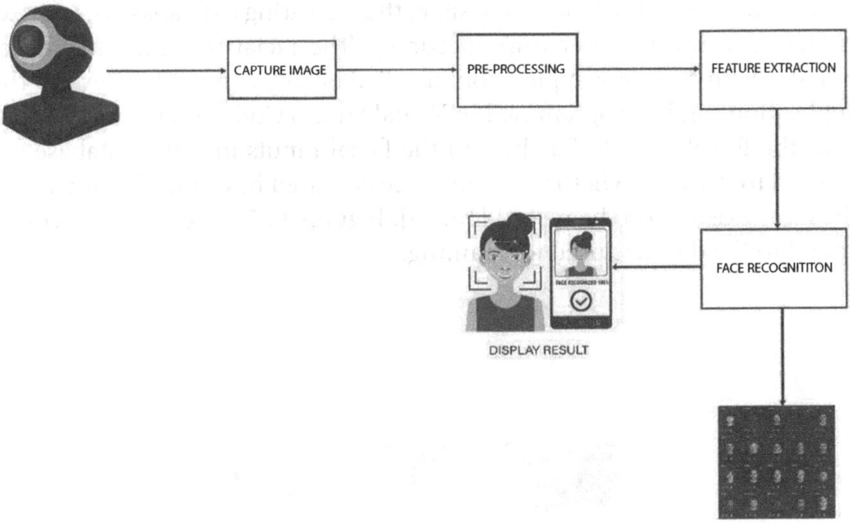

Figure 12.6 Configuration of facial recognition and storing data in the database.

12.6 Building the Database

In building the database, 10-second videos of each individual who is a staff member will be collected. The video will be taken in a standard way to

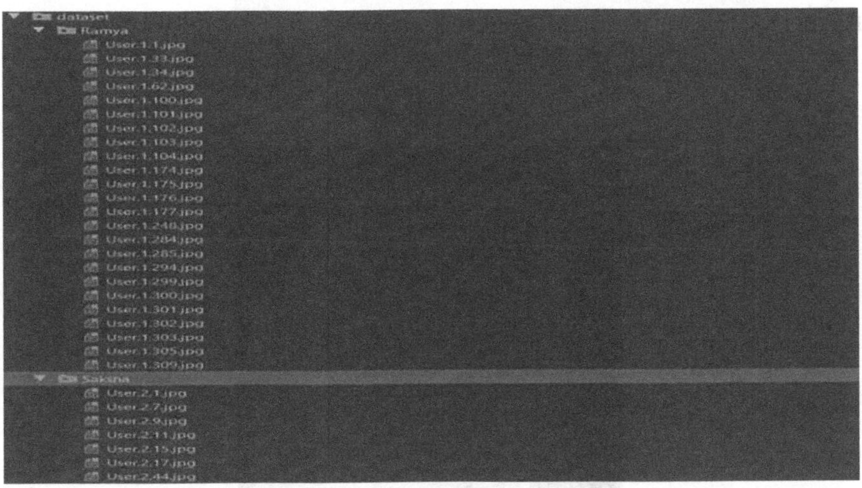

Figure 12.7 Dataset of images with folders corresponding to each person.

ensure all angles of the face are visible; thus, creating a database of approximately 240 images, increasing accuracy. The facial recognition process initiates with a camera application installed on any compatible device. The application can be programmed in Visual Studio Code and works together with the Raspberry Pi. Further, all the facial inputs into the database are utilized to discover whether the new face captured has a match with any of the faces existing on the record [18, 19]. Figures 12.7, 12.8, and 12.9 correspond to the datasets used for training.

Figure 12.8 Image in the dataset/Ramya for training.

Figure 12.9 Image in the dataset/Saksha for training.

12.7 Pi Camera for Extracting Face Features

The extraction of facial features is the method of extracting the components of a face like the eyes, mouth, and nose from an image of a human face. The extraction of facial features is very much vital for the beginning of processing techniques like face recognition, face tracking, or facial expression recognition. A set of images is exposed where they will be fed to a face recognition training algorithm, and the face features will be extracted and it will be stored. Face detection is vital as the image taken via the camera is fed to the system, where the algorithm for face detection is used to recognize the human face(s) in the image. A number of image processing methods are used for facial detection in a picture, as well as determining where the discovered faces are located. The Pi camera module, which is a lightweight portable camera that supports the Raspberry Pi, connects with the Pi using the MIPI camera serial interface protocol. Its most typical uses are in machine learning, image processing, and surveillance projects. The camera module in the Raspberry Pi is a custom-designed Raspberry Pi add-on that communicates to the board via one of the two connectors on the upper surface. The CSI interface is used specifically for interfacing the cameras. The CSI bus is capable of exceptionally high data rates and is designed to carry pixel information only. Capturing a face in real time is shown in Figure 12.10.

Figure 12.10 Capturing a face in real time.

12.8 Real-Time Testing on Raspberry Pi

The Recat, which consists of a Raspberry Pi module and a Pi camera, will be pre-loaded with face features and a detection algorithm that will recognize you by name and prompt you to check your temperature upon recognition. The memory will be used to keep track of your attendance and temperature. Raspberry Pi is a low-cost and compact computing device with a variety of functions that can be used in a variety of applications like mobile robots with embedded control hardware and the implementation of computer motion and vision control. However, the default OS of the Raspberry Pi causes certain limitations in its performance in real-time.

12.9 Contactless Body Temperature Monitoring

Since the corona outbreak, determining whether a person is infected or not has been extremely difficult. Temperature gadgets are frequently used to solve this problem by measuring body temperatures. These gadgets are equipped with non-contactable infrared sensors for temperature measurement, which could detect the temperature of the body without requiring any contact physically. The IoT is altering our lives in today's world by establishing a variety of technologies that can be monitored and controlled remotely. As a result, we have integrated it into the Recat. We will use a Raspberry Pi, an MLX90614, and a Pi camera to create a monitoring system for temperature along with an alert system via email to demonstrate the power of IoT.

12.9.1 MLX90614 Interfaced with the Raspberry Pi

Initially, the library and packages are extracted, then the MLX90614 is connected to the computer. Step 1: From the Raspberry Pi settings, enable I2C and go to the interfacing options after typing sudo raspi-config. Select the I2C option and turn it on. This corresponds to Figure 12.11.

Then, using the commands below, we must install the essential packages. Using cd PyMLX90614- 0.0.3/, the folder that is extracted is navigated and the command is executed via run.

```
>> sudo python setup.py install
```

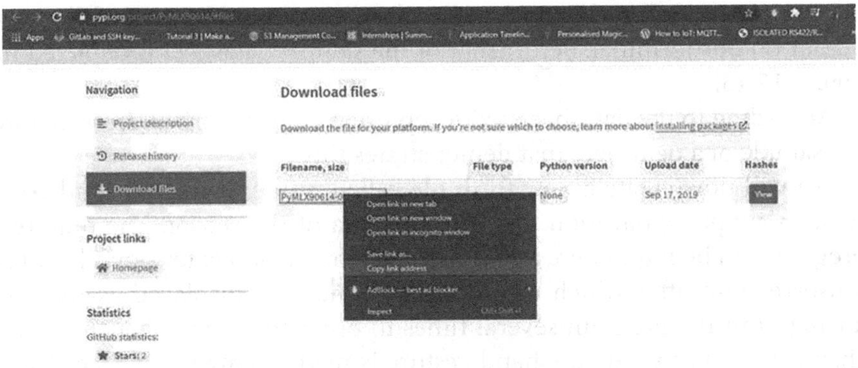

Figure 12.11 Acquiring the MLX90614 package/library.

After completing the instructions, we use the circuit below to build connections with the MLX90614 sensor and the module of the Raspberry Pi. The fritzing software was used in creating the Raspberry Pi and the MLX90614 circuit shown in Figure 12.12.

If the installation and connections are correct, we can use the command i2cdetect -y 1 to see whether the sensor address value on the I2C bus is

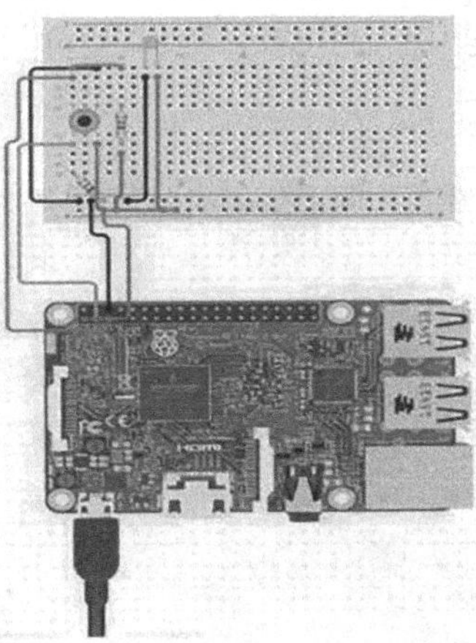

Figure 12.12 Connecting the MLX90614 sensor to the Raspberry Pi.

SDA capacitance	C_{SDA}				10	pF
Slave address	SA	Factory default		5A		hex

Figure 12.13 Screenshot of a sample datasheet.

obtained and, if everything goes accordingly, we should get the following output on our terminal. Screenshot of the sample datasheet is depicted in Figure 12.13.

According to the datasheet, 0x5A represents the sensor's address. Below is a sample of a datasheet that demonstrates this.

We will now create a new fresh file called mlxsensorread.py and construct a simple application to verify the data of the sensor. The required program can be found here as follows: Run the python extension file, mlxsensorread.py, after which the results obtained are displayed as the following. Run the program several times in order to see whether there is a change in values when any hand gesture is made facing the device. Thus, we were able to successfully connect the MLX90614 to our Raspberry Pi.

12.10 Raspberry-Pi Setting Up an SMTP Email

The Python software reads the data from the sensor, compares the data to the set limit value, authorizes the email, and dispatches the email along with an image attachment. Initially, essential libraries, as well as the sender's and recipient's email declarations, are extracted. The Smtplib could send an email, but we have to organize it with a topic, body, and additional attachments, which the multipurpose Internet mail extensions (MIME) can provide, simply by running the Python program on Pi after the hardware and

Figure 12.14 Reading the temperature values from the sensor.

Figure 12.15 Receiving an email from Raspberry Pi.

software are ready. As illustrated Figure 12.14, the temperature value is obtained from the sensor.

Check whether the temperature of the object rised beyond the set value, after the Python application captures a picture via the camera, and the data is saved on the Raspberry Pi, further dispatched through email. An example of an email sent from a Raspberry Pi and the results obtained is shown in Figure 12.15. Connecting the Pi camera to the Raspberry Pi is depicted in Figure 12.16.

The final circuit diagram is depicted in Figure 12.16.

12.11 Uploading to the Database

The code uploads the image to a folder and saves its name in the database. The source code is generated using any code editor. Here, we uploaded the generated datasets to a database named MySQL/Google Firebase, as depicted

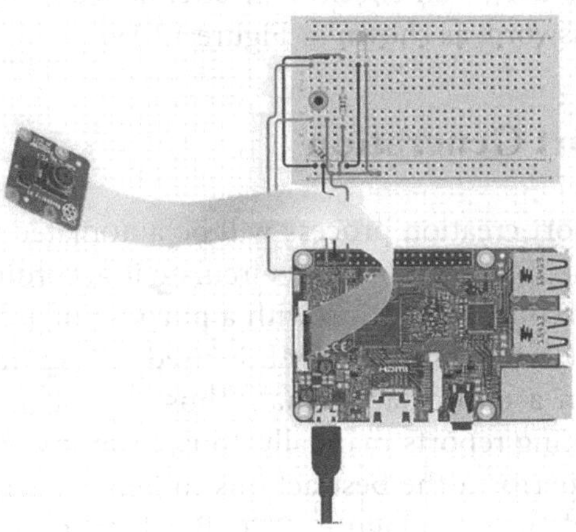

Figure 12.16 Connecting the Pi camera to the Raspberry Pi.

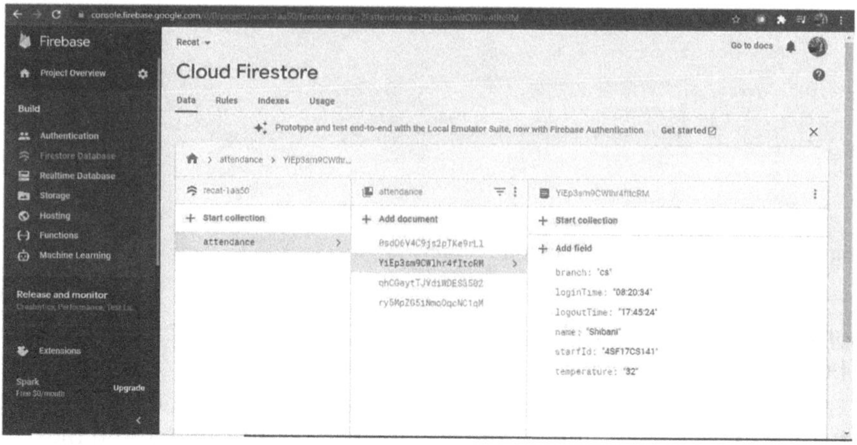

Figure 12.17 Uploading data to the database.

in Figure 12.17. It is employed with the intention of accessing, adding, and managing database contents with speed, reliability, ease of use, and versatility.

12.12 Updating the Website

The data from the device are uploaded to the server, which will be updated on the website. The website has two accesses: one is for the admin and the other is for the users, which are the lecturers in this case. The admin can download the date-wise attendance details of the respective departments and make changes accordingly based on paid leave/holiday, as shown in Figure 12.18. The users can check their attendance using a unique ID or email ID and password, as shown in Figure 12.19.

12.13 Report Generation

The monthly report creation process will be automated where the admin can download the report online and circulate it accordingly. A computer program for the report generation with a purpose of acquiring data from a database source with a stream XML is used. Using this, the document is put together in a particular format. Thus, report generation discards the need for creating reports manually, thus, lowering the error risks and allowing one to perform the best actions to better analyze the data. The sample report is depicted in Figure 12.20. Block diagram of facial recognition system based attendance system is depicted in Figure 12.21.

Figure 12.18 Date-wise attendance details updated on the website.

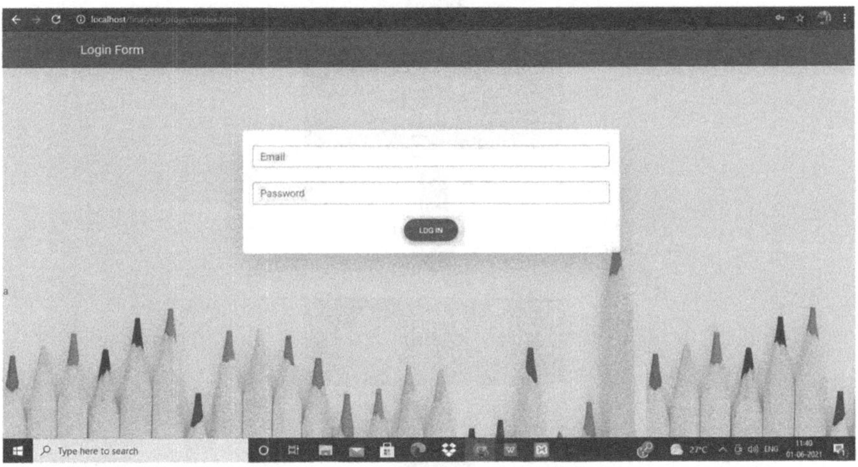

Figure 12.19 Visual of the website.

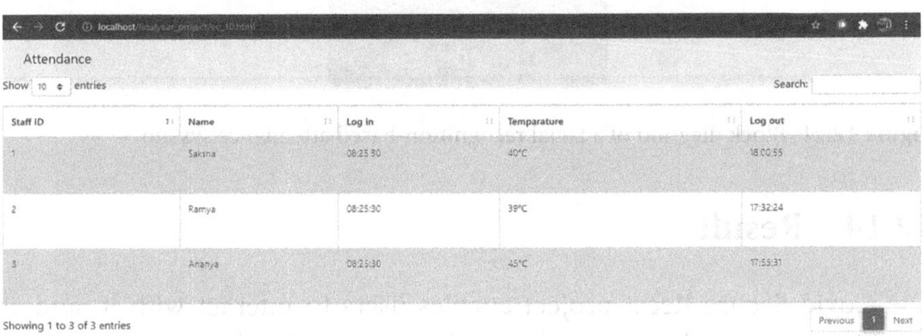

Figure 12.20 Sample report.

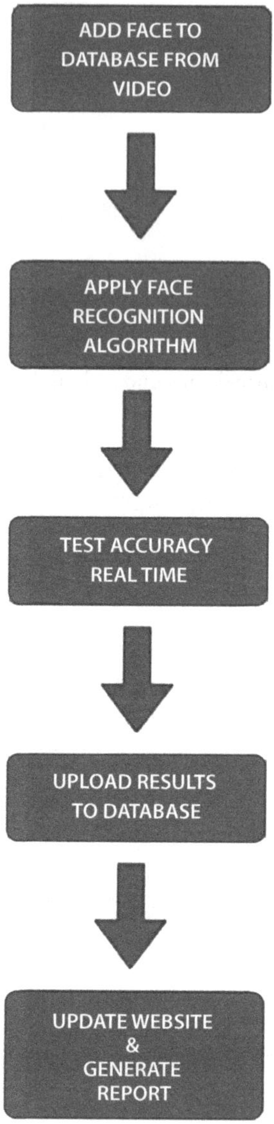

Figure 12.21 Block diagram of a facial recognition-based attendance system.

12.14 Result

The Covid Secure Recat project enables users to interact with it, and at the same time, mark the attendance of every staff/faculty within the campus. This system not only records attendance but also ensures that the attendance is marked while maintaining the anti-touch decorum following the COVID- 19 precaution guidelines and maintaining a hygienic and safer interaction with the system, even when multiple users might interact with the same device. Also, this system not only records the attendance of the user, but also automatically updates the necessary information to a main server, which can be accessed by an admin in general or by individual users to keep track of their own data in the future. Overall, the data are not only of the attendance record but also the temperature update of the individual on a daily basis via a temperature sensor simultaneously measuring his/her body temperature, ensuring that the person is in good health and is safe to be around, especially during this time of the pandemic. From face detection to anti-touch body temperature measurement and auto-updating of vital information to the server, the system ensures a safe and secure way of recording and managing the data related to both attendance and health. The prototype design of the Covid Secure Recat is depicted in Figure 12.22, and its output predictions are shown in Figure 12.23, 12.24, and 12.25.

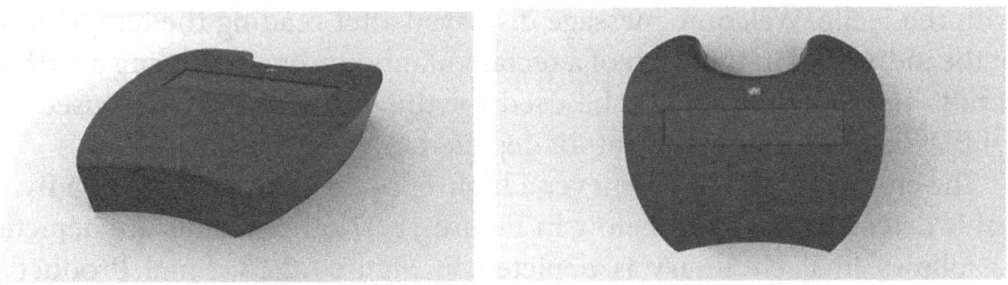

Figure 12.22 Rendered digital design of Covid Secure Recat.

Figure 12.23 Output prediction of Ramya based on trained data.

Figure 12.24 Output prediction of Saksha based on trained data.

For untrained datasets or facial features of the individual, the system does not recognize them. Instead, it recognizes them as 'Unknown.'

The object temperature value is read and displayed as the output, along with the Hello/Welcome message displayed after reading the temperature of the individual in the form of a rectangular bounding box using a Python script. The LCD/OLED can be used for this purpose. We have used the OLED display to show the output depicted in Figure 12.26.

The email is received whenever a high-temperature value of an individual is detected, as shown below in Figure 12.27 other results are depicted as follows final circuitary is depicted in Figure 12.28, Final Product is depicted in Figure 12.29, Admin Page is depicted in Figure 12.30, Data Wise Attendance is depicted in Figure 12.31.

Figure 12.25 Output prediction of no recognition.

Figure 12.26 Displaying 'Hello' message on the OLED.

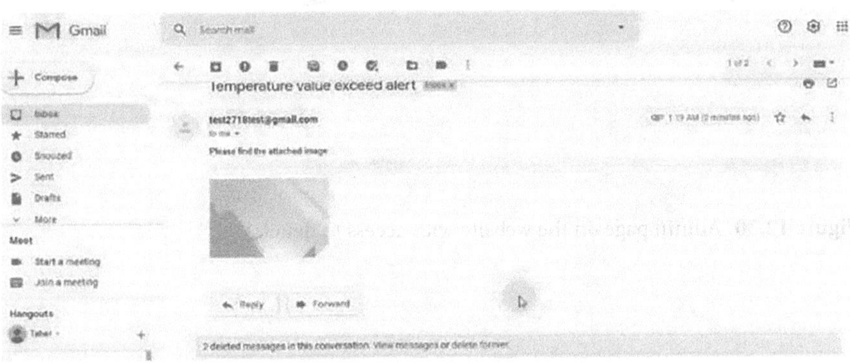

Figure 12.27 Email received from the Raspberry Pi for a person with high temperature.

Figure 12.28 Final circuitry consisting of the Raspberry Pi, Pi camera, and MLX90614 sensor.

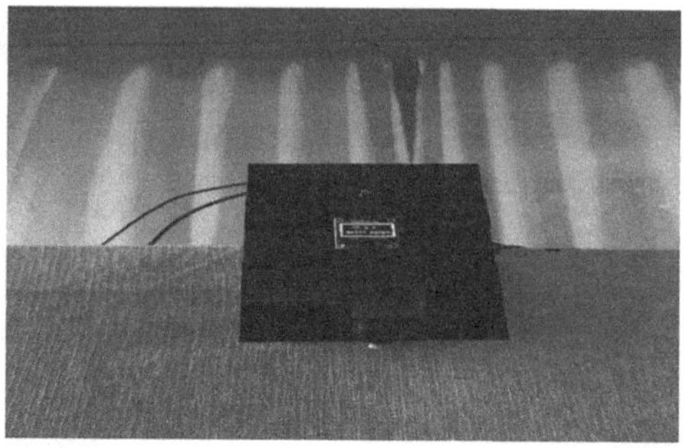

Figure 12.29 Final product displaying the name and temperature of the person.

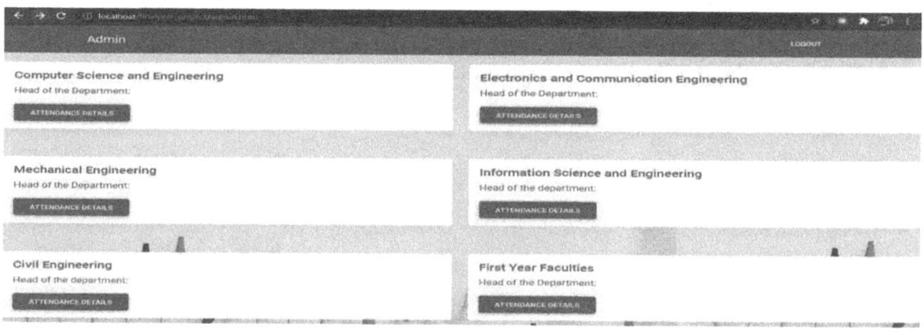

Figure 12.30 Admin page on the website with access to details.

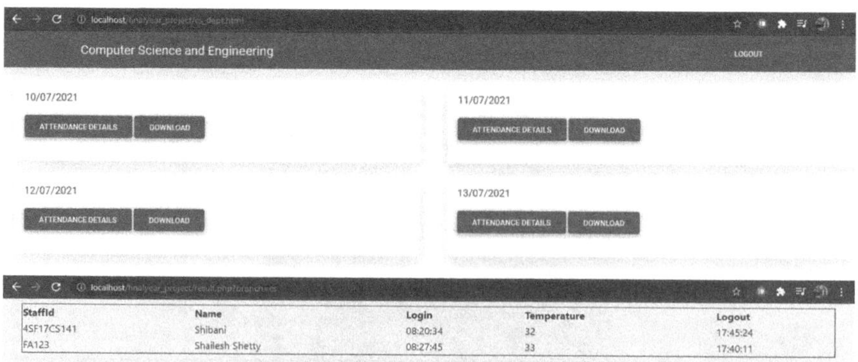

Figure 12.31 Date-wise attendance details.

Figure 12.32 User page on the website with access to details.

12.15 Discussion

While running the system for testing purposes, the facial recognition algorithm running within the product recognizes the face of the individual and then reads the temperature of the individual. The temperature is recorded in the database and if any temperature crossing the threshold value is detected, then an email notification is sent to the user. If the person has no record in the database, then it detects the individual as an unknown person. The details obtained are further recorded in the database and the same are utilized for the updates of the website. Updating the website is done periodically at regular intervals, and both the admin and the user can have access to the concerned data accordingly through our website. Overall, recording the attendance with accuracy and precision, recording and observing the temperature variations, and finally giving privacy-maintained access to the admin and the user to be aware of the details that are recorded over the website. User page on website with details is depicted in Figure 12.32.

Facial recognition has a huge impact even in the future where this technology is expected to transform at an enormous rate by changing the lives of people in helping them get things done faster or use it for any specific applications impacting both personal and professional living. Security and surveillance are the major fields, along with other sectors like banking, private industries, and public buildings, which are and will be influenced deeply. Facial recognition will contribute to generating large revenues in the coming years, and at the same time, continue to provide solutions in accomplishing impractical tasks in the future.

12.16 Conclusion

The smart system for attendance management is created to address problems with existing systems that work with manual support and are based on physical contact. We used the concept of face recognition to mark students' attendance and make the program better. The system works satisfactorily in a variety of conditions and ways. In the future, this program needs to be improved because this program sometimes fails to see students from far away, and we have certain processing limitations. Working with a high-level processing program can have a very positive effect on this program. An automated attendance management system is a necessary tool in any learning management school. Most existing programs are time-consuming and require manual labor by a teacher or other staff. Our approach aims to resolve issues by integrating facial recognition into the process. Although this program has problems, such as getting a large number of faces, there is a lot of room for improvement.

References

1. Kaur, Amandeep, and Ashish Jasuja. "Health monitoring based on IoT using Raspberry PI." *2017 International Conference on Computing, Communication and Automation (ICCCA)*. IEEE, 2017.
2. Sollu, Tan Suryani, Muhammad Bachtiar, and Benyamin Bontong. "Monitoring system heartbeat and body temperature using raspberry pi." *E3S Web of Conferences*. Vol. 73. EDP Sciences, 2018.
3. Damale, Radhika C., and Bazeshree V. Pathak. "Face recognition-based attendance system using machine learning algorithms." *2018 Second International Conference on Intelligent Computing and Control Systems (ICICCS)*. IEEE, 2018.
4. Nandhini R, Duraimurugan N, S.P.Chokkalingam. "Face Recognition Based Attendance System" *Int. J. Eng. Adv. Technol. (IJEAT)*, ISSN: 2249 – 8958, Volume-8, Issue-3S, February 2019.
5. Van Natta, Meredith, *et al.* "The rise and regulation of thermal facial recognition technology during the COVID-19 pandemic." *J. Law Biosci.*, (2020).
6. Mundial, Imran Qayyum, *et al.* "Towards Facial Recognition Problem in COVID-19 Pandemic." *2020 4rd International Conference on Electrical, Telecommunication and ComputerEngineering (ELTICOM)*. IEEE, 2020.
7. Okereafor, Kenneth, *et al.* "Fingerprint biometric system hygiene and the risk of COVID19 transmission." *JMIR Biomed. Eng.*, 5.1 (2020): e19623.
8. Carlaw, Stuart. "Impact on biometrics of Covid-19." *Biom. Technol. Today*, 2020.4(2020): 8-9.
9. Costanzo, Sandra, and Alexandra Flores. "A Non-Contact Integrated Body-Ambient Temperature Sensors Platform to Contrast COVID-19." *Electronics*, 9.10 (2020): 1658.

10. S. U. Amin and M. S. Hossain, "Edge Intelligence and Internet of Things in Healthcare: A Survey," in *IEEE Access*, vol. 9, pp. 45-59, 2021, doi: 10.1109/ACCESS.2020.3045115.
11. Deblu Sahu, Bikash Pradhan, Anwesha Khasnobish, Sarika Verma, Doman Kim, Kunal Pal, "The Internet of Things in Geriatric Healthcare", *J. Healthcare Eng.*, vol. 2021, Article ID 6611366, 16 pages, 2021. https://doi.org/10.1155/2021/6611366
12. Kagalkar R.M., Bhajantri L.B. (2021) Internet of Things based Smart Ring System for Providing Human Safety Services during Emergency Situations. In: Smys S., Palanisamy R., Rocha Á., Beligiannis G.N. (eds) *Computer Networks and Inventive Communication Technologies*. Lecture Notes on Data Engineering and Communications Technologies, vol 58. Springer, Singapore. https://doi.org/10.1007/978-981-15-9647-6_90
13. P C Gunawan *et al* Design and development of telemedicine based heartbeat and bodytemperature monitoring tools ,2020 *IOP Conf. Ser.: Mater. Sci. Eng.*, 850 012018
14. S.M.R. Islam, D. Kwak, M.H. Kabir, M. Hossain, K.S. Kwak, The internet of things for health care: a comprehensive survey, *IEEE Access,* Pages: Page(s) (2015) 678–708.
15. Y. Peng, L. Peng, A cooperative transmission strategy for body-area networks in healthcare systems, *IEEE Access,* Page(s) (2016) 9155–9162.
16. Ioan Orha, Stefan Oniga, Automated system for evaluating health status, in: *Design and technology in Electronic Packaging (SIITME), IEEE 19th International Symposium* for, 2013, pp. 219–222.
17. Karkera, T., & Singh, C. (2021). Autonomous bot using machine learning and computer vision. *SN Comput. Sci.*, *2*(4), 1-9.
18. Shetty, S. H., Shetty, S., Singh, C., & Rao, A. (2022). Supervised Machine Learning: Algorithms and Applications. *Fundamentals and Methods of Machine and Deep Learning: Algorithms, Tools and Applications*, 1-16.
19. Singh, C., Sairam, K. V. S. S. S. S., & MB, H. (2018, August). Global fairness model estimation implementation in logical layer by using optical network survivability techniques. In *International Conference on Intelligent Data Communication Technologies and Internet of Things* (pp. 655-659). Springer, Cham.

13

Real-Time Industrial Noise Cancellation for the Extraction of Human Voice

Vinayprasad M. S.*, Chandrashekar Murthy B. N. and Yashwanth S. D.

Department of Electronics and Communication Engineering, JSS Science and Technology University, Mysuru, India

Abstract

The industrial environment has a wide range of stochastic signals. A major portion of these signals include high decibel level noises (> 85 dB) from various machinery. These are very harmful to the human ear. In addition to these signals, there are human voice signals, which are many times essential. Passive noise cancellation refers to cancelling noise without the use of power. This is achieved using various sound-absorbing and reflecting materials. The problem with this approach is that it does not cancel the noise selectively. Moreover, it is not appropriate for cancelling high-frequency signals. Active noise cancellation is needed to process and extract the desired signal (voice signal) from the undesired signal (mixture of voice and noise signals). The signals are all very high amplitude, which reach the ear through multiple directions; if only active noise cancellation is implemented, then it will not lead to effective cancellation. The method used in this paper has a microphone present at the user end and a microphone present at the machine end. The mixture of noise and voice signals is obtained at the user end. The reference signal is obtained from the machine end then wirelessly transmitted to the user end where processing takes place and is played back to the user. The processing and playing use the concept of multiprocessing in processors with each core of the processor performing a different function. The user is protected from surrounding noises by an additional passive attenuator.

***Keywords*:** Active noise cancellation (ANC), wireless communication, multiprocessing

**Corresponding author*: vpms1408@sjce.ac.in

Chandra Singh, Rathishchandra R. Gatti, K.V.S.S.S.S. Sairam, Manjunatha Badiger, Naveen Kumar S. and Varun Saxena (eds.) Modeling and Optimization of Signals Using Machine Learning Techniques, (271–282) © 2024 Scrivener Publishing LLC

13.1 Introduction

Noise is one of the physical factors that affects humans physiologically, indirectly affects humans physically, and causes environmental pollution. Many regulatory measures have been implemented by governments across the globe. The maximum noise exposure to which employees must be limited is about 90 dB for about 8 hours. Continuous exposure to noise levels of above 80 dB may lead to hearing loss, heightened blood pressure, stress, headaches, distractions, annoyances, increased heartbeat acceleration, sleeping disorders, and various other psychological disorders [1, 4].

The noises in industries are highly random. These noises cannot be cancelled using conventional band-pass filters or other equivalent filters. Thus, adaptive filters are more suitable for these applications.

The simplest adaptive finite impulse response (FIR) filter, the least mean square (LMS) algorithm, is based on a filter coefficient update equation in which the difference between coefficients at time k and time k+1 is given by the negative gradient to the function that is to be minimized. This function, denoted as the performance function, is equal to the expected squared error E {e [k]}.

$$w_{k+1} = w_k + \mu(-\nabla_k) \tag{13.1}$$

The constant μ determines the rate of convergence, i.e., the time to reach the optimum Wiener solution that corresponds to the minimum of the performance function. Simplifying the previous expression leads to the noisy gradient estimate LMS algorithm, where instead of $_k$, the instant error e[k] and the input sample vector x[k] are used:

$$wk+1 = wk + \mu ekxk \tag{13.2}$$

When w is an Lth order filter, the applicable interval of μ is given by:

$$0 < \mu < 1/(L+1)(\sigma_k^2) \tag{13.3}$$

If μ exceeds the upper limit, the two convergences of the algorithm are no longer ensured.

The normalized least mean square (NLMS) algorithm is not different from that of the LMS filter. Only now, the step-size parameter μ is normalized to μ, i.e., the power of x[k] is incorporated in the n algorithm, which takes the form:

$$w_{k+1} = w_k + (2\mu e_k x_k)/(L+1)(\sigma_k^2) \tag{13.4}$$

The instant power estimate of x[k] employs an exponential window defined by the forgetting factor and is given by:

$$(\sigma_k)^2 = (1 - -)\sum_{i=0}^{k} a^i X_{k-1}{}^2 \tag{13.5}$$

The major advantage of the NLMS to the LMS algorithm is the fact that a normalized step size enables the comparison between simulations with different input signals and ensures that the maximum allowable actual step size never exceeds the upper limit. The price, however, is a slight increase in computational complexity [2].

Algorithms like the recursive least aquare (RLS) algorithm are also present for adaptive noise cancellation. This algorithm has a higher rate of convergence than the LMS and NLMS algorithms. However, it is suitable for modeling stochastic signals like music signals. In addition to this, the design involved is more complex and costly compared to the LMS algorithms [2, 3].

13.2 Literature Survey

Tirtharaj Sen *et al.* [1] conducted a case study titled "A Study and Comparison of the Noise Dose on Workers in a Small-Scale Industry in West Bengal, India." Most of the workers were exposed to prolonged levels (greater than 5 hours) of high decibel noise (greater than 70 dB) and many of them suffered from temporary or permanent hearing loss.

Nadder Hamdy [2] is the author of the book "Applied Signal Processing," wherein the need for active noise control (ANC) is explained in detail. Conventional methods of noise cancellation involve the use of attenuators. The problem with this approach is that it is not selective. It attenuates every signal including human voice. It is more suitable for filtering methods to eliminate the noise. Adaptive filters are more suitable as the background noise keeps fluctuating and is highly variable. It has given details of even adaptive interference cancellation. The adaptive filter needs a suitable algorithm for the implementation of noise cancellation. The LMS algorithm can be used to implement background noise cancellation. It involves reducing the expected value of the error signal (E(|e(x) |[2]).

Sayed and Lotfizad [3] proposed a comparative study between the various noise cancellation algorithms in their journal article "A Family of Adaptive Filter Algorithms in Noise Cancellation for Speech Enhancement."

They had proposed an algorithm that is highly convergent and relatively simple, which is not needed in our application. However, they had presented the various signal-to-noise ratios obtained in various algorithms. The LMS signal-to-noise has a relatively low signal-to-noise ratio, but with a simple implementation. In our application, we need to eliminate the noise; that is our primary application, and speech recognition is secondary. A low-quality speech signal can be afforded.

Ishika Sharma *et al.* [8] wrote "Adaptive Filter Design for ECG Noise Reduction using LMS Algorithm" (an IEEE paper). This paper presents the design of an adaptive filter using an LMS algorithm for reducing noise in the ECG waveform to generate a waveform with reduced noise levels. The optimum result is produced when the step size is considered to be as small as possible, but it should also be kept in mind that while designing the adaptive filter, the step size should not be taken very small because it will decrease the value of the rate of convergence. There is ambiguity in deciding the optimum convergence rate.

Jyoti Dhiman *et al.* [5] proposed a comparative study of the different algorithms in their paper, "Comparison between Adaptive Filter Algorithms (LMS, NLMS and RLS)." It gives an analysis of the improvement of S/N ratio for the various values of noise variance. The LMS algorithm takes about 1500 iterations for achieving steady-state error (ess). According to the hardware description of Raspberry Pi, it can handle 14.6 loop iterations per microsecond. Note that 1500 iterations take 102.739 microseconds, according to Jernimo Arenas-Garca, Luis A.

Jernimo Arenas-Garca *et al.* [6] have proposed a technique to optimize between the output and the rate of convergence in their paper, "Combination of Adaptive filter, Convergence and Properties." In order to obtain better convergence and tracking in adaptive filters, the cascaded properties of an adaptive filter are obtained. The robustness of the filter is increased, and for environments where the signals are unknown, it is suitable.

Stefan Liebich *et al.* [7] proposed a technique that shows how the rate of convergence varies according to various external parameters in their paper "Active Noise Cancellation in Headphones by Digital Robust Feedback Control,". The enhancement of the performance of adaptive filtering by considering the parameters of ear position, enclosed volume, material of the earpiece, etc. was discussed. A mixed sensitivity control method, which deals with the addition of extra coefficients without additional audio signals, was used.

13.3 Methodology

This chapter provides details regarding the various algorithms that are used in the design of the technique and the flow of the functionality.

13.3.1 Design of Processing System

The processing system is the flow in which the functionality of the technique is achieved. Here, the system that has to be designed is the adaptive noise cancellation system [9, 10]. The block diagram of the proposed adaptive noise cancellation system, as seen in Figure 13.1, is as follows: Audio is captured for each second. A microphone captures the audio from the users' perspective for 1 s. The user hears both the industrial and the machine noise. The input signal can be considered as follows:

$$x[n] = \textit{industrial noise} + \textit{human voice} \tag{13.6}$$

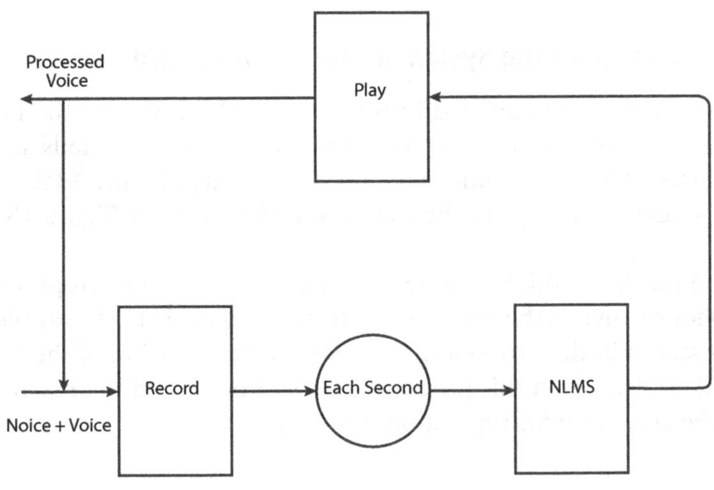

Figure 13.1 The basic flow of the processing system.

13.3.2 The NLMS Algorithm

For this block, there are two inputs; one is the x[n] obtained from one of the microphones. The other one is the machine noise (d[n]) obtained from another microphone.

$$E[n] = d[n]x[n] \tag{13.7}$$

Where e[n] is the error signal.

$$w(n+1) = w(n) + 2ue(n)x(n) \tag{13.8}$$

Where w(n+1) is the updated weights and u is the convergence factor. The convergence factor u is calculated using the equation:

$$U(n) = \sigma/(c + abs(x(n)2)) \tag{13.9}$$

The filtered output is played to the user via a headphone which is padded with a passive attenuator. The play process is designed to function in parallel using multiprocessing.

This process is repeated for every second of the signal, thus obtaining the output.

13.3.3 Design of the System at the Machine End

The external design is divided into two parts: the design at the machine end and the design at the user end. This section provides details about the system present at the machine end and is depicted in Figure 13.2.

The sequence of steps at the machine end as seen in Figure 13.2 is as follows:

At the machine end, the noise is recorded through a microphone. This microphone converts the sound to an electrical signal [11]. A .wav file of the noise is stored in the processor. The wireless transmission of the machine noise takes place from this processor to the processor at the machine end where the algorithm for separation is implemented.

13.3.4 Design of the System at the User End

A system is designed to be implemented at the user end as seen in Figure 13.3. This is where the processing of the algorithm takes place.

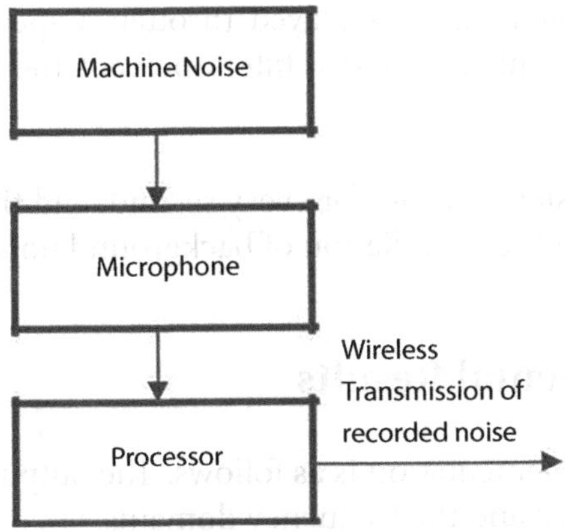

Figure 13.2 Design of the system at the machine end.

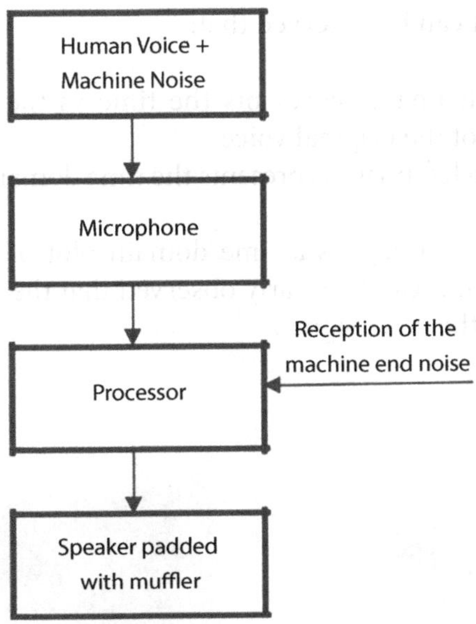

Figure 13.3 Design of the system at the user end.

The sequence of steps at the user end as seen in Figure 13.3 is as follows:

1) The processor at this end does all the processing with the algorithm as mentioned in the design of the processing system. It receives the reference signal from the other processor wirelessly.

2) The processed signal is played through a speaker padded with passive filters. Passive filters prevent the background noise [12].

The sequence of steps repeats for every second, and thus the extraction of human voice with the cancellation of background noise is achieved.

13.4 Experimental Results

The result of the implementation is as follows. The output obtained is analyzed in both the time and the frequency domains.

13.4.1 Time Domain Analysis of the Signals

From Figure 13.4, it can be observed that:

1) The top left image represents the time vs the normalized amplitude of the original voice.
2) The bottom left image represents the time domain plot of the noise signal.
3) The top right image is a time domain plot of the noise + voice signal. It can be clearly observed that the noise dominates over the voice signal.

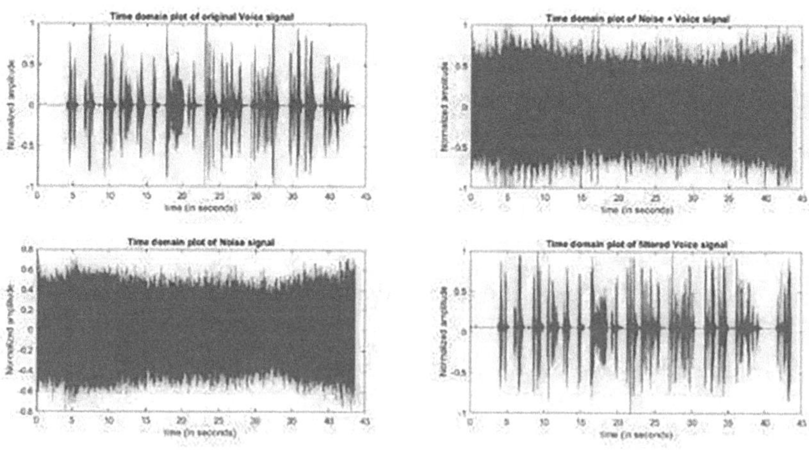

Figure 13.4 Time domain analysis of the signals.

4) The bottom right image represents the time domain plot of the filtered voice signal. From this image, it can be inferred that the presence of noise in the mixture of voice and noise was filtered out and the original voice signal was retrieved almost perfectly. This can be verified with the top left image.

13.4.2 Frequency Domain Analysis of the Signals

The obtained results were also analyzed in the frequency domain.

From Figure 13.5, it can be observed that:

1) The top left image represents the frequency domain plot of the original voice signal. It can be observed that the voice signal was not spread throughout the frequency range. The normalized power was concentrated at the central range of frequencies. This is because the human voice has a frequency range of 100 Hz to 2 kHz, which is a very low frequency compared to the machine noise.
2) The frequency domain plot of noise is seen in the bottom left image. From this image, it can be inferred that the noise is spread throughout the spectrum. Both low- and high-frequency noises are present in the machine noise.
3) The top right image represents the frequency domain plot of the voice + noise signal. It can be observed that the

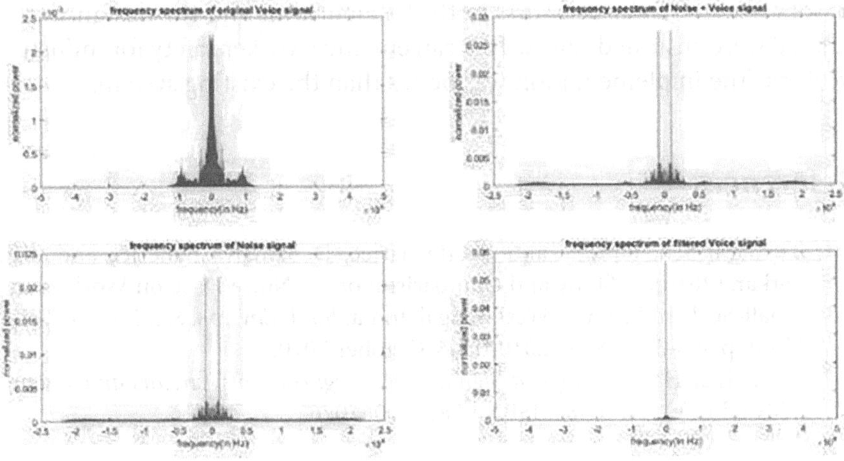

Figure 13.5 Frequency domain analysis of the signals.

normalized power is throughout the frequency spectrum. From this image, it is difficult to distinguish between the noise and voice signals.

4) By using adaptive filtering techniques, the noise was filtered out from the mixture of voice and noise signals. However, there was a slight aliasing of the sound wave. This is because of a lower sampling frequency used to retain the speed at which it is replayed to the user.

13.4.3 Performance of the Algorithm on Hardware

The algorithm is tested on both a fast processor, like that of a computer, and a slow processor, like that of a Raspberry Pi. When the algorithm was run on a computer, the time taken per 1 s of both the signals was 1.33 s. The amount of processor speed used was 3.2 GHz. All four cores of the processor were used.

When the algorithm was run on a Raspberry Pi, the time taken per 1 s of both the signals was 9 s. The amount of processor speed used was 1.2 GHz. All four cores of the processor were used.

13.5 Conclusion

The algorithm was successfully implemented using the NLMS technique, multiprocessing, etc. The time delay was obtained on a fast processor like that of a computer. The delay can be further minimized by using faster processors like 4-GHz processors. Both the algorithm and the design implemented were custom designs. This can enhance worker safety for industrial workers. The implementation was better than the existing system.

References

1. Tirtharaj Sen, Pijush Kanti Bhattacharjee, Debamalya Banerjee and Bijan Sarkar, (2010), A Study and Comparison of the Noise Dose on Workers in a Small-Scale Industry in West Bengal, India, *Int. J. Environ. Sci. Dev.*, in Vol. 1, No. 4, p.366-367, ISSN: 2010-0264, October 2010.
2. A. Nadder Hamdy, *Applied Signal Processing: concepts, circuits and systems*, 2009 edition, p.480-484, ISBN 9781420067026.

3. Sayed. A. Hadei, M. Lotfizad, (2010), A Family of Adaptive Filter Algorithms in Noise Cancellation for Speech Enhancement, *Int. J. Comput. Electr. Eng.*, in Vol. 2, No. 2, p.307, April 2010.
4. E. Atmaca1, I. Peker1, A. Altin, (2005), Industrial Noise and Its Effects on Humans, *Pol. J. Environ. Stud.*, in Vol 14, No.6, p.721-726, 2005.
5. Jyoti Dhiman, Shadab Ahmad, Kuldeep Gulia, (2013), Comparison between Adaptive Filter Algorithms (LMS, NLMS and RLS), *Int. J. Sci. Eng. Technol. Res. (IJSETR)*, in Volume 2, Issue 5, ISSN: 2278 7798, p.11002-11003, May 2013.
6. Jernimo Arenas-Garca, Luis A. Azpicueta-Ruiz, Magno T.M. Silva, Vtor H. Nascimento, Ali H. Sayed (2016), Combination of Adaptive filter, convergence and properties, *IEEE Signal Process Mag.*, p. 120-140, January 2016.
7. Stefan Liebich, Carlotta Anemuller, Peter Vary, Peter Jax, Daniel Ruschen, Steffen Leonhardt (2016), Active Noise Cancellation in Headphones by Digital Robust Feedback Control , *IEEE 2016 24th European Signal Processing Conference (EUSIPCO)*, January 2016.
8. Ishika Sharma1, Rajesh Mehra, Monika Singh (2015), Adaptive Filter Design for ECG Noise Reduction using LMS Algorithm, In *Reliability Infocom Technologies and Optimization (ICRITO)*, IEEE, 2015.
9. S. C, C. M L and M. A. Anusuya, "Noise Cancellation and Noise Reduction Techniques: A Review," *2019 1st International Conference on Advances in Information Technology (ICAIT)*, 2019, pp. 159-166, doi: 10.1109/ICAIT47043.2019.8987262.
10. J. R. Mohammed, "A New Simple Adaptive Noise Cancellation Scheme Based On ALE and NLMS Filter," *Fifth Annual Conference on Communication Networks and Services Research (CNSR '07)*, 2007, pp. 245-254, doi: 10.1109/CNSR.2007.4.
11. W. S. Gan and S. M. Kuo, "An integrated audio and active noise control headset," in *IEEE Trans. Consum. Electron.*, vol. 48, no. 2, pp. 242-247, May 2002, doi: 10.1109/TCE.2002.1010128.
12. M. Abdollahpouri, A. Khaki-Sedigh and H. Khaloozadeh, "A New Method for Active Noise Cancellation in the Presence of Three Unknown Moving Sources," *2008 Second Asia International Conference on Modelling & Simulation (AMS)*, 2008, pp. 1006-1011, doi: 10.1109/AMS.2008.61.

14

Machine Learning-Based Water Monitoring System Using IoT

T. Kesavan[1]*, E. Kaliappan[1], K. Nagendran[2] and M. Murugesan[3]

[1]Department of Electrical and Electronics Engineering, Easwari Engineering College, Chennai, India
[2]Department of Artificial Intelligence and Data Science, Saveetha Engineering College, Chennai, India
[3]Department of Electrical and Electronics Engineering, Karpagam College of Engineering, Coimbatore, India

Abstract
One of the most important components on Earth is water, and it is needed for living things to survive. There is no life on Earth without water. Water is used for a number of purposes besides drinking. Water that is fit for consumption is referred to as potable water. Polluted water is one of the most critical challenges for green globalization. In order to ensure the safety of water, it must be monitored in real-time. The proposed method is used to determine various water parameters by monitoring water quality using data collected by sensors submerged in a water tank. A machine learning-based water monitoring system is proposed and, using multiple sensors, this system can collect a variety of data from the water, including temperature, pH, and water clarity.

***Keywords*:** Internet of Things (IoT), machine learning, water quality management, water resource management

14.1 Introduction

Water has become a valuable resource that is used in almost all human activities. The availability of fresh water is diminishing by the day, and

**Corresponding author*: t.kesavan87@gmail.com

Chandra Singh, Rathishchandra R. Gatti, K.V.S.S.S.S. Sairam, Manjunatha Badiger, Naveen Kumar S. and Varun Saxena (eds.) Modeling and Optimization of Signals Using Machine Learning Techniques, (283–294) © 2024 Scrivener Publishing LLC

the efficient utilization of fresh water is urgently required. Water and its dissolved solids, as well as contaminants, must be monitored continuously [1]. We can monitor the quality of water and efficiently use it for various purposes before it is squandered in saltwater or evaporated into the sky thanks to the availability of modern sensors and valves connected via the Internet of Things (IoT) [2]. This technology allows for the remote monitoring and management of water resources via IoT at any moment of the day [3]. The system is made up of three wireless sensor sub-systems [4].

Wirelessly communication is used to deliver the data to a gateway which is attached with computer hosting the Graphical User Interface (GUI) [5]. Data delivery is often not assured due to wireless communication technologies. There is a chance that data will be lost [6]. The water level sensor will show how much water is in the tank, and the water motor will automatically switch up to the maximum based on the water level [7]. The data are shown on an Android app [8].

Because there is no quality control, the water will be delivered regardless of whether there is water in the tank. It is necessary to download and update the application on a regular basis [9]. The study [10] looks at how a water flow sensor and a water control valve may aid with uniform water distribution and adequate water supply. The amount of water in the tank is unclear since the model lacks a water level sensor. The average populace will be unconcerned about the scarcity of water [11]. Our approach overcomes the problems of previous water management models by storing sensor data in the cloud, which addresses the inadequacies of prior models [12]. This protects data and prevents it from being lost. A water level sensor is also included in our model, which offers real-time data on the water level in the water tank [13]. This technology enables remote water resource monitoring and management via the IoT at any time of the day.

14.2 Smart Water Monitoring System

In the proposed system, depicted in Figure 14.1, we are utilizing various water quality sensors and agriculture sensors to collect data. The collected data are processed by the Raspberry Pi and transferred to the cloud, Azure SQL Database, through wireless local-area network (WLAN) connectivity. By partitioning the data into training and testing groups, the machine learning algorithm gets the data from Microsoft Azure and then

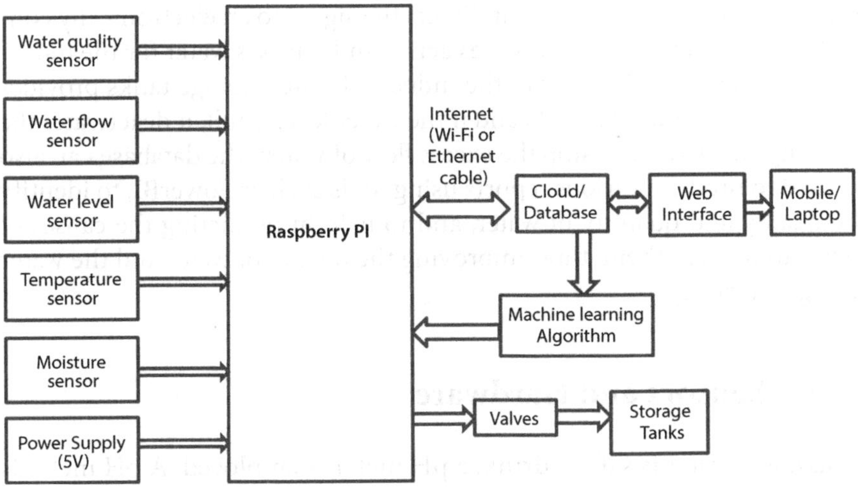

Figure 14.1 Block diagram of a smart water monitoring system.

utilizes it to create a decision tree. Various machine learning models are examined to determine their quality, and the most accurate prediction is used to inform all subsequent models. The machine learning algorithm analyzes the data pattern and suggests suitable actions to be taken by directing the control to various control valves. If the machine learning algorithm analyzes the water quality as good, it provides the signal to an electronically controlled drinking water valve, which becomes active and releases water for drinking water purposes. The water level sensor in the drinking water storage tanks provides the input to the Raspberry Pi and if the water level is full, it deactivates the drinking water valve to stop the excess flow of water.

If the machine learning algorithm analyzes the water quality as moderate, it directs the signal to an electronically controlled water valve, which becomes active and releases water for the agriculture fields. If the moisture sensor provides data that the agriculture fields are wet and do not require water, then the input data from the moisture sensor flow to the Raspberry Pi, which deactivates the agriculture water control valve and stops the water flow. If the machine learning algorithm analyzes the water quality as low, it directs the signal to an electronically controlled water valve, which becomes active and releases water for the industries. The water level sensor in the industrial water storage tanks provides the input to the Raspberry Pi, and if the water level is full, it deactivates the drinking water valve to stop the excess flow of water. If the machine learning algorithm analyzes the water quality as bad and toxic, it directs the signal to an electronically controlled water valve, which becomes active and releases water for the industries. The water level sensor in the industrial water storage tanks provides the input to the Raspberry Pi, and if the water level is full, it deactivates the drinking water valve to stop the excess flow of water. The database can also be used to produce various reports using tools such as PowerBI, to identify the quality and quantity of water, and to help in predicting the causes of toxins to prevent them, thus, improving the quality of water and the water storage efficiency.

14.3 Sensors and Hardware

To assess if water is safe to drink, a pH meter is employed. A pH meter is a scientific instrument that measures the hydrogen-ion activity to detect the acidity or alkalinity of water-based solutions. It is critical for human health to maintain a pH level that is about equal to 7. It measures pH in the whole range from 1.0 to 14.00. It has two reading modes: single and continuous. The YF-S201 is a water level monitor. The temperature varies from -25°C to 80°C, which is ideal for our needs. The time it takes for the echo to return after the scanner emits an ultrasonic pulse is used to calculate the distance to the target. The sensor sends out a frequency between 20 and 200 kHz, with the apex at six degrees. A temperature probe is a piece of equipment that transforms temperature readings into electrical impulses. It monitors the amount of heat energy or even the coldness produced by a product or system, allowing us to "feel" or detect any physical changes in the temperature via an analog or digital output. They have the ability to detect solids, liquids, and gases across a wide temperature range.

Soil moisture sensors are devices that detect the volume of the water content in the soil. Soil water potential sensors, for example, monitor a different aspect of moisture in soils called water potential. In agricultural applications, soil moisture measurement is crucial because it helps farmers to better regulate their irrigation systems.

In the water control valve, a fundamental magnetic device was used to update the movement energy from the electrical energy. The solenoid-based movement valve has a built-in solenoid. A solenoid valve is a electromagnet-based operated valve.

In the Raspberry Pi-based microcontroller used in this hardware circuit, 26 inputs and outputs are available, as well as a cord to attach sensors. In Azure, there is an IoT hub, which allows for extremely secure and

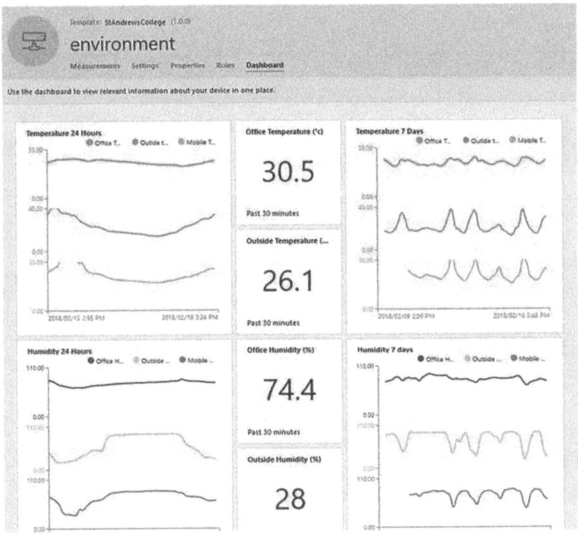

Figure 14.2 Azure IoT hub.

dependable communication between your IoT software and the devices it manages. The Azure IoT hub is a cloud-based solution for connecting practically any device. You can extend your solution from the cloud towards the Edge per the authentication, integrated configuration utility, and scalable configuration management. The Azure IoT hub is shown in Figure 14.2.

14.3.1 Machine Learning Algorithm

The market available software weka was created by New Queensland's University Ottawa, was released with a GNU license. Information coal mines include data reprogramming, clustering, designation, structural equation modeling, visualization, and feature selection, to name a few. The Microsoft Azure is a cloud-based collection of structured query language (SQL) data. Scalability, backup, and high availability are all taken good care of by managed database services. The Azure SQL Database supports a variety of Relational Database Management System) (RDBMS) including non-relational data structures, such as non structured query language (NOSQL). An SQL database management system (DBMS) was implemented in this project to store and handle relational data in tables in a normalized style. A stream dataset is saved in the Azure SQL Database as the source data, as depicted in Figure 14.3.

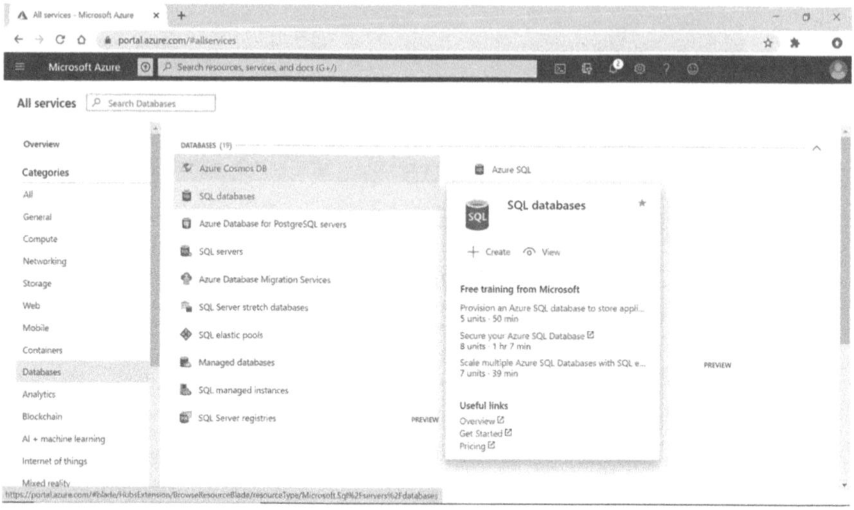

Figure 14.3 Azure SQL database.

14.4 PowerBI Reports

PowerBI (business intelligence) offers a variety of reports, including:

- The water quality variation during each time period;
- How much percentage of water quality is high, low, medium, or toxic;
- The amount of water being used for drinking, agriculture, industries, and sewage; and
- The place where the water turns toxic or the water quality decreases.

The following steps are involved in a smart water monitoring system. The PowerBI report of water resource monitored and distributed level at various stages are presented in Figures 14.4 and 14.5, respectively.

14.4.1 Reading of Data from the Sensors

Once the data are available in storage such as dams and lakes, various sensors such as the water quality sensor, pH meter, water flow sensor, temperature sensor, water level sensor, and tabular data stream (TDS) sensors will be placed in dams and lakes. The water quality sensor will provide water

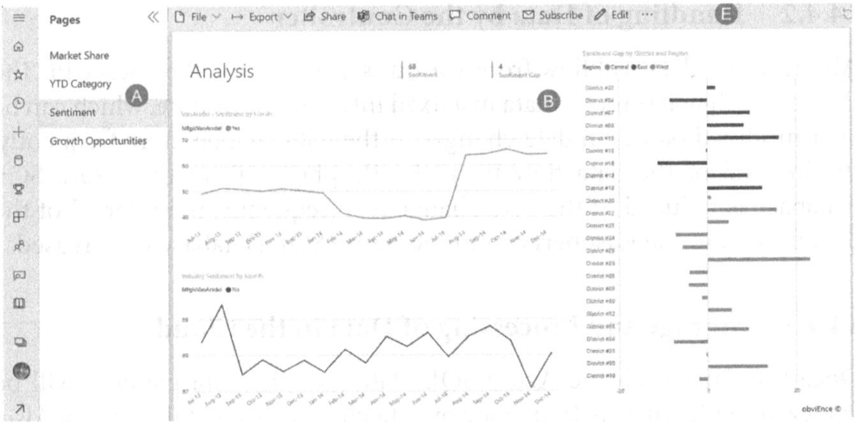

Figure 14.4 PowerBI report generation.

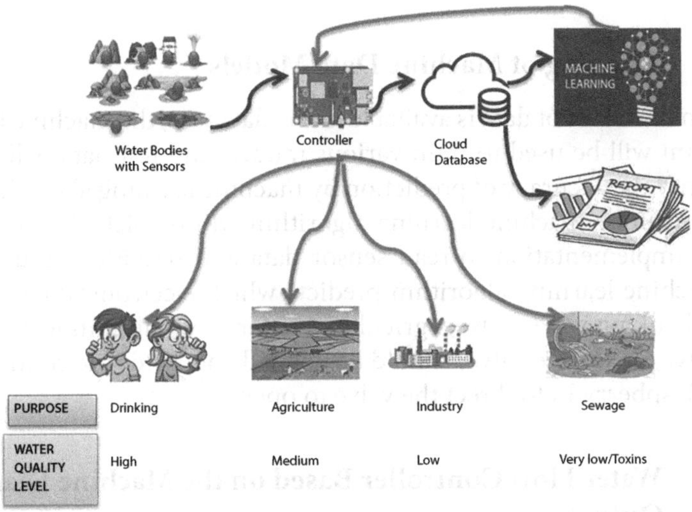

Figure 14.5 Water from the resources is monitored, distributed, and used for various purposes.

quality readings; the pH meter will provide pH readings; the water flow sensor will provide water velocity readings; the water temperature sensor will provide the water temperature; and the TDS sensor will provide the amount of the total dissolved solids in the water.

14.4.2 Handling of Data by the Controller

All the above data will flow from various sensors to the Raspberry Pi. The Raspberry Pi will send the data in a fixed interval of 1 minute, which can be later modified based on data changes. If the data change more frequently, the interval of the data flow from the Raspberry Pi to the Azure SQL Database is reduced. If the data change less frequently, the interval of the data flow from the Raspberry Pi to the Azure SQL Database is increased.

14.4.3 Storage and Processing of Data in the Cloud

Once the data reach the Azure SQL Database, the data patterns will be analyzed to identify the frequency of data change using SQL queries. Also, reporting tools such as PowerBI are used to form graphs and patterns to analyze the water impurity level over a period of time and to determine the factors that increase or decrease water flow levels and water quality levels.

14.4.4 Training of Machine Data Models

Once an initial set of data is available in the database, the machine learning algorithm will be used to train various models and test data will be used to identify the accuracy of prediction by machine learning algorithms. The highly accurate machine learning algorithm will be picked and used for further implementation to read sensor data and provide output signals. The machine learning algorithm predicts which electronically controlled valve (drinking water valve, agriculture water flow valve, industrial water flow valve, or sewage valve) should be opened and sends the control signal to the Raspberry Pi to direct the valve to open.

14.4.5 Water Flow Controller Based on the Machine Learning Output

If the machine learning algorithm analyzes the water quality as high, it provides the signal to an electronically controlled drinking water valve, which becomes active and releases water for drinking water purposes. If the machine learning algorithm analyzes the water quality as medium, it provides the signal to an electronically controlled agriculture valve, which becomes active and releases water for agriculture purposes. If the machine learning algorithm analyzes the water quality as low, it provides the signal to an electronically controlled industrial valve, which becomes active and releases water for industrial purposes. If the machine learning algorithm analyzes the water quality as very low or with a high level of toxicity, it provides the signal to an electronically controlled sewage valve, which becomes active and releases water into the sewage. This reduces the polluted water that gets mixed into drinking water, spoils agriculture fields, and spoils industrial machines. This major action is required to avoid harmful effects on humans, soils, and industries. The machine learning algorithm identifies the time or pattern when the quality of water is high, medium, low, or toxic, and which parameter infuses the quality of water. The machine learning algorithm identifies the time or pattern when the quality of water is high, medium, low, or toxic, and which parameter infuses the quality of water. The machine learning algorithm identifies the time or pattern when the quality of water is high, medium, low, or toxic, and which parameter infuses the quality of water. The proposed machine learning algorithm is used to improve the quality of water. In addition, water level sensors will be placed in drinking water storage tanks, industrial water storage tanks, and sewage water storage tanks. These sensors will give signals back to the Raspberry Pi to stop or close the electronic valves when the tanks are full. This is to avoid water overflow in storage, thus reducing water wastage. The moisture sensors in agriculture fields help to identify the level of water in agriculture fields. During the wet condition, the control is send from the moisture sensors to the Raspberry Pi to close the electronically controlled valve to stop the flow of water into agricultural fields. This helps in avoiding the flooding of agricultural fields during rainy seasons.

14.4.6 Analysis of the Water Data Reports

The Water Data Report helps to identify the following patterns and help the government or water control management with the following observations: the water quality variation during each time period; how much percentage of water quality is high, low, medium, or toxic; the amount of water being used for drinking, agriculture, industries, and sewage; and the place where water turns toxic or the water quality decreases.

14.5 Conclusion

The above model can be used to improve both the quality of water resources and the efficiency of using water resources to avoid water wastage. The advancements in IoT have enabled the recording of the various parameters of water and their integration into a single database for analysis. The cloud storage used for data can be increased or decreased over a period of time so that we can efficiently use the system and avoid huge capital costs and maintenance. The PowerBI provides various reports that help the management and the authorities to monitor and effectively improve the water quality and reduce water wastage. The machine learning algorithm predicts the quality of water and water requirements for various purposes and directs the water accordingly as per the needs. With the decrease in freshwater around the world, the optimum use of fresh water has become more valuable than ever before. Water quality improvement is needed for healthy humans and to maintain soil and land resources. Thus, this paper focuses on improving the quality and quantity of freshwater resources.

References

1. Yamuna, A.; Kesavan, T.; Sivashankari, V; Harmonic compensation in residential distribution system with MPPT. *Int. J. Appl. Eng. Res.*, 2015, 10(20), 15737–15741.
2. Lakshmi, K..; Kesavan, T.; Kavin, R.; Senthilkumar, M.; Gomathy, V.; Quick search optimization algorithm-based implementation of virtual power plant for distribution network. *Advances in Intelligent Systems and Computing*, 2021, 1163, pp. 261–272.
3. Subhashini, N.; Gnanamalar, S.; Geethamani, R.; Gomathy, V.; Kesavan, T.; Effectiveness of pitch control scheme in load balance of WECS. *J. Adv. Res. Dyn. Control Syst.*, 2018, 10(11 Special Issue), pp. 542–545.
4. Kesavan, T.; Sheebarani Gnanamalar, S.; Sivaranjani, S.; Radhakrishnan, G.; Sitharthan, R; SVM Based Reduction of Input Current Harmonics in Three Phase Rectifier. *IOP Conf. Ser.: Mater. Sci. Eng.*, 2020, 937(1), 012058
5. Kesavan, T.; Sheebarani, S.; Gomathy, V.; Kavin, R.; Sivaranjani, S.; Renewable Energy Based on Energy Conservation and Crossover System. *2020 6th International Conference on Advanced Computing and Communication Systems*, 2020, pp. 155–157, 9074157.
6. Kavin, R.; Kesavan, T.; Sheebarani Gnanamalar, S.; Rameshkumar, K.; Optimal Charging and Discharging Planning for Electric vehicles in Energy saving system. *2019 5th International Conference on Advanced Computing and Communication Systems*, 2019, pp. 976–978, 8728477.
7. Ruman, M. R.; Barua, A.; Rahman, W.; Jahan, K. R.; Jamil Roni, M.; Rahman, M. F.; IoT Based Emergency Health Monitoring System. *International Conference on Industry 4.0 Technology*, 2020.
8. Yeri, V.; D C, S.; IoT based Real Time Health Monitoring. *Second International Conference on Inventive Research in Computing Applications*, 2020. doi:10.1109/icirca48905.2020.9183194

9. Yuan-Hsiang Lin.; I-Chien Jan.; Ko, P. C.I.; Yen-Yu Chen.; Jau-Min Wong.; Gwo-Jen Jan. ; A wireless PDA-based physiological monitoring system for patient transport. *IEEE Trans. Inf. Technol. Biomed.*, 2014, 8(4), 439–447.
10. Ding, S.; Wang, X.; Medical Remote Monitoring of Multiple Physiological Parameters Based on Wireless Embedded Internet. *IEEE Access,* 2020, 8, 78279–78292.
11. Tirkey, A.; Jesudoss, A.; A Non-Invasive Health Monitoring System for Diabetic Patients. *International Conference on Communication and Signal Processing,* 2020.
12. Liau, J. C.; Ho, C. Y.; Intelligence IoT(Internal of Things) Telemedicine Health Care Space System for the Elderly Living Alone. *IEEE Eurasia Conference on Biomedical Engineering, Healthcare and Sustainability,* 2019.
13. Amin, P.; Anikireddypally, N. R.; Khurana, S.; Vadakkemadathil, S.; Wu, W; Personalized Health Monitoring using Predictive Analytics. *IEEE Fifth International Conference on Big Data Computing Service and Applications,* 2019, doi:10.1109/bigdataservice.2019.00048.

15

Design and Modelling of an Automated Driving Inspector Powered by Arduino and Raspberry Pi

Raghunandan K. R.[1]*, Dilip Kumar K.[2], Krishnaraj Rao N.S.[3], Krishnaprasad Rao[2] and Bhavya K.[4]

[1]*Department of Computer Science and Engineering, NMAM Institute of Technology, Affiliated to Nitte University, Nitte, India*
[2]*Department of Mechanical Engineering, NMAM Institute of Technology, Affiliated to Nitte University, Nitte, India*
[3]*Department of Information Science and Engineering, NMAM Institute of Technology, Affiliated to Nitte University, Nitte, India*
[4]*Department of Mathematics, NMAM Institute of Technology Nitte (Deemed to be University), Nitte, India*

Abstract

The 2019 novel coronavirus (COVID-19) has spread quickly among people and there are 26,27,630 cases worldwide, according to the statistics of the European Centre for Disease Prevention and Control. In this proposed work, an effort was made to secure a way of inspecting a test drive during the COVID-19 pandemic using new technologies for societal benefits. A secured and automated driving inspector powered by Arduino and Raspberry Pi is a new technology, which can be embedded in test cars to produce eligible drivers by assisting the inspector during a driver's test. The product will be placed in the car and will read the car's diagnostic information such as the acceleration, clutch, gear, and braking. During the driving session, tagged driving events will be recorded using data collected by one or more sensors. The description of each driving event, its weight or severity in relation to a scoring system, and the timing of the driving event will all be included in the tagged driving events. The tagged driving events will be used to construct the driving session report, which may contain a driver's feedback score that is weighted based on the severity of each of the marked driving events. Tagged

**Corresponding author*: raghunandan@nitte.edu.in

Chandra Singh, Rathishchandra R. Gatti, K.V.S.S.S.S. Sairam, Manjunatha Badiger, Naveen Kumar S. and Varun Saxena (eds.) Modeling and Optimization of Signals Using Machine Learning Techniques, (295–312) © 2024 Scrivener Publishing LLC

events are reported and will be analyzed by a respective authority. Based on the overview of all the parameters, the efficiency of the driver is identified by the program in the form of a report, and if the driver passes the minimum requirements, he/she will be eligible for a license. An analyzed report of the tagged driver will be converted to a performance grading. This result will help in the approval of the license for the tagged driver. After the completion of the test, the inspector is to update the database at the Regional Transport Office (RTO).

Keywords: Arduino, Raspberry Pi, driving, cryptography, on-board diagnostics (OBD), rotations per minute (RPM), RSA

15.1 Introduction

This idea deals with artificial assistance to the driver's test and a technical approach for identifying the driver's efficiency. It also provides brief information about the vehicle's diagnostic information and the various sensors used for the project. The methodology employed in designing and fabricating a computing device and selecting the sensors is discussed. A secured and automated driving inspector powered by Arduino and Raspberry Pi is a new technology that can be embedded in test cars to produce eligible drivers by assisting the inspector during a driver's test. The device will be placed in the car and will read the car's diagnostic information such as the acceleration, clutch, gear, and braking. Based on the overview of all the parameters, the efficiency of the driver is identified. Artificial assistance is used to monitor the actions of the driver throughout the duration of the test. The different parameters are monitored using different kinds of sensors and are analyzed carefully to produce the results. The driving inspector can monitor the test closely through the website designed, which displays the tagged events. The errors committed by the driver during the test are categorized into minor errors, medium errors, and major errors, and marks are deduced accordingly. A detailed report containing the results of the driver's test will be automatically generated. This device consists of an Arduino Uno, a Raspberry Pi, a voltage regulator IC and heat sink, a battery supply, ultrasonic sensors, IR proximity sensors, single- and multi-strand wires, a clad board, male and female Berg connectors, and a cooling fan.

15.2 Literature Survey

An Indian driving license is mandatory to drive a two-wheeler or a four-wheeler safely on the roads. But it is not necessary to obtain a new driving license automatically. For that, there is a different method. A person wishing to drive some form of motor vehicle in India must first get a license from his/her learner. A license for instruction is given to a learner. Within a month of the license being given to the learner, he/she will appear in front of a Regional Transport Office (RTO) authority for the test and, after careful inspection, may decide whether he/she passed the exam. Throughout India, the Regional Transport Authority (RTA) or the state's RTO issues a driving license. The Motor Vehicles Act of 1988 specifies that nobody is permitted to drive a motor vehicle in a public venue without a driving license [1].

If the applicant has obtained the license for the learners, he/she can present for the practical driving test in any period after one month. When you feel comfortable enough to give the test, he/she may pick a slot. On the day of the test, an inspector scrutinizes the documents. The first applicant would have to proceed to the field test, which is simply driving down the designated 'H' without contacting the edge poles. This is specifically to test his/her ability to control the engine, reverse, switch gears, and focus. If this is flunked, applicants will be given another date to retake the exam. When the field test is completed, they are eligible for the road test, which seems to be the final step. Here, the inspector must follow and test his/her driving ability and how knowledgeable they will be in the effective implications of traffic laws and regulations [2].

The Ministry of Road Transport and Highways claims that 77% of injuries occur due to the negligence of the driver [3]. Nineteen percent of motor vehicle victims elaborate on drivers with untrue licenses [4]. Drivers with unfounded licenses comprised 13% of every one of the drivers involved in significant crashes [5]. A survey shows that out of every 10 applicants, an average of 6 people get their license without being tested. The survey was done across 10 cities, including five metropolitan cities, which incidentally have the highest vehicle population [6].

Across the UK, the test is passed only by 40–43% of people applying for a driving license. As a result, inexperienced drivers are required to attend certified trainers' training and spend on average the Indian equivalent of some ₹80,000 [6]. We have to remember that just 12% of drivers in Agra genuinely got their licenses. Eighty-eight percent of respondents had no test at all. The same was the case for 72% of drivers in Jaipur, 64% in Guwahati, about 54% in Delhi, and 50% in Mumbai [7].

The question is more complicated by the fact that licenses do not have a single database. Therefore, there is no clarification or formal credential for driver education, approved pedagogy, instructional content, or any indication of a procedure capable of ensuring that the driver of the vehicle is certified to do so. Two of the hallmarks of Swedish road safety campaigners and community groups are careful planning, fences dividing traffic areas, minimum speed limits, and stringent enforcement. To be able to reproduce even any of this in India, the driver's test for his/her car would be a sensible starting point: his/her license. The Regional Transport Officer's approval to grant a driving license will, as required by regulation, ensure that the applicant follows traffic laws and knows how to drive. Nevertheless, with more than one crore licensing given annually and increasing by 7–10% annually, the Center has not invested enough in assets to test any person requesting a license, thus allowing untrained citizens to do so and unsafe drivers to handle vehicles.

A Road Traffic Accidents Public Safety Review states more than 1.3 Lakh people have died in motor vehicle accidents. This has made India top on the worldwide road accident list. Two years ago, India was ranked second behind China by the International Road Federation. Yet China has reduced rates of injuries from around 90,000 to 50,000, and the situation in India has worsened. India continues to account for 10% of road deaths despite just having 1% of the world's cars, up from 8% at the last count. In India, our policymakers are exacerbating the problem by poor enforcement of traffic laws and myopic policies [8].

Gujarat has launched an automatic monitoring program focused on sensors for drivers who apply for licenses. The pass rate has fallen dramatically from 100 to 40, following the launch of this automated test. Equally interesting was the domino effect in this process. Those who failed to complete the exam had no option but to improve their expertise. To refine their driving and retake the test, many participated in good standard driving schools. In addition, the driving schools were required to follow improved training methods. Although there are no definitive statistics currently connecting this with traffic fatalities, the initiative has prompted license applicants to take driving training professionally and made driving schools more efficient and credible in their interactions [9].

It has been proved that people who appeared for the permanent driving license test have experienced that the driving test on an advanced track is the undoing of 52 out of the 152 aspirants. More than 125 candidates may very well have accomplished it in the case of the old testing system. The RTO conducted the driving test on the specialized track set up at the Institute of Driving Training and Research (IDTR) near Nashik Phata for the permanent Department of Car Licenses. The first day of the test was attended by 152 participants, of whom 100 successfully completed all three tests while driving the car in an 8-shape, an H-shape, and on a gradient.

Improvements introduced during the driving test also made the process more rigorous, and greater accuracy was obtained. In order to pass the three examinations, applicants will now have to train even harder, claims sources to the RTO and others at the motor driving schools. The RTO will check 202 applicants in one day [10].

Various sensors can be used to monitor the actions of the driver during the test. Ultrasonic sensors can be used to measure the angle of inclination of the acceleration, brake, and clutch pedals. An ultrasonic sensor comprises a transmitter and a receiver. Trigger pulses are sent by the Arduino to the ultrasonic sensor, which transmits ultrasonic waves. Whenever an obstacle is detected, the transmitted ultrasonic wave bounces back and is received by the receiver of the sensor [11]. The IR proximity sensors can also be used to detect obstacles. The light-emitting diodes (LED) emit IR rays, which bounce back from the obstacle in its path and onto the photo sensors [12].

The RSA algorithm can be used for the safe transmission of data along a channel. The RSA security relies upon the capacity to factorize n by an unapproved individual. One of the biggest drawbacks of the RSA algorithm is the integer factorization process to split (n, e) while obtaining the corresponding (p, q) and decryption key d. If the key is extremely small, Wiener's attack may be used to identify the private key d. Computational time increases when long key sizes are used [13]. Dual RSA, which is a modified variant of RSA, can be used to resolve the disadvantages faced by RSA. This approach produces two different sets of keys with similar public and private keys [14].

Observations from the survey: From the literature survey, it is observed that the highest rate of road accidents and vehicle fatalities are due to faulty drivers and drivers with invalid licenses. A considerable number of drivers obtain licenses without even attending a driver's test. Most drivers lack knowledge of the key aspects of road safety, such as flaws, safe spacing, and the three-second law, and thus essentially risking lives. Every software research program has produced good outcomes. The modified format makes the test more comprehensive by testing the applicants' driving skills with greater precision. For automating the purpose of monitoring the actions of the driver, various sensors like ultrasonic sensors and IR proximity sensors can be used. Having just a normal public key cryptography is not secure enough. A modified algorithm mentioned in the paper gives a higher level of security.

15.2.1 Objectives

This study aimed to design and fabricate a product that will reduce the work of an inspector by monitoring the candidate using different sensors and artificial intelligence. A few of the important parameters to be monitored were considered, and the required sensors to monitor these parameters were acquired and tested. These sensors, along with other components, were then assembled in the required manner to fabricate the final product.

15.2.2 Preliminaries Used

A variety of sensors must be used to monitor different parameters of the car like the acceleration, brake, clutch, gear, and indicators. Arduino UNO was selected as the microcontroller to make the sensors operational. The aim of the project was to eliminate the use of a computer in this process.

Arduino does not have enough computing ability. To overcome this, we used a Raspberry Pi 3B, which is a microprocessor with 1 GB of RAM and Bluetooth and Wi-Fi availability. The Raspberry Pi stores all the data from the Arduino in the database and computes the result. This result is then sent to the server via Wi-Fi. A wide range of distance measurement sensors are available in the market like optical sensors, sound sensors, and resistance sensors. After a detailed study of all the different possibilities, ultrasonic sensors were chosen for the measurement of distance. It uses sound waves to measure the distance. The main reason for choosing this sensor is that it is contactless, affordable, and easily available. The sensor has a small size and can be easily mounted. Four of these sensors were attached to the acceleration, brake, and clutch pedals and in a free spot near the tire for determining the turns and plotting the path of the test drive. The distance between the ultrasonic sensor and the obstacle can be calculated using the following formula:

$$Test\ distance = high\ level\ times \times \frac{Velocity\ of\ Sound}{2}$$

Where the velocity of the sound in air is about 343 m/s.

Use hall effect sensors to track the location of the gear level. It is a transducer, which varies its output voltage in response to a magnetic field, which leads to the switching action. Using this, the location of the gear lever can be tracked and hence helps in the measurement of the gear. Two important parameters, rotations per minute (RPM) and speed, are readily available from a tool called onboard diagnostics, also known as OBD. This tool fits into the OBD port available in all the cars sold after 1996 and gives different diagnostic information about the car when programmed via Bluetooth. A transistor in switching a circuit is used as an indicator sensor. It is cost-efficient and switching a circuit is simple in construction and is easy to use. The Arduino is coded for an indicator sensor in such a way that when the output is high, the indicator is not active, and when the output is low, the indicator is active. All the sensors chosen must be mounted at different locations of the car. Mounts and fixtures were designed and fabricated for this purpose. The motherboard was etched, and the Raspberry Pi was placed on the device, which is the housing for the circuit.

15.2.3 Method Proposed

The process begins with the inspector and the test driver taking their respective positions in the car. The driver will be behind the wheel, and the inspector will proceed by logging in using a public key, which is confidential to the department (RTO). This will be followed by a selection of the driver from the drivers list. The details of the test candidates for that day will be stored in the database of the virtual inspector. The specification and standard diagnostic values will already be pre-fed in the database from which the test car will be selected.

The inspector begins the test by clicking on the "START SESSION" button from the application, followed by which the driver begins his/her test. All the major actions mentioned above will be monitored using different sensors. By default, 100 points will be awarded to each test driver in the beginning, and for each mistake/error performed by the driver, a certain number of points will be deducted from the denomination. At the end of the session, the total points remaining with the test driver will be used to judge his/her efficiency. The test can either take place under normal operating conditions or on the test track. A few of the skills checked are listed below:

i. Ability of the driver to efficiently operate the accelerator, brake, and clutch (ABC) pedals;
ii. Ability of the driver to change gears at the right instance;
iii. Ability of the driver to apply the indicators at the right time;
iv. Ability of the driver to drive in a straight path;
v. Ability of the driver to maintain the required distance behind another vehicle;

vi. Ability of the driver to park;
vii. Ability of the driver to drive according to road conditions, etc.

From the above, the first five important operations were considered. It was studied that the application of the ABC pedals can easily be monitored by measuring the distance between the firewall and the pedal. The turning of the car can again be obtained by monitoring the distance between a wheel surface and a fixed point. The RPM and speed can be easily obtained as it is already provided by the vehicle through the OBD. An architecture diagram of the sensors installed in the car is demonstrated in Figure 15.1.

Driving events tagged during the driving session will be registered using the information collected from one or more sensors and/or entered by the inspector. Tagged driving events should provide a summary of each driving incident and the corresponding frequency or weight in relation to a rating mechanism, as well as the driving time. The tagged driving will include the car's diagnostic information such as acceleration, braking, and clutch. With the values received from the OBD, the RPM vs. TIME graph will be plotted. Also, hand signals applied by the driver are evaluated. Tagged events are reported and will be analyzed by the respective authority. The analyzed report of the tagged driver will be converted to a performance grading. This result will help in the approval of the license for the tagged driver. The process depicting the performance measurement of a driver is indicated using a flow diagram, which is depicted in Figure 15.2.

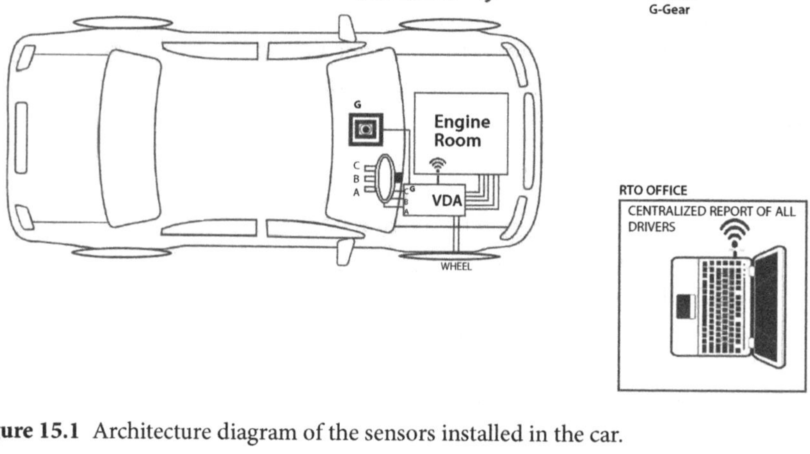

Figure 15.1 Architecture diagram of the sensors installed in the car.

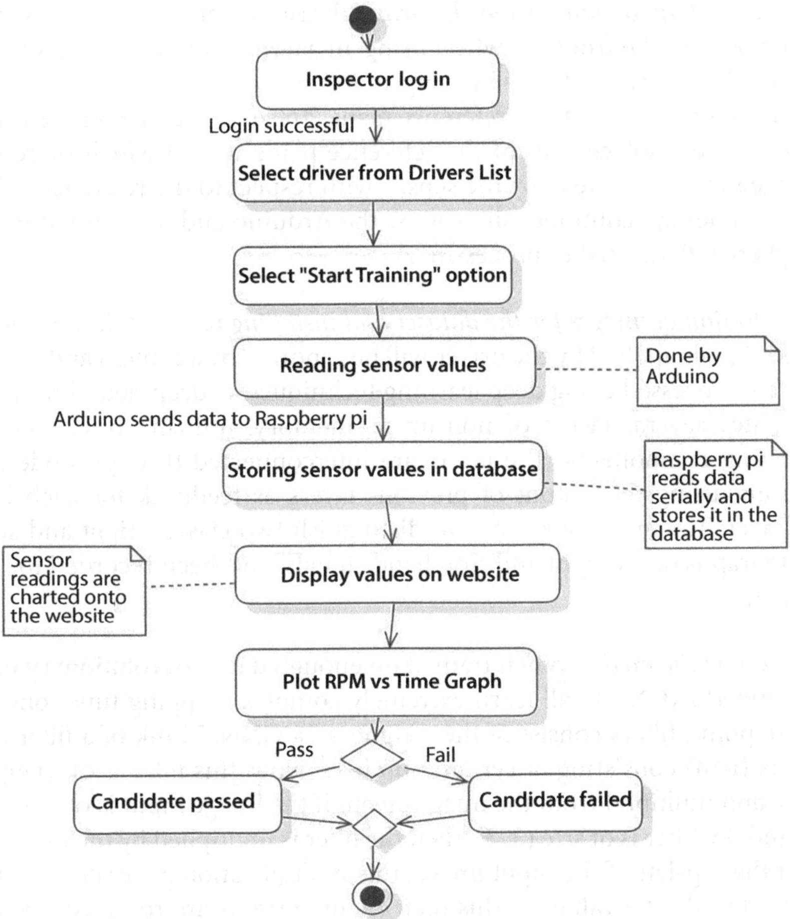

Figure 15.2 A block diagram depicting the process flow.

A. Sensors are set up to detect acceleration, brake, and clutch actions: Ultrasonic sensors are fixed to the ABC pedals. These sensors take the plane opposite to it as a reference plane. The displacement of the ABC pedal leads to the displacement of the sensors due to which there is a change in the value read by the sensor with respect to the reference value. These values are continuously read by the Arduino and then stored in the Raspberry Pi for further processing. Every pedal has unique mounts and fixtures for fixing ultrasonic sensors to it. The values received from these sensors will be displayed on the website as a meter.

B. Sensor set up to monitor vehicle turns: Ultrasonic sensors are fixed parallel opposite to the front left wheel using an acrylic mount and sheet metal fixtures. The sensor is fastened using an M8 bolt and nut. The sensor takes the wheel opposite to it as a reference plane. The displacement of the wheel leads to the displacement of the reference plane due to which there is a change in the value read by the sensor with respect to the reference value. These values are continuously read by the Arduino and then stored in the Raspberry Pi for further processing.

C. Collection of images for the dataset and installing required dependencies: Hand signals applied by the driver will be captured by a camera and appropriately processed using deep learning techniques. A deep neural network integrates several layers of non-linear memory, utilizing basic parallel operating components. The layers are interconnected through nodes, or neurons, using the output of previous layers as feedback for each hidden layer. The model is trained to distinguish two classes: right and stop. Photographs of the right and stop hand signals have been recorded for the dataset.

D. Training the model: When trained on enough data, coevolutionary neural networks (CNN) will learn extremely complex mapping functions. At a base point, filters consist of the weights of a CNN. Think of a filter as a matrix (n*n) consisting of certain numbers. Now, this filter is overlapped (slide and multiply) from the image given. If the image input is of size (10, 10) and the filter is of size (3, 3), then the filter is multiplied by the nine pixels at the top-left of the input image, this multiplication generates another matrix (3, 3). The values of this matrix's nine pixels are rounded up, and this value is a single-pixel representation on the CNN's top-left layer two. MobileNet is a pre-trained model that has been used for transfer learning as it is lightweight (17 MB) in its architecture. The steps that are followed in transfer learning are as follows:

i. Importing and building the required model;
ii. Loading the training data into the image data generator, which is inbuilt in Keras and helps us to train the model; and
iii. Training the model on the dataset by compiling and training. The model has been trained for 20 epochs.

Figure 15.3 Images collected to form the dataset.

Images were captured from different subjects in various environmental and light conditions. A total of 540 images for each of the two hand signals were collected, which is shown in Figure 15.3.

E. Automated result generation: After the completion of the test, a report containing the results will be generated using the decisions made by considering the values obtained by the various sensor data. The script contains various testing conditions against which data from the sensors are fed. Say the driver takes a turn without applying an appropriate indicator, which is considered a fault in driving. We take similar consequences into consideration, and the score will be deducted appropriately depending on the severity of the fault. The device fitted inside the car is depicted in Figure 15.4.

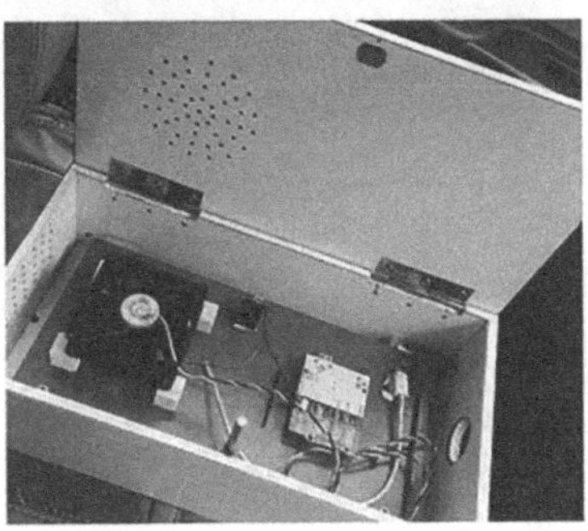

Figure 15.4 The device to be fitted inside the car.

F. Security: An analyzed report of the tagged driver will be converted to a performance grading. This result will help in the approval of the license for the tagged driver. After the completion of the test, the inspector is to update the database at the RTO office. This is a crucial stage because hackers may try to modify the results. To provide an excellent level of security while transferring results from the database to the RTO server, a public key cryptographic approach is used, which uses the concept of Pell's key generation scheme [14–18].

15.3 Results

The work provides a very good implementation of sensors to compare theoretical values with practicality to interpret the skill of the driver virtually on a display, making the work easier for the inspector without having any close contact. Every action of the driver will be closely monitored from the results obtained. After the completion of the driving test, the inspector updates the database at the RTO office.

The inaccuracy in the outputs from the hall sensors led to wrong results, and an alternative for the hall sensor was to be found. A fixed acceleration mount with an ultrasonic sensor is depicted in Figure 15.5. The IR proximity sensors were used for monitoring the position of the gear, which gave positive results. When the obstacle is present, the output of the sensor will be low, which can be detected by a microcontroller. Using this, the location of the gear lever (Figure 15.6) can be tracked and hence helps in the measurement of the gear.

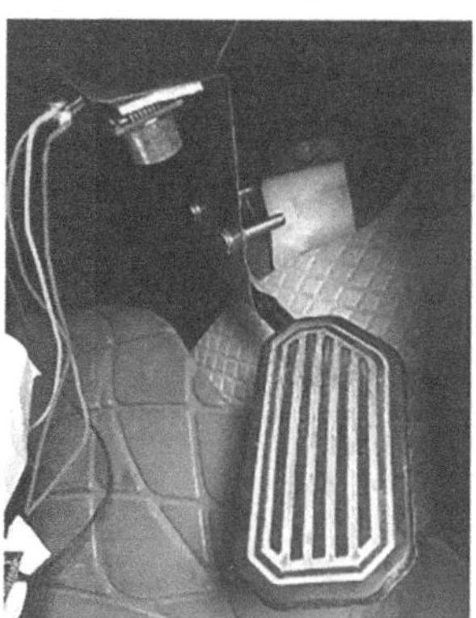

Figure 15.5 Fixed acceleration mount with ultrasonic sensor.

Figure 15.6 Gear mount placed around the gear lever.

Figure 15.7 showcases the results of the sensors attached to the ABC pedals, from which the training set is going to get the data in terms of graphs.

Figure 15.8 represents the results of the sensors attached to the gear and the values from switching a circuit for the indicators. To notify this, a flag is used in the model.

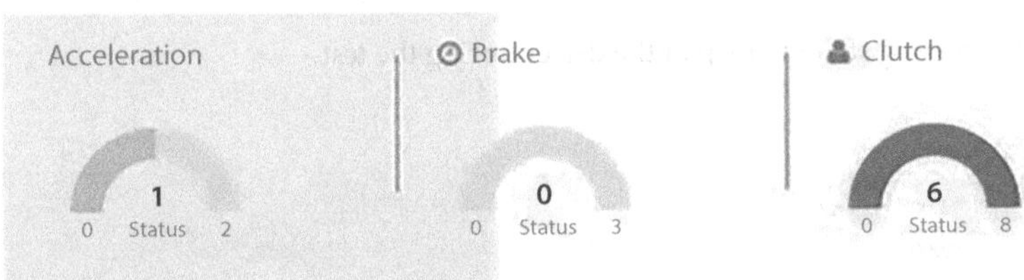

Figure 15.7 The result of the sensors attached to the accelerator, brake, and clutch.

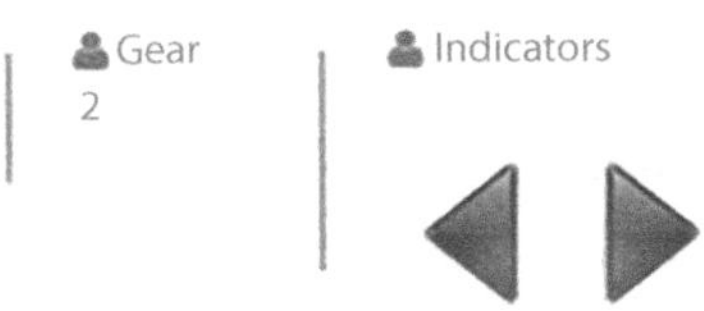

Figure 15.8 The result of the sensors attached to the gear and the values from switching a circuit for the indicators.

A driver's steering stability is one of the essential parameters that is considered during the driving test. Through this result, the driver will gain full knowledge of all the errors committed during the test. Figure 15.9 displays the driver's steering stability graph during the study.

The model's performance data are correlated with the actual driving test datasets depicted in Figure 15.10. The results obtained from the experiment are shown in the figures using blue lines, and the results of the datasets taken from existing models are shown in red lines. The graph reveals that the model provides an accuracy result of 88% relative to existing driving approaches and has the following advantages:

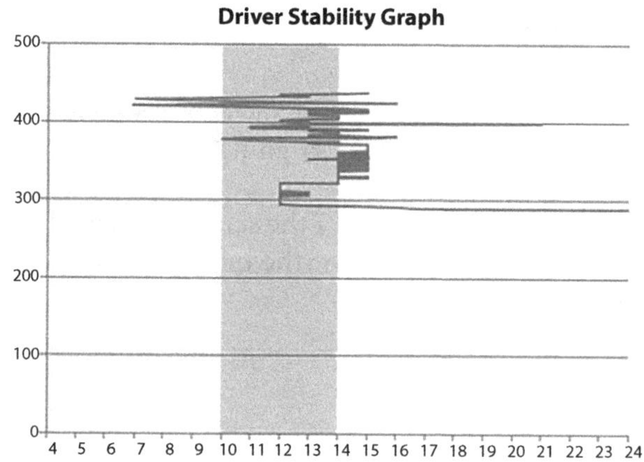

Figure 15.9 Steering stability map of the driver during the test.

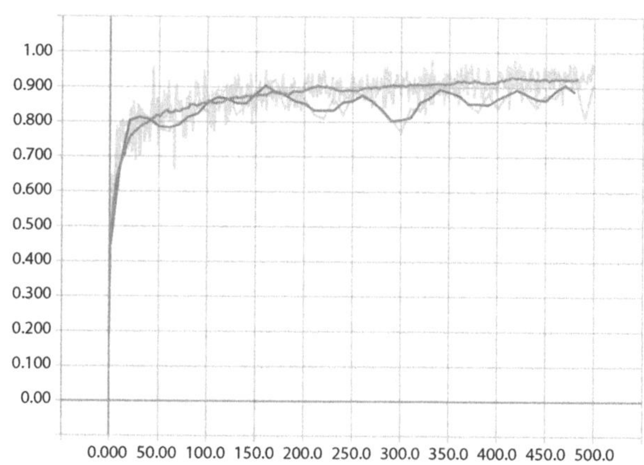

Figure 15.10 Comparisons of driving skills with the dataset using MobileNet.

i. The driver will have knowledge about all the mistakes committed during the test;
ii. Due to the complete automation of the inspection process, the effort of the inspector will be greatly reduced and secured during the COVID-19 pandemic; and
iii. An old-fashioned way of the driver's test can be replaced with a very technical approach.

The proposed work provides a higher level of security as compared to the traditional way of encryption to ensure that no middleman attack can happen between the client system and the server.

15.4 Conclusion

In the work, it has been demonstrated how a relatively new technology is embedded in test cars to produce eligible drivers and, therefore, making it easier for the inspector by assisting him/her by maintaining social distancing. This was achieved by successfully designing, fabricating, and programming a virtual driving inspector using sensors and a microprocessor along with a successful implementation of artificial assistance to save the time and efforts of the inspector. There are several benefits to this method because it differs from existing approaches in that it replaces the old-fashioned driver's test with a very technical approach, ensuring the quality examination of the driver and the reported documentation of the same. This can be very useful against corruption along with reducing the chance of unethical issues of the driving license, thus improving road safety by approving quality drivers only.

References

1. Thierry Delvaulx, Bernard Goffart, Kris Jooris, Bram Kerkhof, Pierre Pourveur, "Obtaining a Driver's License in India: An Experimental Approach to Studying Corruption", *Q. J. Econ.*, Volume 122, Issue 4, Pages 1639–1676, 01/ 11/2007, DOI: 10.1162/qjec.2007.122.4.1639.
2. "Driver's license test procedure in Karnataka", an article by Bankbazaar.com, 2018.
3. Mahesh Rajoria, "Making driving safer in India- Issuing licences should be made more stringent", The Hindu Business Line, 05/2017.

4. Research Note-DOT HS 812 046, "Driver License Compliance Status in Fatal Crashes". NHTSA'S National Centre for Statistics and Analysis 1200 New Jersey Avenue SE. Washington, DC 20590, 10/2014.
5. "34% fail new format of driving test", Times of India, 17/03/2015.
6. Dipak K Dash, "6 out of 10 get driving licence without test in India", Times of India, Jul 16, 2017.
7. S. Gopala Krishnan, Department of Community Medicine, SRM Medical College Hospital and Research Centre, Tamil Nadu, India, "A Public Health Perspective of Road Traffic Accidents", Article Literature Review in *J. Family Med. Prim. Care,* 1(2):144-150, March 2012, DOI: 10.4103/2249-4863.104987.
8. Press Trust of India, "Gujarat introduced a sensor-based automated testing system for drivers applying for licences.", Business Standard, March 23, 2015.
9. Massimo Banzi, Michael Shiloh, *Getting Started with Arduino: The Open Source Electronics Prototyping Platform*, Maker Media, 3rd Edition, 2009.
10. News, C. and News, P. (2019). 34% fail new format of driving test - Times of India. [online] The Times of India. Available at: https://timesofindia.indiatimes.com/city/pune/34-fail-new-format-of-driving-test/articleshow/46591242.cms [Accessed 6 Feb. 2017].
11. Aliyu, Ahmed, *et al.* "An ultrasonic sensor distance induced automatic braking automobile collision avoidance system." Electro-Technology for National Development (NIGERCON), *2017 IEEE 3rd International Conference on. IEEE*, 2017.
12. Kim, Jeong Seok, *et al.* "An IR proximity-based 3D motion gesture sensor for low-power portable applications." *IEEE Sens. J.*, 15.12 (2015): 7009-7016.
13. K. R. Raghunandan, R. Shetty and G. Aithal, Key generation and security analysis of text cryptography using cubic power of Pell's equation. In: *2017 International Conference on Intelligent Computing, Instrumentation and Control Technologies (ICICICT)*, Kannur, 2017, pp.1496-1500.
14. R. K.R., G. Aithal and S. Shetty, "Comparative Analysis of Encryption and Decryption Techniques Using Mersenne Prime Numbers and Phony Modulus to Avoid Factorization Attack of RSA," *2019 International Conference on Advanced Mechatronic Systems (ICAMechS)*, Kusatsu, Shiga, Japan, 2019, pp. 152-157
15. Rao, Raghunandan, Aithal, Ganesh and Shetty, Surendra. Secure RSA Variant System to Avoid Factorization Attack using Phony Modules and Phony Public Key Exponent. In: *Int. J. Innov. Technol. Explor. Eng. (IJITEE)*, 2019, vol 8.
16. Raghunandan K.R., Dsouza R.R., Rakshith N., Shetty S., Aithal G. Analysis of an Enhanced Dual RSA Algorithm Using Pell's Equation to Hide Public Key Exponent and a Fake Modulus to Avoid Factorization Attack, In: Chiplunkar N., Fukao T. (eds) *Advances in Artificial Intelligence and Data Engineering.* Advances in Intelligent Systems and Computing, Springer, Singapore. doi.org/10.1007/978-981-15-3514-7_60, vol 1133.

17. Raghunandan K. R., Ganesh Aithal, Surendra Shetty and Bhavya K. Image encryption scheme in public key cryptography based on cubic Pells quadratic case, In: *Indones. J. Electr. Eng. Comput. Sci.*, Vol. 20, No. 1, October 2020, pp. 385~394
18. Rao, Raghunandan & Ganesh, Aithal & Surendra, Shetty & Kallapu, Bhavya. (2020). Key Generation Using Generalized Pell's Equation in Public Key Cryptography Based on the Prime Fake Modulus Principle to Image Encryption and Its Security Analysis. *Cybern. Inf. Technol.*, 20. 86-101. 10.2478/cait-2020-0030.

16

Kalman Filter-Based Seizure Prediction Using Concatenated Serial-Parallel Block Technique

Purnima P. S.[1]* and Suresh M.[2]

[1]*SSAHE, Agalakote, Tumakuru, Karnataka, India*
[2]*Department of ECE, SSIT, Tumakuru, Karnataka, India*

Abstract

Present state-of-the-art seizure prediction approaches use machine learning techniques and neural networks, which include K-means clustering, discriminant analysis, radial basis function neural networks (RBFNN), and backpropagation neural networks for efficient results, but most of them are not real-time approaches. We propose to implement an alternate real-time approach based on the Kalman filtering of electroencephalogram (EEG) time-series signal with a novel 'serial-parallel block concatenation' method, which predicts multiple signal samples ahead of time, and hence allows real-time processing. We also propose a novel probability density function-based (PDF-based) method to estimate signal data points for the adaptive filter updates. This technique applies to 250 different EEG records, each lasting 23.6 seconds. The receiver operating characteristic (ROC) analysis is used for evaluating the performance of the proposed recognition system. The proposed work achieves approximately a 94% success rate. The proposed concatenation approach, along with the PDF method, gave more accurate real-time predictions than a standard adaptive filter.

Keywords: Kalman filter, seizure detection, EEG signal, series-parallel concatenation

**Corresponding author*: purnima.p.s.2015@gmail.com

Chandra Singh, Rathishchandra R. Gatti, K.V.S.S.S.S. Sairam, Manjunatha Badiger, Naveen Kumar S. and Varun Saxena (eds.) Modeling and Optimization of Signals Using Machine Learning Techniques, (313–326) © 2024 Scrivener Publishing LLC

16.1 Introduction

Brain abnormal electrical activity can be considered as seizures, which can be classified as a very high or very low entropy activity. Consequently, the subject can either go into a dazed state or a state of uncontrollable fits or convulsions. They need quick medical attention and, presently, there are no complete end-to-end treatment methods. In a certain part of the brain, adult seizures usually occur, may or may not spread throughout, and are called focal seizures. A condition of such frequent seizures is called epilepsy.

A seizure is divided into three parts: pre-ictal, ictal, and post-ictal stages. The onset of a seizure until it subsides is defined as the ictal state; the states preceding and post-ceding the ictal stage are defined as pre-ictal and post-ictal states, respectively. A subject takes some time to recover after a seizure. Thus, both the ictal and post-ictal states require medical attention and treatment. Currently, epilepsy affects about 50 million people worldwide; out of which 80% stay in developing nations. The United States alone currently spends $15.5 billion towards epilepsy treatment. Hence, there has been a lot of research focusing on treating seizures.

An electroencephalogram (EEG) is used for monitoring the brain by measuring the brain's electrical activity. An EEG can be used to monitor and detect seizures. If seizures are accurately predicted, the recovery time and the cost of treatment, can be reduced. A review by Ramgopal *et al.* [7] gives a highly exhaustive study of seizure detection methods and products available, which gives insight into the state-of-the-art detection methods, as well as emphasizes the importance of seizure prediction. An EEG is used most frequently on account of its low noise and millisecond time-frame detection.

16.2 Prior Work

There is a lot of research directed toward studying seizures, and there have been some successful approaches to detecting seizures. Some of the early work dates back to the 1980s, when Gotman published an automatic seizure detection method and an improvement over it in 1990 [1] using background wave characterization and thresholding, which had about a 75% accuracy as a whole, whereas 41% of the seizures were not notified to the caregivers. His research was extended to infants by Gotman and the team, employing a three-way approach of spectral analysis of overlapping epochs, spike detection, and low energy spectral evaluation in 1997 [2].

Ghosh-Dastidar *et al.* proposed the use of classifiers and neural networks to classify EEG data into seizures [3]. The classifiers used were K-means clustering, discriminant analysis, radial basis function neural networks (RBFNN), and backpropagation neural networks. The overall accuracies of these methods typically ranged from about 60% to 90%. The data were decomposed using wavelet transforms. The method proposed by Tzallas *et al.* [4] used spectrograms (STFT) of the EEG data to extract features and apply artificial neural networks (ANNs) to it. There are many more approaches using wavelet transforms [5], principal component analysis (PCA) [6], etc. However, these methods detect and classify seizures that undoubtedly help in understanding seizures, but do not provide the advantages that we would hope for by predicting a seizure in advance.

There has also been a lot of research interest in predicting seizures. There have been statistical, algorithmic, and concept-based proofs of the prediction methods proposed. Most methods consider a direct variation in the time-series characteristics. Lehnertz *et al.* and Aschenbrenner-Scheibe *et al.* [8–10] used the correlation dimension approach for prediction, whereas Martinerie *et al.* [11] used correlation density. With these approaches, they achieved 90% to 95% sensitivity and 3–12 minutes in advance prediction of seizures. The use of the similarity index was proposed [12–17]. The similarity index methods claim very high sensitivity (96% typically) and a large time in advance prediction of seizures (some up to 80 minutes). Deep learning to know [18] is another paradigm in this regard, which is able to process huge quantities of sign statistics generated with the aid of using wearable Internet of Things (IoT) sensor gadgets together with EEG headsets for epilepsy [19]. Algorithms primarily based on deep getting-to-know strategies triumph over the restrictions of conventional device getting-to-know algorithms with the aid of using a decreasing processing time and presenting the cap potential to technique huge statistics in multi-channel biomedical indicators [20]. Therefore, those strategies play a promising function in presenting real-time healthcare solutions. To offer a wise answer for epilepsy patients, a two-layer, long-time period, reminiscence model (LSTM) [21] was proposed based totally on spectral traits for the automatic prediction of epileptic seizures. This method makes use of a long-time period Children's Hospital Boston Massachusetts Institute of Technology (CHBMIT) [22] EEG database of 24 instances of epilepsy amassed at a children's sanatorium in Boston. First, 23 channels of uncooked EEG indicators recorded from the patient's scalp are preprocessed, filtered, and segmented into brief segments starting from 5 to 50 seconds. Divided parts are then converted into a frequency domain with the use of the fast fourier transform (FFT) to appropriately interpret the purposeful and behavioral traits of the complicated shape of the mind during an epileptic attack. It is split into bands [23, 24].

16.3 Proposed Method

The breakdown of the proposed solution is as follows:

- Phase 1: Use data points 1 to M to predict the $(M + 1)^{th}$ data point using a Kalman filter (KF).
- Phase 2: Implement a serial-parallel block concatenation of these filters to predict future points.
- Phase 3: Attempt to normalize the seizure-predicted signals to subside the seizure.

16.3.1 Phase 1

In Phase 1 (Figure 16.1), M data points are used to predict the $(M + 1)^{th}$ data point. This is a standard prediction procedure and the focus here was the –Proof of Concept||, which has been described in Section 16.3. From the predictions on the various data sets of M points to test the robustness of our filter design, we safely concluded that the adaptation we use is a highly accurate one.

16.3.2 Phase 2

In Phase 2, more data points can be predicted with the help of the serial-parallel block concatenation approach, as shown in Figures 16.2 and 16.3. This phase is also called filter block processing, as the whole processing of serial and parallel filters that we implemented happens after the $(i + M)^{th}$ sample is achieved and before the $(i + M + 1)^{th}$ sample is obtained,

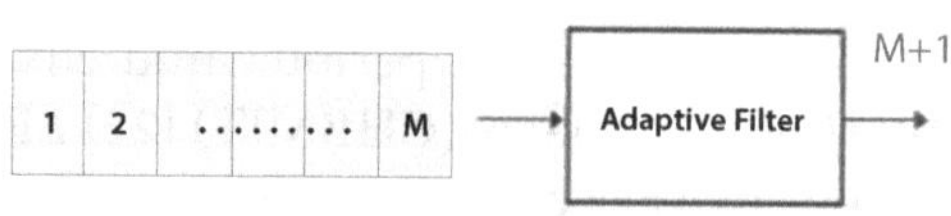

Figure 16.1 Standard adaptive filter.

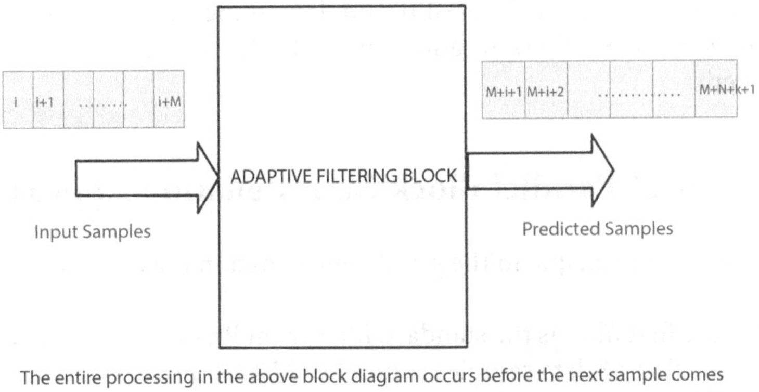

Figure 16.2 Adaptive filter block processing.

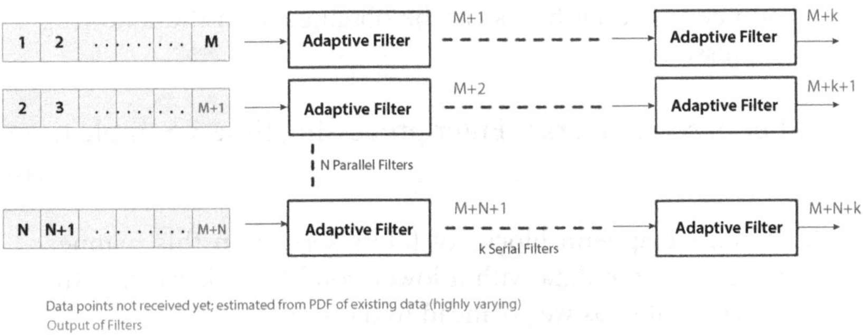

Figure 16.3 Serial-parallel block concatenation approach.

where i is the instance at which we start to predict. This method helps us achieve a prediction of N + k samples ahead in one single block processing, i.e., in one sample interval. This is defined in detail in the next section.

16.3.3 Phase 3

If the prediction in Phase 2 classifies a signal as a seizure, we intend to subside the seizure by using a nerve stimulation technique to bring the brain back to its normal state. One of these techniques is the vagus nerve stimulation procedure. This involves a device implantation that sends electrical impulses to the vagus nerve. In this method, a device is implanted under the skin connected to the vagus nerve, and electrical signals are sent to the brainstem along the vagus nerve, which sends signals to specific areas of the brain to calm the lesion. Since Phase 2 of the project is intensive on adaptive filtering, we focused on implementing particle and recursive least squares (RLS) filters in addition to the KF for further testing and assessment.

16.4 Serial-Parallel Block Concatenation Approach

The underlying principle in the serial concatenation is as follows:

1. The first filter is the standard filter from Phase 1, which gives the first M data samples using the $(M+1)^{th}$ prediction.
2. Using this $(M+1)^{th}$ prediction that we got in the previous step, we now predict the M + 2 data sample.
3. We append serial filters in this manner, and a bound on the number of serial filters can be obtained with the following expression:

$$\text{Number of serial filters} \times \text{Filter processing time} < \text{Sample time} \tag{16.1}$$

4. When we append blocks of filters serially in this manner, we predict the data with a lower confidence level than the previous filter as we go ahead in the series.

While serial concatenation includes a delay due to filter processing, parallel concatenation gives us the prediction without any delay.

The underlying principle in parallel concatenation is as follows:

1. The first filter is the standard filter from Phase 1, which gives us the $(M+1)^{th}$ prediction using the first M data samples.
2. The second filter in parallel uses two to M data samples and the $(M+1)^{th}$ data sample from the PDF method and so on.
3. We append parallel rows in this manner; however, every parallel row predicts data with a lower confidence level than the previous one.

This way, we achieve a prediction of the next N + k data samples even before the next actual data sample is observed. We tested the block processing rigorously with different adaptive filters such as the Kalman, RLS, and particle filters; with the KF being our primary focus. Figure 16.3 shows the 'serial-parallel block concatenation' approach.

16.4.1 PDF Method

In order to implement the filter block processing explained in Phase 2, to predict the $(i+M+1)^{th}$ sample before the occurrence of the $(i+M)^{th}$ sample, we need to generate the signal's PDF distribution, which helps us to predict the M + N + k samples in a single block of processing, using serial-parallel concatenation of KF blocks. The PDF basically establishes a near-accurate relationship with the test data, to obtain x(M + 1), x(M + 2), x(M+ 3), x(M + n) from the data x(1), x(2)...x(M), much ahead of the occurrence of it. It achieves this by using prior knowledge of data collected (in this case, the data set) and generating a conditional PDF distribution of each of the points x(M + 1) to x(M + n) given x(1) to x(M). A number of criteria could be followed to bind the generation of the PDF function.

16.5 Algorithm

Step 1: The input test data, x(1), x(2), x(3)..x(M), where M is equal to the number of filter taps, is fed as the input.
Step 2: The error minimum is initialized as ∞.
Step 3: From all the training data sets available, a PDF distribution is created.
Step 4: Since the neurotic signals are bivariate, a separate $n \times L$ matrix is formed for neuron X and neuron Y, where n is the number of data present in each file and L is the total number of files.
Step 5: When the PDF directory is generated, consisting of all the data sets organized in matrix form, the test data are matched with each of the files.
Step 8: If the average error is less than the error minimum, the error minimum is replaced, and the best fit is established.
Step 9: This process is repeated to generate the x best fit and the y best fit, where the x best fit and the y best fit contain the data points generated by the PDF function.
Step 10: The x best fit and the y best fit are then passed over to predict the samples using the serial-parallel concatenated adaptive filter block processing unit.

We observe that both the minimum mean error and the minimum mean square error approach give nearly the same estimate for the given set of data, and hence we selected the minimum mean square error approach for our paper. Even though there are other methods to generate the best-fit data points, the minimum mean square approach yields the most accuracy and hence is adopted in this paper.

16.6 Kalman Filter

State parameter models are used by the KF to obtain estimation and prediction. It also takes into account the measurements over time, statistical noise, and sensor noise to generate estimates that are more accurate than a single measurement/source of data. The KF uses both the observations and measurements to come up with the best estimate. The two most important applications of the KF are smoothing and prediction. Mathematically, the KF is a set of five equations that include 'prediction and update' steps.

The effectiveness of the KF comes from the fact that it does not assume errors to be Gaussian.

The KF requires a clear definition of the system model, initial conditions, and error sources. The Kalman Gain (K) is crucial in determining how effectively it performs. It performs a trade-off between the observations and predictions by weighing the observations or predictions based on their respective noise covariance. In this project, we will be using the KF to predict the signal states of EEG data, and in turn, use the predicted data to classify a seizure with a given level of confidence. In this project, we have employed an adaptation of the KF, since the KF requires an exact description of a state model, and it is not accurately available in literature for the brain EEG data. Thus, we used a filter that is heavily based on the KF in its framework and update yet makes use of the knowledge we have of the EEG data. The derivation goes as follows:

$$X=\{x(0),x(1),\ldots..x(M-1)\}\text{—Data frame of M points}$$

$$W=\{w(0),w(1),w(2),\ldots.w(M-1)\}\text{---Filter tap weights}$$

$$S=E((w-E(w)(w-E(w))^{T})\text{—covariance matrix}$$

Predict equations:

$$\begin{aligned} Y &= wx^{T} \\ K &= \frac{Sx^{T}}{(w)+xSx^{T}} \end{aligned} \tag{16.2}$$

Update equations:

$$w^{T}_{\text{new}} = w^{T} + (y_{est} - y)$$

$$S_{new} = (I = Kx) \tag{16.3}$$

These equations justify why the filtering is termed as an adaptation of the KF. We have two equations in the prediction stage to update the estimate as we have no exact state model, and since we are predicting by more than one sample, we have no current measurement data. Combining them, a simplified KF is obtained. As the update step is no longer with respect to the measurement data, the y_{est} here is obtained from a novel PDF approach. This parameter replaces the measurement obtained and is defined as the estimated state. We use a method that takes into account a set of data obtained in the past and obtains the best estimate of what the measurement could be. W and S are updated in every iteration. The convergence is defined by a lower bound error between y_{est} and y_{actual} or by the maximum number of iterations in case the error bound is not met as a way of safe exit. This process predicts the signal state one step ahead. We can now use the block concatenation approach to successively predict more points and subsequently, seizures.

16.7 Results and Discussion

For experimental analysis, more than 250 data set records are considered, where every data set consists of 4096 samples. The five profile types and their respective pass rates are tabulated in Table 16.1. This procedure applies to both healthy and epileptic subjects and assesses the presence of false positives. When an algorithm is applied to the healthy records, the overall success of the 88% (44/50) method was achieved in open-eye situations and 100% in closed-eye situations.

A group of healthy volunteers with open eyes showed a correspondingly low recognition rate. These false alarms are believed to be the result of a lack of a regular alpha rhythm that is directly related to the eye. Background noise is modeled by the KF, not the signal itself, because the open eye suppresses the dominant alpha frequency. This reduces the success rate of filter modeling.

Table 16.1 Data sets and recognition rate.

Data sets	Healthy/epileptic	Condition	Recognition rate
Set 1	Healthy normal	Eyes open	44/50=88%
Set 2	Healthy normal	Eyes closed	50/50=100%
Set 3	Epileptic	Inter-octal (Epileptic focus)	47/50=94%
Set 4	Epileptic	Inter-octal (Hippocampal region)	48/50=96%
Set 5	Epileptic	Ictal state	46/50=92%

The overall success rate for these cases is 94% (47 of 50) in the epilepsy focus, 96% (48 of 50) in the hippocampal region, and 92% (46 of 50) after seizures. The conditions are reported in Table 16.1.

The ROC analysis was used to evaluate the performance of the method. The ROC shows the overall performance of the discovery process, regardless of the decision limits. Figure 16.4 shows the results of the ROC analysis of the system. The sensitivity and specificity values of the ROC are calculated as follows:

$$\begin{aligned}\text{Sensitivity} &= \text{True positives/total positives}\\ &= 46/50\\ &= 0.92\end{aligned} \tag{16.5}$$

$$\begin{aligned}\text{Specificity} &= \text{False positives/total negatives}\\ &= 11/200\\ &= 0.055\end{aligned} \tag{16.6}$$

The success rate obtained was assessed as the area under the ROC in Figure 16.4. The large area below the ROC indicates that the system can be very sensitive. The success rate related to the sensitivity is given

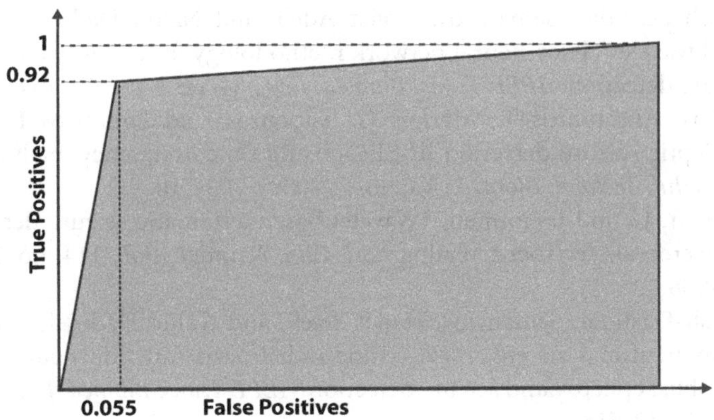

Figure 16.4 ROC result.

by Equation 16.5, the square of 0.936 units. This result confirms the test results shown in Table 16.1.

16.8 Conclusion

In this paper, seizure detection and prediction were implemented and tested using a serial-parallel concatenated KF. The innovative PDF method adopting mean square error was implemented to predict additional N + k states, where k is limited based on the efficiency of the accuracy parameters. On extensive testing of the KF, the accuracy parameters improved with serial implementation while the parallel concatenation had no significant improvement. The KF was used in the proposed work to predict the seizure, which does not include the learning stage, which makes it easy to apply with a high detection rate. A simple and effective method implemented for epileptic seizure detection can be considered a motivation for advanced improvements in this field.

References

1. Gotman, Jean. "Automatic seizure detection: improvements and evaluation." *Electroencephalogr. Clin. Neurophysiol.*, 76, no. 4 (1990): 317-324.
2. Gotman, J., D. Flanagan, J. Zhang and B. Rosenblatt. "Automatic seizure detection in the newborn: methods and initial evaluation." *Electroencephalogr. Clin. Neurophysiol.*, 103, no. 3 (1997): 356-362.

3. Ghosh-Dastidar, Samanwoy, Hojjat Adeli, and Nahid Dadmehr. "Mixed-band wavelet-chaos-neural network methodology for epilepsy and epileptic seizure detection." *IEEE Trans. Biomed. Eng.*, 54, no. 9 (2007): 1545-1551.
4. Tzallas, Alexandros T., Markos G. Tsipouras, and Dimitrios I. Fotiadis. "Epileptic seizure detection in EEGs using time–frequency analysis." *IEEE Trans. Inf. Technol. Biomed.*, 13, no. 5 (2009): 703-710.
5. Khan, Y. U. and J. Gotman. "Wavelet based automatic seizure detection in intracerebral electroencephalogram." *Clin. Neurophysiol.*, 114, no. 5 (2003): 898-908.
6. Ghosh-Dastidar, Samanwoy, Hojjat Adeli, and Nahid Dadmehr. "Principal component analysis-enhanced cosine radial basis function neural network for robust epilepsy and seizure detection." *IEEE Trans. Biomed. Eng.*, 55, no. 2 (2008): 512-518.
7. Ramgopal, Sriram, Sigride Thome-Souza, Michele Jackson, Navah Ester Kadish, Iván Sánchez Fernández, Jacquelyn Klehm, William Bosl, Claus Reinsberger, Steven Schachter, and Tobias Loddenkemper. "Seizure detection, seizure prediction, and closed-loop warning systems in epilepsy." *Epilepsy Behav.*, 37 (2014): 291-307.
8. Lehnertz, Klaus, and Christian E. Elger. "Can epileptic seizures be predicted? Evidence from nonlinear time series analysis of brain electrical activity." *Phys. Rev. Lett.*, 80, no. 22 (1998): 5019.
9. Lehnertz, Klaus, Ralph G. Andrzejak, Jochen Arnhold, Thomas Kreuz, Florian Mormann, Christoph Rieke, Guido Widman, and Christian E. Elger. "Nonlinear EEG Analysis in Epilepsy: Its Possible Use for Interictal Focus Localization, Seizure Anticipation, and." *J. Clin. Neurophysiol.*, 18, no. 3 (2001): 209-222.
10. Aschenbrenner-Scheibe, Richard, Thomas Maiwald, Matthias Winterhalder, Henning U. Voss, Jens Timmer, and Andreas Schulze-Bonhage. "How well can epileptic seizures be predicted? An evaluation of a nonlinear method." *Brain*, 126, no. 12 (2003): 2616-2626.
11. Martinerie, Jacques, Claude Adam, M. Quyen, Michel Baulac, Stéphane Clemenceau, Bernard Renault, and Francisco J. Varela. "Epileptic seizures can be anticipated by non-linear analysis." *Nat. Med.*, 4, no. 10 (1998): 1173-1176.
12. Le Van Quyen, Michel, Jacques Martinerie, Michel Baulac, and Francisco Varela. "Anticipating epileptic seizures in real time by a non-linear analysis of similarity between EEG recordings." *NeuroReport*, 10, no. 10 (1999): 2149-2155.
13. Le Van Quyen, M., C_ Adam, J. Martinerie, M. Baulac, S. Clemenceau, and F. Varela. "Spatio-temporal characterizations of non-linear changes in intracranial activities prior to human temporal lobe seizures." *Eur. J. Neurosci.*, 12, no. 6 (2000): 2124-2134.
14. Le Van Quyen, Michel, Jacques Martinerie, Vincent Navarro, Paul Boon, Michel D'Havé, Claude Adam, Bernard Renault, Francisco Varela, and

Michel Baulac. "Anticipation of epileptic seizures from standard EEG recordings." *Lancet*, 357, no. 9251 (2001): 183-188.
15. Navarro, Vincent, Jacques Martinerie, Michel Le Van Quyen, Stéphane Clemenceau, Claude Adam, Michel Baulac, and Francisco Varela. "Seizure anticipation in human neocortical partial epilepsy." *Brain*, 125, no. 3 (2002): 640-655.
16. De Clercq, Wim, Philippe Lemmerling, Sabine Van Huffel, and Wim Van Paesschen. "Anticipation of epileptic seizures from standard EEG recordings." *Lancet*, 361, no. 9361 (2003): 971.
17. Winterhalder, M., T. Maiwald, H. U. Voss, R. Aschenbrenner-Scheibe, J. Timmer and A. Schulze-Bonhage. "The seizure prediction characteristic: a general framework to assess and compare seizure prediction methods." *Epilepsy Behav.*, 4, no. 3 (2003): 318-325.
18. LeCun, Yann, Yoshua Bengio, and Geoffrey Hinton. "Deep learning." *Nature*, 521, no. 7553 (2015): 436-444.
19. Faust, Oliver, Yuki Hagiwara, Tan Jen Hong, Oh Shu Lih and U. Rajendra Acharya. "Deep learning for healthcare applications based on physiological signals: Areview." *Comput. Methods Programs Biomed.*, 161 (2018): 1-13.
20. Mohammadi, Mehdi, Ala Al-Fuqaha, Sameh Sorour, and Mohsen Guizani. "Deep learning for IoT big data and streaming analytics: A survey." *IEEE Commun. Surv. Tutorials*, 20, no. 4 (2018): 2923-2960.
21. Hochreiter, Sepp, and Jürgen Schmidhuber. "Long short-term memory." *Neural Comput.*, 9, no. 8 (1997): 1735-1780.
22. Shoeb, Ali Hossam. "Application of machine learning to epileptic seizure onset detection and treatment." PhD diss., Massachusetts Institute of Technology, 2009.
23. Kumar, J. Satheesh and P. Bhuvaneswari. "Analysis of Electroencephalography (EEG) signals and its categorization–a study." *Procedia Eng.*, 38 (2012): 2525-2536.
24. Usman, Syed Muhammad, Muhammad Usman, and Simon Fong. "Epileptic seizures prediction using machine learning methods." *Comput. Math. Methods Med.*, 2017 (2017).

17

Current Advancements in Steganography: A Review

Mallika Garg[1]*, Jagpal Singh Ubhi[2] and Ashwani Kumar Aggarwal[3]

[1]*Department of Electronics and Communication Engineering, Indian Institute of Technology, Roorkee, Uttarakhand, India*
[2]*Electronics and Communication Engineering Department, Sant Longowal Institute of Engineering and Technology, Longowal, Punjab, India*
[3]*Electrical and Instrumentation Engineering Department, Sant Longowal Institute of Engineering and Technology, Longowal, Punjab, India*

Abstract

In today's world, communication over the Internet has increased to a great extent. With this increased communication, enormous data are to be kept and communicated secretly and safely. Many techniques have been developed to ensure the secret transfer of data over the internet. Steganography is used to protect data from unauthorized persons. This chapter provides a review of various traditional, convolutional neural network (CNN)-, and generative adversarial network (GAN)-based steganographic techniques. Details of the different types of steganography based on the host and domain, its applications, its disadvantages, and various evaluation parameters are also illustrated. This paper also presents the details of the datasets used for steganography and contains some basic details of the counterpart of steganography called steganalysis.

Keywords: Image security, convolutional neural network, information hiding, StegoExpose, steganography

Corresponding author: mallika@ec.iitr.ac.in

Chandra Singh, Rathishchandra R. Gatti, K.V.S.S.S.S. Sairam, Manjunatha Badiger, Naveen Kumar S. and Varun Saxena (eds.) Modeling and Optimization of Signals Using Machine Learning Techniques, (327–348) © 2024 Scrivener Publishing LLC

17.1 Introduction

Steganography's primary goal is to wrap and hide secret data in a host data from some unauthorized party. The main intention behind performing wrapping is to encapsulate sensitive information. The second is the prevention of withdrawal from the cover file without disturbing the cover object. The secret information may be easy to identify in steganography, but suspecting the sensitive message is the least possible.

In steganography, we do not modify the form in which the secret information is present, but we wrap it within some cover image. It takes into consideration that humans are not able to distinguish or identify the concealed information. Human senses are not capable of suspecting and recognizing the information hidden inside a host image. Steganography is beneficial for securely storing secret data, such as a hiding system in military communication [29]. Steganography requires a cover object, a secret message, a stego object, and an embedding algorithm for encapsulation, as shown in Figure 17.1.

1. Cover Object: It contains the information used to enclose the sensitive information in it. It is also referred to as the host object as it can hide sensitive data.
2. Message: The sensitive and secure information to be communicated over the network is the message data. This information is usually concealed within the host information.
3. Stego Object: The host within which the sensitive data are embedded makes the stego object. Then, the obtained data are transferred over the network to reach the receiving end.

Cryptography and steganography can be jointly used to enforce layers of secrecy on the secret message. There are many schemes in the literature that extract the features of both steganography and cryptography to enhance security. Edward [1] proposed a multilayer steganographic and

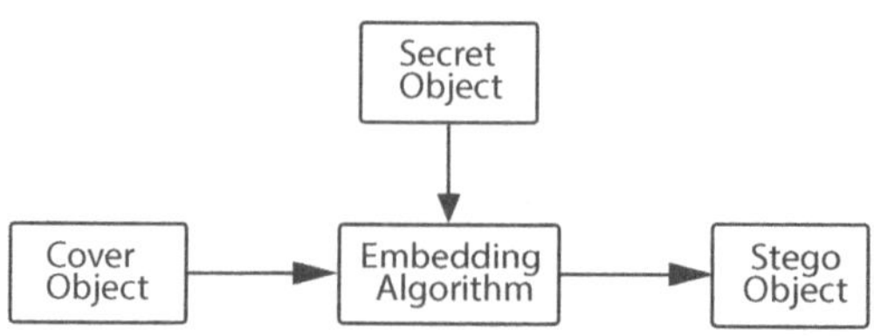

Figure 17.1 Common terms used in steganography.

cryptographic scheme using genetic algorithm (GA) and residual numbers. GA is employed for image encryption and decryption for hiding texts.

17.2 Evaluation Parameters

The parameters used to evaluate steganography are the following:

1. Imperceptibility: It is the noticeable changes in the stego object that should not be visible to human eyes. The higher the imperceptibility, the more secure the data are. The stego image should resemble the host image to maintain imperceptibility. After the secret image has been embedded, the imperceptibility of the stego image will be enhanced.
2. Capacity Payload: The maximum piece of information that the host image can embed within itself so that the distortions are not observable depends on the capacity payload of the host image. It can be measured in units of information. The greater the capacity payload, the more efficient the steganographic technique will be. The quantity of data to be embedded in the carrier depends on the dimensions and size of the host and the type of steganography used to embed them for safe and secure communication. There is a trade-off between the capacity payload and the imperceptibility; as with the increase in capacity payload, the algorithm becomes more imperceptible and vice versa.
3. Computational Complexity: It is the amount of resources and time required for embedding and extracting secret information. The lower the computational complexity, the more efficient the algorithm is.
4. Robustness: Robustness depends on the extent to which the secret information can withstand the changes. The stego image can undergo some transformations such as image cropping, rotation, compressing, and changing the resolution and exposure to different types of noises.
5. Undetectability: The measure of undetectability depends on the extent to which the presence of the secret data is undetectable whenever the stego object is statistically tested.
6. Accuracy: The extent to which the hidden information is correctly or precisely extracted from the host information affects the accuracy of the algorithm.

7. Secrecy: The security of the data is ensured if the statistical tests are not capable enough to differentiate between the cover image and the stego image [8].

Conflicts between various parameters
The major objective of steganography is to optimize the performance of steganographic systems based on various performance evaluation parameters. Often, enhancing one particular parameter can negatively influence the others, like one cannot maximize the imperceptibility and the capacity payload at once. The alteration done in the host image by concealing secret information is directly dependent on the content and the amount of the secret data. So, there must be a trade-off between these parameters to optimize the steganographic algorithm.

17.3 Types of Steganography

Steganography can be classified based on different factors.

17.3.1 Host

Steganography can be implemented on various hosts such as text, audio, image, video, and protocol, as shown in Figure 17.2.

1. Image Steganography: In this, the secret pixel bit is embedded in each pixel of the image. This is most commonly used because it considers the limitations of human vision. The images can be seen as a collection of pixels placed at

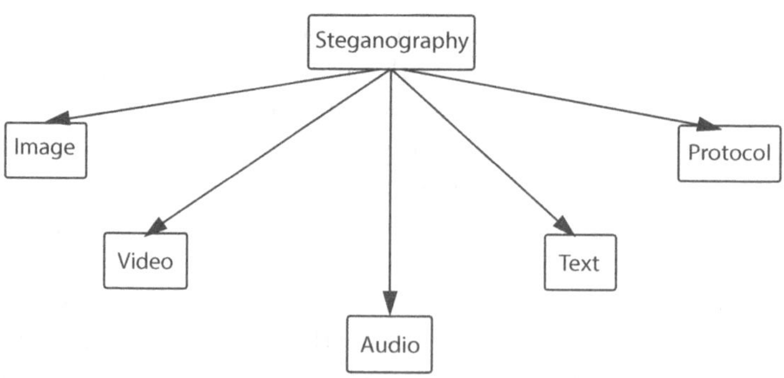

Figure 17.2 Types of steganography based on the host.

different locations. Each pixel is 8 bits or 24 bits depending on whether the image is grayscale or colored. The secret bit replaces a bit in each pixel to hide it in the cover image.

2. Video Steganography: In this steganography, the confidential information is encapsulated in a video stream. While embedding secret information in video, usually the discrete cosine transform (DCT) method is used. Secret data is embedded in each frame of the video in the transfer domain in a way that any change is not noticed by subjective analysis.
3. Text Steganography: It uses text as a cover to hide the secret data. It can be categorized as format-based, random generation, statistical generation, and linguistic method.
4. Audio Steganography: This exploits the psychoacoustical occurrence of the human auditory system (HAS). The psychoacoustical or auditory masking property arises because of the low differential range of the HAS. Audio steganography transmits hidden information within some audio files to preserve the security of the information. It is used to hide data in various audio file formats such as AIFF, MP4, WAV, FLAC, AU, and MP3.
5. Protocol Steganography: This is a new steganographic technique that uses network protocols such as TCP, IP, UDP, and ICMP, to embed a message in the carrier. The message is hidden in the carrier of a TCP/IP packet in some fields that are either optional or are never used.

17.3.2 Domain

According to the domain in which embedding is done, steganography can be of various types, as shown in Figure 17.3.

1. Spatial Domain: In this transform steganography, secret pixel bits are directly replaced by cover pixel bits. Spatial domain steganography applies bit-wise insertion and manipulation of cover bits. These are computationally efficient techniques as the insertion or manipulation of bits takes place in the spatial domain only.
2. Transform Domain: In the transform or frequency domain, domain transformation is applied on both images first. Then, the secret information is encapsulated in the transform domain information of the cover image. After the

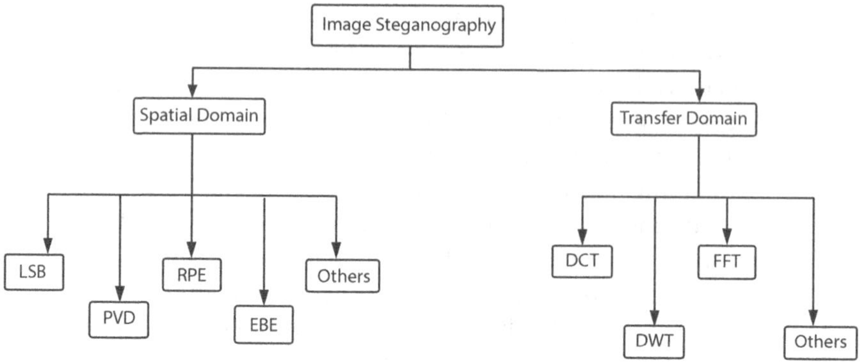

Figure 17.3 Types of steganography based on the domain.

embedding is completed, the embedded information is transformed back to the spatial domain. The various frequency domain steganographic methods are:

(a) Discrete Fourier transformation (DFT),
(b) Discrete cosine transformation (DCT),
(c) Discrete wavelet transformation (DWT),
(d) Lossless or reversible technique, and
(e) Embedding in coefficient bits.

Frequency domain steganography is more robust than spatial domain embedding. Although, when embedding is done for a high capacity, spatial transform is preferred over frequency transform steganography.

17.4 Traditional Steganographic Techniques

There are various types of steganography and in all of them, cover bits are directly altered for embedding the secret image bits. The summary of different traditional steganographic techniques is given in Table 17.1. Spatial domain techniques are classified based on the way the secret bits are embedded in the following types.

17.4.1 Least Significant Bit (LSB) Steganography

In this technique, the LSBs of the cover image pixels are replaced by the secret data bits [29]. We can replace up to four positions of a cover pixel

Table 17.1 Summary of traditional steganographic techniques.

Method	Technique	Imperceptibility	Capacity payload	Summary
[19]	LSB	Yes	No	Very less distortion
[18]	LSB with secret key	Yes	No	Secure and complex
[24]	LSB and RRCF	Yes	Yes	Secure and less complex
[39]	PVD	No	Yes	High hidden capacity
[32]	LSB and PVD	No	Yes	High hidden capacity and high PSNR
[46]	LSB and PVD	Yes	Yes	Robustness
[2]	EBD	Yes	Yes	Less distortion
[25]	RPE	Yes	No	Security
[6]	RPE	Yes	No	High security
[40]	PMM	Yes	Yes	Not an optimal mapping
[11]	PMM	Yes	No	Optimal mapping and fast
[45]	PMM	Yes	No	Data cannot be easily extracted

with the bits of the secret message. Replacing more bits can cause distortion in the cover as the color changes and pixel intensity changes become noticeable. However, it embeds the secret bits in the LSBs of the cover image without affecting the perceptual quality of the cover image.

LSB steganography has a limitation in that its sample value changes asymmetrically [8]. If the LSB of a host is similar to the secret image bit that is to be embedded, then no modification is done to that bit of the cover image. Else, the even value is modified to an odd value or the odd pixel intensity value is changed to an even intensity value. This asymmetry can be detected by simple steganalysis techniques [33].

A new method, which is the enhanced least significant bit (ELSB) method, has been proposed for color images [8], which is found to have less distortions than the simple LSB algorithm. ELSB embeds in the blue channel bit so it has less distortion than simple LSB which is negligent to the human eye. Ankita Gangwar *et al.* [18] provided an improved RGB-LSB-based image steganography. A secret key is considered while hiding secret information in the LSB of RGB cover pixels. The unknown intruder cannot reconstruct the secret image without knowing the secret key.

Moustafa M. Kurdi *et al.* [24] used LSB and random right circular shift (RRCF). The RRCF is utilized to change the order or arrangement of bit sequences. Various performance evaluation parameters, such as the peak signal-to-noise ratio (PSNR) and the mean squared error (MSE), are used for testing and evaluating the algorithm. The watermarked image is also attacked by changing the format, cropping, and adding noise.

The advantages of LSB are the following:

1. Difficulty in detecting the secret (i.e., embedded) data
2. More capacity to hide secret information
3. Less chances of data modification
4. Simple implementation and less complexity
5. Very little degradation of the original image.

The disadvantages of LSB are the following:

1. Loss of hidden information with image manipulation
2. Easy recovery of confidential information (in case of detection)
3. Difficulty in recovering secret information in the presence of simple attacks
4. Sensitivity to image manipulations.

17.4.2 Pixel-Value Differencing (PVD)

The stego object obtained by PVD is highly imperceptible to human eyes when compared to the cover object. Chung-Ming Wang *et al.* [39] proposed a PVD scheme in which two consecutive pixels are selected and a quantization range table is designed. This range table determines the payload between the consecutive pixels' differences in values. These pixel values having more differences between them have larger capacity payloads than the ones having less differences between them. This signifies that the less amount of secret messages can be embedded at the edges than in smoother areas.

Gandharba Swain *et al.* [32] develops a combination of LSB substitution and PVD. The image is broken into 2*2 non-overlapping image patches. For every image patch, LSB substitution is used to embed the upper-left pixel with k-significant bits. Edges from both horizontal and vertical directions are used to hide the secret data bits. Higher PSNR and capacity payload values are obtained by this method on the Signal and Image Processing Institute (SIPI) image database. Cheng-Hsing Yang *et al.* [46] also combined LSB substitution and PVD not only to enhance the capacity payload and imperceptibility but also to reduce the risk of RS-steganalysis detection.

17.4.3 Edge-Based Embedding (EBE)

This method uses the edge information of the cover pixels for embedding the secret data bits. The edges are more capable of shielding the confidential data bits with less distortion in the stego image. BrahmaTeja, K.N. *et al.* [9] proposed a method in which the LSB of the cover image is masked first to form the masked image. The edge pixels are used to hide the secret data in the LSB bits of the edge pixels only. Thus, this algorithm ensures better security against attackers.

Junlan Bai *et al.* [2] proposed a novel steganography approach based on the combination of the LSB substitution mechanism and edge detection. First, the edge pixels were detected. These pixels can embed a greater number of secret bits, thus increasing the capacity payload of the cover image. The key idea behind edge steganography is based on the fact that edges are more tolerable to the embedding bits than smooth areas. Canny and Sobel edge detectors are used to detect the edge pixels.

17.4.4 Random Pixel Embedding (RPE)

Random pixels from the cover image are selected for embedding the bits of the secret image. Laskar *et al.* [25] proposed a method in which the secret bits are placed in the randomly selected pixels on the basis of the random sequence generated by the pseudorandom number generator (PRNG). So, the secret bits are placed at random positions which makes it difficult for the third party to reconstruct the original bits accurately. Rupali Bhardwaj *et al.* [6] present a technique for providing three levels of security. This method outperforms the basic LSB method. The PSNR is determined between the stego and cover images to compare the proposed algorithm with the inverted LSB, simple LSB, and random LSB.

17.4.5 Pixel Mapping Method (PMM)

The pixel mapping method uses pixel intensity-based mathematical functions to select the cover image pixels for embedding [7]. These pixels are then embedded with the bits of the secret image and the capacity payload of each pixel is determined based on its intensity and the intensity of its 8 neighborhood pixels. Ran-Zan Wang *et al.* [40] hid the secret image by a novel method using optimal LSB substitution and GA. The GA utilized for finding the optimal result may take a huge computation time for hiding important information in the rightmost k LSBs. Lastly, the features of human perception are exploited for hiding large-scale LSB in the four most significant bits (MSBs) and two LSBs for an improved hiding technique. This improves the high-quality embedding in the host image.

Chang, C.C. *et al.* [11] found the optimal solution of LSB substitution that is computationally faster. This algorithm takes less processing time as the optimal solution can be generated very fast using GA and the exhaustive LSB substitution along with the method proposed. Xu Huan *et al.* [45] proposed a pair-wise LSB matching scheme. The pair-wise LSB includes the alteration of two cover pixels at the same time. This system uses Mielikainen's method to distinguish the matching process and then adjusts the order of secret bits inserted in the cover image. Therefore, the visual and perceptual quality of the stego image is improved.

17.5 CNN-Based Steganographic Techniques

With the tremendous growth in the traditional steganographic techniques, deep learning frameworks are also proposed which can enhance the security of the secret image. These frameworks include an encoder-decoder approach that first performs steganography and then performs the reverse process of extracting the secret image. The encoder is fed with two images, cover and secret, to generate the stego image which is then inputted to the decoder to reconstruct the original image. After the neural networks are used in steganography, a sudden rise in this field is noted in the literature. A detailed summary of CNN-based techniques is shown in Table 17.2.

Baluja [3] proposed a deep learning-based steganographic framework that takes the secret image as an input distributed among all the bits of the cover image. Hidden information of the same size as that of the host was hidden in the proposed network. The system consists of three networks, which are the preparation network that prepares the secure image, the hiding network that embeds the secret bits in the cover image, and the

Table 17.2 Summary of CNN-based steganographic techniques.

Method	Model	Dataset	Remarks
[4]	CNN	ImageNet and Corel Dataset [36]	1:1 ratio of hidden-to-host information was hidden.
[37]	U-Net	Imagenet	Speeds up the training process by improvising the learning rate. Imperceptibility is maintained.
[30]	Encoder-decoder	COCO	Training time is reduced by using different loss functions.
[43]	Fully connected dense network	ImageNet	High capacity.
[16]	Improved FC-DenseNet	ImageNet	High capacity of 23.96 bits per pixel.

reconstruction network that regenerates the secret image. Baluja has made efforts to detect where the secret bits are stored in the cover image and found that the secret image bits are distributed in the surrounding pixels of the host image and the different color channels of the same pixel. He also found that the representation of each secret pixel is distributed in up to seven surrounding pixels in the host image. Also, the spread of the pixel bits relies directly on the size of the neural network.

In [31], the encoder and decoder are trained by reducing the error between the cover-stego pair and secret-recover pair jointly, as given by Equation 17.1.

$$L(c, c^0, s, s^0) = ||c - c^0|| + \beta||s - s^0|| \tag{17.1}$$

where c is the cover image, c^0 is the stego image, s is the secret image, and s^0 is the regenerated secret image from the revealed network. β is a hyper-parameter that regulates the amount of secret images to be restored. This algorithm uses the Adam optimizer to minimize this loss function and uses an extra layer of security by using an encrypted secret image instead of a real secret image for image hiding in the cover image.

As mentioned earlier, the neural network used for steganography includes three networks that are trained jointly, so they need more time to train, and the training process is very complex. So, Rehman [30] proposed a network that reduces the training time by using different optimization strategies, as given by Equation 17.2.

$$L(c, c^0, s, s^0) = \alpha||c - c^0||^2 + \beta||s - s^0||^2 + \lambda(||W_e||^2 + ||W_d||^2) \tag{17.2}$$

where W^e and W^d are the trained weights of the hiding and reconstruction networks, respectively, and all the others are the same as in Equation 17.1.

Van *et al.* [37] further sped up the training process of the network without even sacrificing accuracy. The speed of training is reduced to half as compared to [4] by reducing the learning rate when the improvement in learning is reduced. The author of this paper proposed two independent optimization techniques for each network. The proposed network in [37] uses an H-net for hiding the secret image in the cover image and an R-net for retrieving the secret image. This idea of networks was based on the U-net.

Zhong *et al.* [49] proposed a steganographic strategy for stylized images. Two styles of images were obtained from one original image using different parameters of the stylized image. The first was used as the host image and the second as the reference image. Finally, the message was encapsulated in the host image pixels with the embedding cost which is defined by the detection risk. A fully connected dense network (FC-DenseNet) [43] can also be used to perform steganography with advanced visual effects and high steganographic capacity since this network can overcome the problems of gradient dissipation and gradient explosion, and a large number of features are fused to construct the stego image.

An improvement is proposed by using an improved FC-DenseNet [16]. A steganographic scheme (SteganoCNN) [15] is also proposed to embed two secret images in a carrier image and reconstruct them. This scheme can effectively embed the two images with a capacity payload of 47.92 bits per pixel while avoiding detection.

17.6 GAN-Based Steganographic Techniques

With the development and enhancement in steganography, GAN-based steganography has also gained much attention. The security of secret images is the basis for evaluating steganography algorithms. A good

Table 17.3 Summary of the GAN-based steganographic methods.

Method	Architecture	Dataset	Remarks
[38]	Info-GAN	MNIST and CelebA	Key-based embedding adds additional security
[48]	Auxiliary classifier generative adversarial network (ACGANs)	MNIST	High robustness and high capacity payload
[12]	Encoder-decoder	ImageNet and Photo2Monet	Generates imperceptible and high-frequency signals
[23]	Kerckhoffs' principle-based GAN	MNIST	High security and capacity

steganographic algorithm must be capable of resisting the detection of the stego image. Adversarial training algorithms can be used to synthesize a robust image that can compete with the other steganographic techniques. The steganography without embedding (SWE) presented in [21] does not alter the carrier image. This algorithm is based on generative adversarial networks. A detailed summary of GAN-based techniques is shown in Table 17.3.

Zhang [47] shows that invisible steganography via generative adversarial networks (ISGAN) hides only in the Y-channel of the carrier to withstand the stego images' color distortion [3]. The author has strengthened the security of secret information by reducing the divergence between the empirical probability distributions of the two images. A mixed loss function was introduced to associate imperceptibility with the human visual system to generate a more realistic stego image. GANs [20] can perform steganography and steganalysis at the same time using the concept of Alice, Bob, and Eve in steganography. Alice synthesizes the stego image, Bob is used for extracting secret bits from the generated image, and Eve is used as a steganalysis network.

Volkhonskiy *et al.* [38] presented steganographic generative adversarial network (SGAN) which creates an adaptive container generator based on deep convolutional generative adversarial network (DCGAN). Both networks are learned using the optimization loss given in Equation 17.3.

$$L_{GAN}(G, D) = E_x[\log D(x, z)] + E_z[\log(1 - D(x, G(z)))] \quad (17.3)$$

where Ex is the expected value over every training data sample, D(x, z) is the discriminator's estimate that x is the original image, E^z is the expected value over all generated samples G(z), and D(G(z)) is the discriminator's estimate that the instance is generated by the generator.

Tang *et al.* [35] proposed an automatic steganographic distortion learning (ASDL- GAN) system by utilizing GANs that generate artificial samples with a unified framework. The generator produces stego images by the adversarial training as indicated by the insignificant inserting mutilation determined on the change probabilities. The learned change probabilities are adopted for minimal-distortion embedding. Adversarial embedding (ADV-EMB) [34] generates an adversarial stego image by back-propagating a gradient from the target CNN steg-analyzer.

Yan Ke *et al.* proposed a steganographic framework with Kerckhoffs' principle (GSK) [23]. In GSK, the two generators are trained according to the Kerckhoffs' principle. The secret information is synthesized by the generator only if the extraction key and the host image are given as inputs to the receiver. No meaningful results are obtained without the knowledge of these two. CycleGAN also hides bits of the secret image into the generated stego images. Casey Chu *et al.* [12] connect the phenomenon of cyclic consistency while generating a realistic image of adversarial attacks by viewing CycleGAN's [51] training procedure. The model generally learns to fool by encoding information imperceptibly and adversarially, even though the cyclic consistency loss is intended to force the network into encoding information about a source image semantically into the generated image.

17.7 Steganalysis

Along with steganography, an important area of research that comes into existence is steganalysis, the counterpart of steganography that detects the presence of steganography. Tapan [22] proposes a neural network-based steganalysis that distinguishes between cover and stego images. The algorithm has been designed for the detection of spatial domain steganography like LSB, MSB, and PVD, and after various experiments, it has been found that it performs best for LSB steganography. Butaro [10] found that truncating the DCT coefficients during JPEG compression has consequences for steganography and its detection.

Many deep neural networks are also trained to detect the presence of steganography [27, 44]. So, more robust neural networks are to be proposed, which not only embed the secret information with high imperceptibility but are also robust to steganalysis. An end-to-end model [50] was proposed that uses intermediate representations to embed information at any pixel location of the cover image. This algorithm uses a discriminator to minimize the loss. The model has a 50% accuracy when trained without noise and shows an improvement of 35% when trained with a simulated JPEG noise. High-capacity convolutional video steganography [41] and light field messaging with deep photographic steganography [42] are proposed for hiding information in videos to minimize the detectability of the message embedding while maximizing the accuracy of secret data recovery. All these neural networks use basic image feature extraction for data hiding while maintaining less distortion.

17.8 Applications

Steganography can be used by any two users between whom a secret data transfer is carried out during communication. Depending on numerous areas where secrecy in communication is needed, there are different ways by which secure data transfer can take place. One such way is steganography. It can be used by political organizations to share confidential plans or strategies among themselves, or even terrorists or criminals can use steganography to plan terrorist operations secretly. Steganography can also be used in copyright protection to avoid misuse, in smart identity cards, in maintaining secrecy in the patients' information, and in military and police communications. Confidential information may also be shared in the business sector and online transaction methods.

17.9 Dataset Used for Steganography

There are many datasets that can be used for steganography. These datasets are important for researchers to develop various algorithms for steganography. The algorithm is first trained on the training dataset and tested on the testing dataset. The evaluation of the algorithm is then done on the test dataset. Some of the publicly available images are used for traditional steganography like Baboon, Lena, Cameraman, and Pepper, which are shown in Figure 17.4.

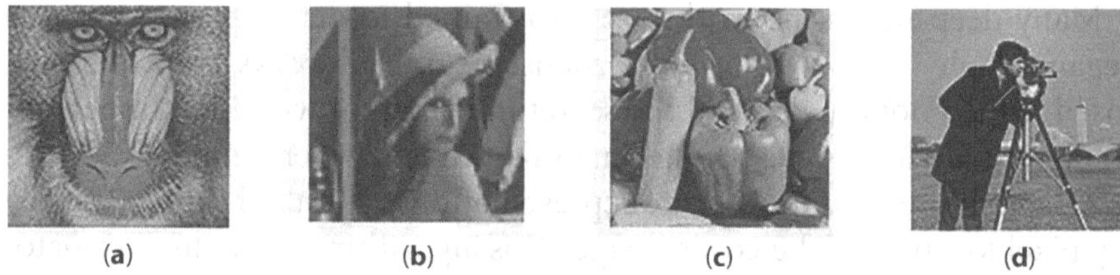

Figure 17.4 Images used for traditional steganography: (a) Baboon, (b) Lena, (c) Pepper, (d) Cameraman.

Table 17.4 Details of the datasets used in literature for steganography.

Dataset	Image	Dimension of image	Task
ImageNet	14M	Random	Computer vision
Pascal VOC	11,530	500×500×3	Object detection and image segmentation
CelebA	More than 200K	-	Face recognition
COCO	330K	640×640×3	Object detection and image segmentation
MNIST	70K	28×28×1	Computer vision
BOSS	11K	512×512×1	Steganography

A few of the most frequently used datasets for steganography are Break Our Steganographic System (BOSS) [5], Imagenet [13], MNIST handwritten digits [14], Pascal VOC [17], CelebA, and COCO [26]. Details of the datasets are described in Table 17.4.

17.9.1 BOSS

The Break Our Steganographic System (BOSS) is a challenge conducted to design an efficient steganalysis algorithm that can break the steganographic images that are highly undetectable stego images [28]. The dataset consists of image samples with the steganographic algorithm to synthesize the stego images. The training set contains 10,000 original samples with size 512 ×512. The testing set contains 1,000 samples of the same size as that of the training set.

17.9.2 Pascal VOC

The PASCAL visual object classes (Pascal VOC) is also a publicly available dataset for object detection and image segmentation tasks. VOC datasets are highly irregular considering the object size, position, orientation, pose, illumination, and occlusion. It has images with 20 image classes and all images were taken from the flickr's website. The total number of annotated images is 11,540 with 5,717 train images and 5,823 validation images. The total number of annotated objects is 27,450. So, it contains an average number of 1,372 objects per class.

17.9.3 ImageNet

The ImageNet dataset is a publicly available large dataset that can be used for any computer vision task. In ImageNet, classes are arranged according to the WordNet hierarchy. The dataset consists of about 14 million images, with 21,841 synsets. The images vary in dimensions and resolutions. Each node consists of more than 1,000 images. It does not have the images itself but provides the URL for the original images in each synset.

17.9.4 COCO

The Common Objects in Context (COCO) dataset was first introduced for object detection, segmentation, localization, and other computer vision applications. It contains about 330,000 images with image annotations. Image annotations include category labeling, instance segmentation, and instance spotting. The size of each image is 640 × 640 × 3. This dataset contains more object instances per image than the ImageNet and Pascal VOC datasets.

17.9.5 MNIST

The Modified National Institute of Standards and Technology (MNIST) database is a standard dataset used for various computer vision applications. It consists of images from handwritten digits 0 to 9 of size 28 × 28 pixels. The images are normalized and are grayscale images. The training set has 60,000 images and the testing set has 10,000 images.

17.10 Conclusion

This chapter presents different steganographic methods for embedding information within text, image, and audio/video, traditional steganography, CNN and GAN-based steganography, steganalysis, datasets used, and their applications. Steganography is an effective method that has gained importance rapidly in the search for new and better algorithms. The reliability of these techniques is determined based on the evaluation parameters according to the requirements. Most traditional steganography is reliable and robust to steganalysis, but with the recent advancements in deep learning algorithms, steganography has gained much attention. The encoder-decoder-based architecture has prevailing results and more capacity payload as compared to the traditional methods.

More recently, GANs have proven to be significant for the reconstruction of the image. Since there is no standard benchmark dataset for image steganography, almost all computer vision datasets can be used for this purpose. In the future, many new reliable algorithms can be developed with high security and less computation. So, it can be concluded that there can be tremendous growth in this area with respect to the development of deep learning frameworks and GAN models.

References

1. Edward Yellakuor Baagyere, Peter Awon-Natemi Agbedemnab, Zhen Qin, Mohammed Ibrahim Daabo, and Zhiguang Qin. A multi-layered data encryption and decryption scheme based on genetic algorithm and residual numbers. *IEEE Access*, 8:100438–100447, 2020.
2. Junlan Bai, Chin-Chen Chang, Thai-Son Nguyen, Ce Zhu, and Yanjun Liu. A high payload steganographic algorithm based on edge detection. *Displays*, 46:42– 51, 2017.
3. Shumeet Baluja. Hiding images in plain sight: Deep steganography. In *Advances in Neural Information Processing Systems*, pages 2069–2079, 2017.
4. Shumeet Baluja. Hiding images within images. *IEEE Trans. Pattern Anal. Mach. Intell.*, 2019.
5. Patrick Bas, Tomat́sı Filler, and Tomat́sı PevnŒy. " break our steganographic system": the ins and outs of organizing boss. In *International workshop on information hiding*, pages 59–70. Springer, 2011.
6. Rupali Bhardwaj and Vaishali Sharma. Image steganography based on complemented message and inverted bit lsb substitution. *Procedia Comput. Sci.*, 93:832–838, 2016.

7. Souvik Bhattacharyya, Aparajita Khan, Aunkita Nandi, Aveek Dasmalakar, Somdip Roy, and Gautam Sanyal. Pixel mapping method (pmm) based bit plane complexity segmentation (bpcs) steganography. In *2011 World Congresson Information and Communication Technologies*, pages 36–41. IEEE, 2011.
8. Rajkumar L Biradar and Ambika Umashetty. A survey paper on steganography techniques. *J. Innov. Res. Comput.Commun. Eng.*, 4(1), 2016.
9. K Naveen BrahmaTeja, DG Madhumati, and K Rama Koteswara Rao. Data hiding using edge based steganography. *Int. J. Emerg. Technol. Adv. Eng.*, 2(11):285–290, 2012.
10. Jan Butora and Jessica Fridrich. Steganography and its detection in jpeg images obtained with the" trunc" quantizer. In *ICASSP 2020-2020 IEEE International Conference on Acoustics, Speech and Signal Processing (ICASSP)*, pages 2762–2766. IEEE, 2020.
11. Chin-Chen Chang, Ju-Yuan Hsiao, and Chi-Shiang Chan. Finding optimal least-significant-bit substitution in image hiding by dynamic programming strategy. *Pattern Recognit.*, 36(7):1583–1595, 2003.
12. Casey Chu, Andrey Zhmoginov, and Mark Sandler. Cyclegan, a master of steganography. arXiv preprint arXiv:1712.02950, 2017.
13. Jia Deng, Wei Dong, Richard Socher, Li-Jia Li, Kai Li, and Li Fei-Fei. Imagenet: A large-scale hierarchical image database. In *2009 IEEE conference on computer vision and pattern recognition*, pages 248–255. Ieee, 2009.
14. Li Deng. The mnist database of handwritten digit images for machine learning research [best of the web]. *IEEE Signal Process Mag.*, 29(6):141–142, 2012.
15. Xintao Duan, Nao Liu, Mengxiao Gou, Wenxin Wang, and Chuan Qin. Steganocnn: Image steganography with generalization ability based on convolutional neural network. *Entropy*, 22(10):1140, 2020.
16. Xintao Duan, Liu Nao, Gou Mengxiao, Dongli Yue, Zimei Xie, Yuanyuan Ma, and Chuan Qin. High-capacity image steganography based on improved fc- densenet. *IEEE Access*, 8:170174–170182, 2020.
17. Mark Everingham and John Winn. The pascal visual object classes challenge 2012 (voc2012) development kit. Pattern Analysis, Statistical Modelling and Computational Learning, Tech. Rep, 8, 2011.
18. Ankita Gangwar and Vishal Shrivastava. Improved rgb-lsb steganography using secret key. *Int. J. Comput. Trends Technol.*, - volume 4 Issue 2-2013, 2012.
19. Shilpa Gupta, Geeta Gujral, and Neha Aggarwal. Enhanced least significant bit algorithm for image steganography. *IJCEM Int. J. Comput. Eng. Manage.*, 15(4):40–42, 2012.
20. Jamie Hayes and George Danezis. Generating steganographic images via adversarial training. In *Advances in Neural Information Processing Systems*, pages 1954–1963, 2017.

21. Donghui Hu, Liang Wang, Wenjie Jiang, Shuli Zheng, and Bin Li. A novel image steganography method via deep convolutional generative adversarial networks. *IEEE Access*, 6:38303–38314, 2018.
22. Tapan Jain. Spatial domain steganography techniques and neural network based steganalysis with differential storage. In *2020 International Conference on Emerging Trends in Information Technology and Engineering (ic-ETITE)*, pages 1–4. IEEE, 2020.
23. Yan Ke, Min-qing Zhang, Jia Liu, Ting-ting Su, and Xiao-yuan Yang. Generative steganography with kerckhoffs' principle. *Multimedia Tools Appl.*, 78(10):13805–13818, 2019.
24. Moustafa M Kurdi, Imad A Elzein, and Akram M Zeki. Least significant bit (lsb) and random right circular shift (rrcf) in digital watermarking. In *2016 12th International Computer Engineering Conference (ICENCO)*, pages 111–116. IEEE, 2016.
25. Shamim Ahmed Laskar and Kattamanchi Hemachandran. Steganography based on random pixel selection for efficient data hiding. *Int. J. Comput. Eng. Technol.*, 4(2):31–44, 2013.
26. Tsung-Yi Lin, Michael Maire, Serge Belongie, James Hays, Pietro Perona, Deva Ramanan, Piotr Dollatŕ, and C Lawrence Zitnick. Microsoft coco: Common objects in context. In *European conference on computer vision*, pages 740–755. Springer, 2014.
27. Zinan Lin, Yongfeng Huang, and Jilong Wang. Rnn-sm: Fast steganalysis of voip streams using recurrent neural network. *IEEE Trans. Inf. Forensics Secur.*, 13(7):1854–1868, 2018.
28. Tomatśı PevnŒy, Tomatśı Filler, and Patrick Bas. Using high-dimensional image models to perform highly undetectable steganography. In *International Workshop on Information Hiding*, pages 161–177. Springer, 2010.
29. Birgit Pfitzmann. Information hiding terminology–results of an informal plenary meeting and additional proposals; *information hiding*, lncs 1174, 1996.
30. Rafia Rahim, Shahroz Nadeem, *et al.* End-to-end trained cnn encoder-decoder networks for image steganography. In *Proceedings of the European Conference on Computer Vision (ECCV)*, pages 0–0, 2018.
31. Kartik Sharma, Ashutosh Aggarwal, Tanay Singhania, Deepak Gupta, and Ashish Khanna. Hiding data in images using cryptography and deep neural network. arXiv preprint arXiv:1912.10413, 2019.
32. Gandharba Swain. A steganographic method combining lsb substitution and pvd in a block. *Procedia Comput. Sci.*, 85:39–44, 2016.
33. Gandharba Swain and Saroj Kumar Lenka. Classification of image steganography techniques in spatial domain: a study. *J. Comput. Science Eng. Technol. (IJCSET)*, 5(03):219–232, 2014.
34. Weixuan Tang, Bin Li, Shunquan Tan, Mauro Barni, and Jiwu Huang. Cnn-based adversarial embedding for image steganography. *IEEE Trans. Inf. Forensics Secur.*, 14(8):2074–2087, 2019.

35. Weixuan Tang, Shunquan Tan, Bin Li, and Jiwu Huang. Automatic steganographic distortion learning using a generative adversarial network. *IEEE Signal Process Lett.*, 24(10):1547–1551, 2017.
36. Dacheng Tao, Xiaoou Tang, Xuelong Li, and Xindong Wu. Asymmetric bagging and random subspace for support vector machines-based relevance feedback in image retrieval. *IEEE Trans. Pattern Anal. Mach. Intell.*, 28(7):1088–1099, 2006.
37. Toan Pham Van, Thoi Hoang Dinh, and Ta Minh Thanh. Simultaneous convolutional neural network for highly efficient image steganography. In *2019 19th International Symposium on Communications and Information Technologies (ISCIT)*, pages 410–415. IEEE, 2019.
38. Denis Volkhonskiy, Ivan Nazarov, and Evgeny Burnaev. Steganographic generative adversarial networks. In *Twelfth International Conference on Machine Vision (ICMV 2019)*, volume 11433, page 114333M. International Society for Optics and Photonics, 2020.
39. Chung-Ming Wang, Nan-I Wu, Chwei-Shyong Tsai, and Min-Shiang Hwang. A high quality steganographic method with pixel-value differencing and modulus function. *J. Syst. Software*, 81(1):150–158, 2008.
40. Ran-Zan Wang, Chi-Fang Lin, and Ja-Chen Lin. Image hiding by optimal lsb substitution and genetic algorithm. *Pattern Recognit.*, 34(3):671–683, 2001.
41. Xinyu Weng, Yongzhi Li, Lu Chi, and Yadong Mu. High-capacity convolutional video steganography with temporal residual modeling. In *Proceedings of the 2019 on International Conference on Multimedia Retrieval*, pages 87–95, 2019.
42. Eric Wengrowski and Kristin Dana. Light field messaging with deep photographic steganography. In *Proceedings of the IEEE Conference on Computer Vision and Pattern Recognition*, pages 1515–1524, 2019.
43. Duan Xintao and Liu Nao. Hide the image in fc-densenets to another image. arXiv preprint arXiv:1910.08341, 2019.
44. Guanshuo Xu. Deep convolutional neural network to detect j-uniward. In *Proceedings of the 5th ACM Workshop on Information Hiding and Multimedia Security*, pages 67–73, 2017.
45. Huan Xu, Jianjun Wang, and Hyoung Joong Kim. Near-optimal solution to pair- wise lsb matching via an immune programming strategy. *Inf. Sci.*, 180(8):1201–1217, 2010.
46. Cheng-Hsing Yang, Chi-Yao Weng, Shiuh-Jeng Wang, and Hung-Min Sun. Varied pvd+ lsb evading detection programs to spatial domain in data embedding systems. *J. Syst. Software*, 83(10):1635–1643, 2010.
47. Ru Zhang, Shiqi Dong, and Jianyi Liu. Invisible steganography via generative adversarial networks. *Multimedia Tools Appl.*, 78(7):8559–8575, 2019.
48. Zhuo Zhang, Guangyuan Fu, Jia Liu, and Wenyu Fu. Generative information hiding method based on adversarial networks. In *International Conference on Computer Engineering and Networks*, pages 261–270. Springer, 2018.

49. Nan Zhong, Zhenxing Qian, Zichi Wang, and Xinpeng Zhang. Steganography in stylized images. *J. Electron. Imaging*, 28(3):033005, 2019.
50. Jiren Zhu, Russell Kaplan, Justin Johnson, and Li Fei-Fei. Hidden: Hiding data with deep networks. In *Proceedings of the European Conference on Computer Vision (ECCV)*, pages 657–672, 2018.
51. Jun-Yan Zhu, Taesung Park, Phillip Isola, and Alexei A Efros. Unpaired image-to-image translation using cycle-consistent adversarial networks. In *Proceedings of the IEEE international conference on computer vision*, pages 2223–2232, 2017.

18

Human Emotion Recognition Intelligence System Using Machine Learning

Bhakthi P. Alva*, Krishma Bopanna N., Prajwal S., Varun A. Naik and Lahari Vaidya

Department of Electronics and Communications Engineering, Sahyadri College of Engineering & Management, Mangaluru, India

Abstract

The human voice is versatile and displays substantial emotional variances, allowing for a better understanding of human behaviour. This model is proposed to benefit the visually impaired population to successfully engage with people and socialize by recognizing their emotions. A speech emotion recognition (SER) system is now being developed, which is based on various classification models and feature extraction algorithms. Mel-frequency cepstrum coefficient (MFCC) characteristics are obtained from voice signals and these are utilized to train some classifiers. Feature selection (FS) has been used to obtain a relevant feature subset. A wide range of machine learning (ML) methods have been used for the emotion classification challenge. To initiate, the K-nearest neighbors (KNN) algorithm is used to detect seven distinct emotions. Their results will be compared to support vector machine (SVM) approaches and are commonly used in the domain of emotion recognition for voice signals. The Toronto emotional speech set (TESS) and knowledge extraction based on evolutionary learning (KEEL) databases comprise the experimental dataset. With the objective of more accurately identifying speech percepts based on emotions, this proposed study will investigate a speech emotion detection system that outperforms previous systems on the basis of data, FS, and methodology.

Keywords: Emotion recognition, SVM, KNN, MFCC

**Corresponding author*: bhakthiprashanth@gmail.com

Chandra Singh, Rathishchandra R. Gatti, K.V.S.S.S.S. Sairam, Manjunatha Badiger, Naveen Kumar S. and Varun Saxena (eds.) Modeling and Optimization of Signals Using Machine Learning Techniques, (349–360) © 2024 Scrivener Publishing LLC

18.1 Introduction

Communication is essential for expressing one's ideas. Speech is the most important aspect of human communication and a significant approach to human-computer interaction (HCI). Speech is enriched with emotions like happiness, sadness, disgust, anger, and many more depending upon one's mood. Emotions play a vital role in day-to-day interactions with other people and living organisms. Speech emotion recognition (SER) tries to recognize the emotions from a speech through various techniques and features. The speech signal comprises a huge variety of factors that convey emotional features. One of the major obstacles in emotion identification is determining which characteristics to employ. Many common aspects, such as energy, pitch, and formant, as well as certain spectrum features, have been obtained. More distinct and informative audio can be found in the TESS and KEEL. The model uses two English-language datasets, the TESS and KEEL. It has the potential to significantly cut the running time of learning algorithms. An SER system based on multiple classification algorithms and feature extraction methods was developed to carry out this study. From the speech signals, Mel-frequency cepstrum coefficients (MFCCs) and modulation spectral features are obtained and utilized to train various classifiers. Feature selection (FS) is used to select the most desired feature subset. For the emotion classification tasks, several machine learning (ML) approaches and methods have been used. Nowadays, SER is used widely for HCI, lie identification, automobile systems to support steering, smart training systems, sound extraction, protection, mobile communications, and human-machine interaction in homes, hospitals, stores, and others. SER's main purpose is to improve HCI. It would be used to monitor a person's physiological and psychological status in lie detectors. Lately, SER has found uses in clinical and forensic evidence.

18.2 Literature Review

Nishchay Parikh *et al.* (2021) [1] described the capability of recognizing emotions with an accuracy of 78.65% on the RAVDESS dataset with the help of feature extraction techniques that extracts aspects like MFCC chroma and the Mel spectrogram. The model is first trained using the Ryerson audio-visual database of emotional speech and song (RAVDESS) dataset, ML algorithm, and the MLP (multi-layer perceptron) classifier. The system can operate in one of two modes: real-time input or input from a previously stored file. The MLP classifier is used to train and predict emotions like calmness, happiness, fearfulness, and disgust. The real-time functionality was implemented with the help of the PyAudio library, which records audio and extracts and predicts features from it.

Esma Mansouri-Benssassi Juan Ye (2021) [2] described how spiking neural networks (SNNs) have been effectively used for FER and SER tasks using unsupervised learning on different types of information: static pictures in FER and time series data in SER. Through cross-dataset examination, cutting-edge methodologies result in greater recognition accuracy. It displays SNN's generalization capacity as well as subject independence for both SER and FER.

Issa *et al.* (2020) [3] explained the description of various fields where emotion recognition can be used. It also covers the usage of three separate audio datasets: RAVDESS, emotional speech database (EMO-DB), and the interactive emotional dyadic motion capture (IEMOCAP), which are used to compare actual audio signals even without the requirement for pictorial representations. This process also involves feature extraction to provide a better-trained model and various spectral representations for the same. The study also mentioned the accuracy of training data for each dataset and compared it with previous approaches.

Akshat Agrawal and Anurag Jain (2020) [4] noted that statistical classification approaches such as principal component analysis (PCA) and linear discriminant analysis (LDA) are used to extract prosodic and acoustic information from speech in order to identify emotions. For the examination of one's emotional state, the KNN and the Naive Bayes classifier (NBC) algorithms are utilized. With the use of PCA and LDA, the local and global aspects of emotions are retrieved in order to improve the outcomes. After applying KNN and NBC to these attributes, it is clear that both algorithms provide distinct results, although the KNN outperforms the NBC in recognizing emotions (sad, happy, and angry).

N. Vryzas *et al.* (2020) [5] discussed ASEDD in training datasets with the CNN (convolution neural network) architecture. CNN input vectors can be one-dimensional signals as a result of the pulse code modulation or two- dimensional signals as a result of the spectro-temporal transformation and Mel-scale coefficients. The suggested CNN model outperformed the baseline SVM model in terms of accuracy, coming in at 69.2% (60.8%). Data augmentation had little effect on the outcomes, although it did enhance robustness [6–8].

Xianxin Ke *et al.* (2018) [9] described how the characteristics are separated into two categories: acoustic and statistical features, which are generated using an emotional model built from SVM and artificial neural network (ANN). The Chinese Academy of Sciences (CASIA) Chinese emotional corpus is utilized to investigate the major speech technologies and the influence of feature reduction on SER. To minimize the dimensions of characteristics, an approach like the principal component analysis PCA is utilized. The models developed using SVM and ANN are compared with and without PCA. As a consequence, the SVM model outperforms the ANN-based model in terms of accuracy (46.67% vs. 76.67% with PCA). Second, the benefits are shown when feature dimension reduction is performed.

Taherkhani *et al.* (2018) [10] described the existing techniques that can have problems generalizing across datasets, and that they perform poorly on data that have not been seen. In addition to these two issues, most present techniques rely on well-annotated training data to learn distinguishing features that may be used to distinguish between different types of emotions [11].

Wang *et al.* (2015) [12] discovered that the first and second order changes in harmony features have a significant effect in SER. The output from different emotional states is seen by variation in first-person shooter (FPs). The study was also extended to a speaker-independent emotion recognition using SVM and the Bayesian classifier. It shows that the recognition of speech emotion is much more effective using FP and can be improvised by combining FPs and MFCCs.

18.3 Problem Statement

For the visually challenged, an emotion recognition system is being developed. Blind persons are unable to detect the facial emotions of others who interact with them. These individuals might be given a device that recognizes people's emotions. The global population of persons with visual impairments is relatively large, and the majority of them have impaired vision. They have difficulty interacting with others. An innovative system is introduced to aid visually impaired persons in identifying the expression or emotions of the confronting person and therefore enabling better communication. In this proposed method, a blind person uses a device to gather voice signals from a person speaking with them via a microphone, and the identified emotion is returned to the user. This technology recognizes emotions such as happiness, sadness, and surprise, allowing the blind individual to engage in successful social contact. This technology recognizes all of the fundamental emotions and so serves as an effective and helpful aid for the blind.

18.4 Methodology

To implement a speech emotion detection system, the ML model is used. The implementation procedures are similar to other ML projects, with additional precise techniques to enhance the system performance [13, 14].

Figure 18.1 depicts a schematic representation illustrating a visual representation of the speech recognition process. The speech signal comprises a huge variety of factors that convey emotional features. One of the major obstacles in emotion identification is determining which characteristics to employ. This proposed work has obtained many common aspects, such as energy, pitch, formant, and certain spectrum features like linear prediction coefficients (LPC) and MFCCs. The initial stage is to acquire voice samples from the user. Following that, the voice samples that were gathered will go through the feature extraction procedure. In this process of extracting features from speech segments, the MFCC representation method is applied, which takes into account the spectral properties of voice signals. These are best for voice recognition since they consider visual recognition responsiveness to frequencies. The Fourier transform and power spectra of each frame have been computed and converted to the Mel-frequency scale. The Mel log energies DCT is employed and the classifier's MFCC targets were given by the first 12 DCT coefficients. The operation for computing MFCC values is depicted in Figure 18.2.

This feature's primary goal in ML is to lower the count of aspects required to characterize a set of data in order to enhance the performance of a learning algorithm on a specific function. The goal is to maximize

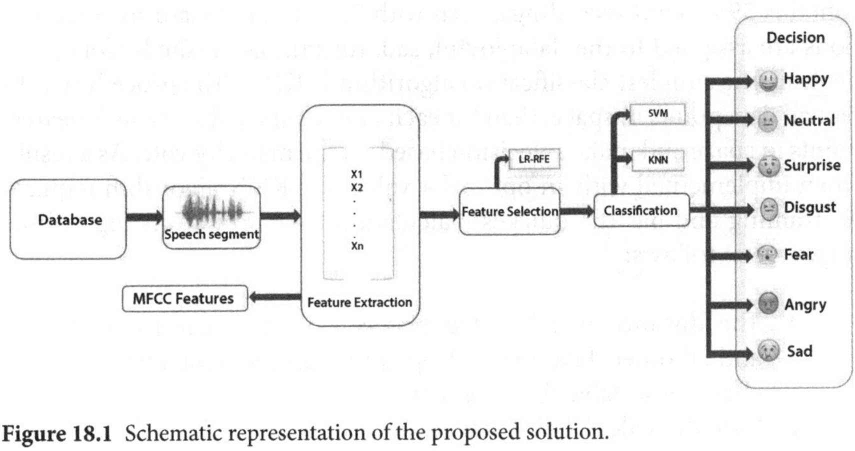

Figure 18.1 Schematic representation of the proposed solution.

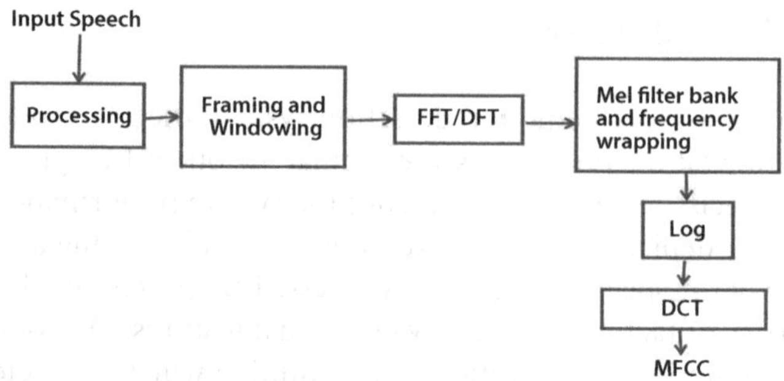

Figure 18.2 Schema of the MFCC extraction.

the classification performance in a particular function for a specific learning process, which decreases the number of features in order to affect the ultimate classification model. The goal of FS is to select a subset of appropriate characteristics from the original set related to a specific relevance assessment criterion, which usually leads to a higher recognition rate. It has the potential to significantly reduce the time required for learning algorithms to run. The retrieved features will be subjected to a selection procedure using the datasets listed as follows. The model employs two English-language datasets, TESS and KEEL, which contain more distinct and realistic audio [15].

The collection includes 2,800 spoken word sound recordings grouped into seven basic emotional categories: joyful, sad, furious, shocked, afraid, nasty, and neutral. It is a tape in which actors aged 64 and 26 execute the dictation. KEEL is a data repository developed by supervised learning professionals from all around the world. The emotion for the speech dataset contains 593 sound recordings, each with 72 extracted features. Six emotions are assigned to the data: joyful, sad, furious, astonished, afraid, and neutral. The simplest classification algorithm is KNN. The procedure is to plot all data points in space, then for each new sample, observe its k nearest points in space and make a decision based on the majority vote. As a result, when implemented with an optimal k value, the KNN algorithm requires no training and has the quickest calculation time. The KNN algorithm's stages are as follows:

1. The distance is calculated between a particular instance and all other data points. Depending on the issue instance, choose a suitable distance metric.
2. Sort the calculated distances in ascending order. Observe the nearest k points depending on the value of k.

3. Determine the majority class among the k points and designate it as the projected class.

Support vector machines (SVMs) are supervised algorithms that may be used to solve classification and regression issues. Support vectors are coordinate points in space that are produced by combining the attributes of a data point.

SVM is best suited for high-dimensional data. Finding the hyperplane is a critical part of implementing the SVM algorithm. When selecting the appropriate hyperplane, two requirements must be satisfied in the order specified. The hyperplane appropriately categorizes the classes. Increase the distance between the hyperplane and the nearest data points. The kernel trick technique in SVM gives extra features for low-dimensional data, allowing the detection of the hyperplane [16, 17]. Following the categorization procedure, a choice will be made based on seven emotions: happy, angry, disgusted, fearful, neutral, surprised, and sad. For all the extreme conditions, certain actions can be carried out to neutralize the thoughts. One such approach is the use of an emotional chatbot which acts according to the emotion of the user. When such feelings are detected the designed chatbot can attempt to convince or console the user. Additional features like the SMS notification can be triggered in case of emergency for the first three contacts of the user. This is one such approach that can be generalized for different scenarios or conditions to deal with harsh emotions.

18.5 Results

The implementation results can be used to draw a number of results. The conclusions are depicted in Figure 18.3 while the confusion matrix is shown in Figure 18.4. The efficiency scores of the various approaches improved. The first technique performed well with a best value of 70% using the SVM algorithm, whereas the new approach performed well with a best score of 96% utilizing the KNN and SVM algorithms.

18.6 Applications

- The suggested approach helps visually impaired persons communicate more effectively during social encounters. Patients might use an emotional analyzer to become more conscious of their emotional states.

```
Epoch 78/1000
71/71 [==============================] - 20s 284ms/step - loss: 0.0563 - accuracy: 0.9806 - val_loss: 0.0780 - val_accuracy: 0.
9603
Epoch 79/1000
71/71 [==============================] - 20s 285ms/step - loss: 0.0450 - accuracy: 0.9837 - val_loss: 0.1699 - val_accuracy: 0.
9484
Epoch 80/1000
71/71 [==============================] - 20s 286ms/step - loss: 0.0721 - accuracy: 0.9718 - val_loss: 0.0720 - val_accuracy: 0.
9722
Epoch 81/1000
71/71 [==============================] - 20s 284ms/step - loss: 0.0632 - accuracy: 0.9780 - val_loss: 0.0721 - val_accuracy: 0.
9762
Epoch 82/1000
71/71 [==============================] - 20s 285ms/step - loss: 0.0670 - accuracy: 0.9757 - val_loss: 0.0519 - val_accuracy: 0.
9802
Epoch 83/1000
71/71 [==============================] - 20s 284ms/step - loss: 0.0598 - accuracy: 0.9788 - val_loss: 0.1404 - val_accuracy: 0.
9524
Epoch 84/1000
71/71 [==============================] - 20s 287ms/step - loss: 0.0557 - accuracy: 0.9806 - val_loss: 0.1819 - val_accuracy: 0.
9325
Epoch 85/1000
71/71 [==============================] - 20s 287ms/step - loss: 0.0673 - accuracy: 0.9780 - val_loss: 0.0497 - val_accuracy: 0.
9841
Epoch 86/1000
71/71 [==============================] - 20s 284ms/step - loss: 0.0433 - accuracy: 0.9859 - val_loss: 0.1786 - val_accuracy: 0.
9484
Epoch 87/1000
71/71 [==============================] - ETA: 0s - loss: 0.0563 - accuracy: 0.9828Restoring model weights from the end of the b
est epoch: 57.
71/71 [==============================] - 20s 285ms/step - loss: 0.0563 - accuracy: 0.9828 - val_loss: 0.0984 - val_accuracy: 0.
9683
Epoch 87: early stopping
Training completed in time: 0:30:17.639995
```

```
results = model.evaluate(X_test, y_test, verbose=0)

for name, value in zip(model.metrics_names, results):
    print(f"{name:10} {value:.4f}")
```

```
loss       0.0800
accuracy   0.9679
```

```
y_proba = model.predict(X_test)
y_pred = np.argmax(y_proba, axis=-1)
y_true = np.argmax(y_test, axis=-1)

np.mean(y_true == y_pred)
```

```
0.9678571428571429
```

Figure 18.3 Evaluation of results.

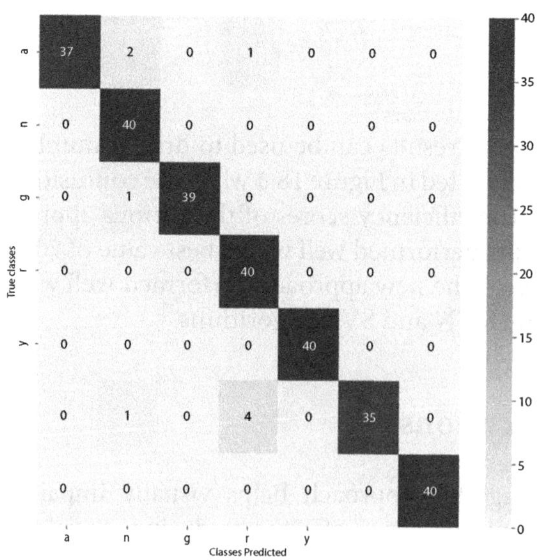

Figure 18.4 The confusion matrix.

- The approach aids in a better understanding of stressful or difficult situations, as well as assisting clinicians in suicide risk prediction and therapeutic care.
- It can help cognitive science departments that rely on human contact.
- It may be used in lie detectors to track a person's psychophysiological condition.

18.7 Conclusion

A new age of automation has emerged as a result of increasing research in the fields of artificial intelligence and ML. The majority of these speech-activated devices are controlled by the person's voice. Substantial advantages may be gained over current methods if computers could read the author's sentiment in addition to recognizing speech. A voice emotion detection approach is commonly used in computer-based training programs, automated help desk dialogues, complete diagnostic tools for treatment, and automatic translation systems. In this chapter, the methodologies for constructing a speech emotion identification system were thoroughly examined, and a variety of tests were conducted to determine the impact of each stage. It was initially difficult to implement a well-trained model due to the limited number of publicly available speech datasets. Following that, several other methods to extract the feature had been published in prior papers, and determining the best method needed doing numerous experiments. It is conceivable to conclude that an integrated feature space would provide a greater recognition accuracy than a particular feature.

18.8 Future Work

The proposed chapter can be optimized in terms of performance, precision, and usefulness for any further advancements. The system can recognize responses like anxiousness and mood swings in addition to emotions. The creation of an emotional robot's sarcasm detection system is a difficult task. Sarcasm would be a more challenging task to solve than emotion recognition as sarcasm cannot be detected just by listening to the speaker's speech or intonation. To detect probable sarcasm, sentiment detection employing language can be combined with speech emotion detection. As a result, several applications of a speech-based emotion identification systems might arise in the future.

References

1. Nishchay Parikh, Khyati Mistry, Yashvi Bhavsar, Abdul Basit Hakimi and Archana Magare, "Real Time Speech Emotion Recognition using Machine Learning", *Int. Res. J. Eng. Technol.,* Volume 08 Issue 06, 2021.
2. Mansour Benssassi, Esma & Ye, Juan, "Generalisation and Robustness Investigation for Facial and Speech Emotion Recognition using Bio-Inspired Spiking Neural Networks", *Int. J. Soft Comput.,* Volume 25 Issue 2, 2021.
3. Issa, D., Demirci, M., & Yazici, A, "Speech Emotion Recognition with Deep Convolutional Neural Networks", *Int. J. Biomed. Signal Process. Control,* Volume 59 Issue 1, 2020.
4. Akshat Agrawal and Anurag Jain, "Speech Emotion Recognition of Hindi Speech using Statistical and Machine Learning Techniques", *J. Interdiscip. Math.,* Volume 23 Issue 01, Jan 2020.
5. N. Vryzas, L. Vrysis, M. Matsiola, R. Kotsakis, C. Dimoulas and G. Kalliris, "Continuous Speech Emotion Recognition with Convolutional Neural Networks", *J. Audio Eng. Soc.,* Volume 68 Issue 1, 2020.
6. Koduru, A., Valiveti, H.B. and Budati, A.K, "Feature Extraction Algorithms to Improve the Speech Emotion Recognition Rate", *Int. J. Speech Technol.,* Volume 23 Issue 1, 2020.
7. Lin, Jiang & Tan, Ping & Junfeng, Yang & Liu, Xingbao & Wang, Chao, "Speech Emotion Recognition using Emotion Perception Spectral Feature", *Special issue of 13th International Conference on Parallel Processing and Applied Mathematics*, Volume 33 Issue 11, 2019.
8. Mustaqeem and Soonil Kwon, "A CNN-Assisted Enhanced Audio Signal Processing for Speech Emotion Recognition", *Int. J. Signal Process.,* Volume 20 Issue 183, 2019.
9. Xianxin Ke, Yujiao Zhu, Lei Wen, and Wenzhen Zhang, "Speech Emotion Recognition Based on SVM and ANN", A Review," *Int. J. Mach. Learn. Comput.,* Volume 8 Issue 3, 2018.
10. Taherkhani, A. Belatreche, Y. Li and L. P. Maguire, "A Supervised Learning Algorithm for Learning Precise Timing of Multiple Spikes in Multilayer Spiking Neural Networks", In *IEEE Trans. Neural Networks Learn. Syst.*, Volume 29 Issue 11, 2018.
11. S. Zhang, S. Zhang, T. Huang and W. Gao, "Speech Emotion Recognition Using Deep Convolutional Neural Network and Discriminant Temporal Pyramid Matching", In *IEEE Trans. Multimedia*, Volume 20 Issue 6, 2018.
12. Wang, K., An, N., Li, B. N., Zhang, Y., & Li, L, "Speech Emotion Recognition using Fourier Parameters", *Int. J. Trans. Affect. Comput.,* Volume 6 Issue 1, 2015.
13. Karkera, T., & Singh, C. (2021). Autonomous bot using machine learning and computer vision. *SN Comput. Sci.*, *2*(4), 1-9.

14. Shetty, S. H., Shetty, S., Singh, C., & Rao, A. (2022). Supervised Machine Learning: Algorithms and Applications. *Fundamentals and Methods of Machine and Deep Learning: Algorithms, Tools and Applications*, 1-16.
15. Singh, C., Sairam, K. V. S. S. S. S., & MB, H. (2018, August). Global fairness model estimation implementation in logical layer by using optical network survivability techniques. In *International Conference on Intelligent Data Communication Technologies and Internet of Things* (pp. 655-659). Springer, Cham.
16. Badiger, M., Kumara, V., Shetty, S. C., & Poojary, S. (2022). Leaf and skin disease detection using image processing. *Global Transit. Proc.*, 3(1), 272-278.
17. Badiger, M., and J. A. Mathew. "Tomato Plant Leaf Disease Segmentation and Multiclass Disease Detection Using Hybrid Optimization Enabled Deep Learning." *J. Biotechnol.* (2023): S0168-1656.

19

Computing in Cognitive Science Using Ensemble Learning

Om Prakash Singh

Department of Computer Science and Engineering, Vidya Vihar Institute of Technology, Purnea, Bihar, India

Abstract

Cognitive science is deeply ingrained in artificial intelligence (AI). The phenomenon of thinking and acting either humanly or logically is how it is defined as an AI. Many academics have been concentrating on human activity recognition using ensemble learning in recent years. The dataset of human activities is unbalanced. This collection is a reflection of all human thoughts and behaviors that effectively forbids the coexistence of cognitive science and AI. The term "unbalanced dataset" refers to a set of data where at least one class has more instances than the other classes combined. The goal of machine learning with such an unbalanced data collection is to continuously learn from the skewed data distribution. Many common machine learning techniques, such as linear regression, k-nearest neighbor, and decision trees, perform poorly in terms of accuracy with this kind of dataset (77%, 62%, and 89% accuracy, respectively). Boosting is a general machine learning technique for raising any learning algorithm's current accuracy. It is built on the simple idea that a group of simple classifiers working together can perform significantly better than a single classifier working alone. The foundation of ensemble learning is accelerated learning techniques. Simple classifiers are weak learners, while ensemble learning uses weak learners that have been purposefully assembled and created to address a specific computational intelligence challenge. The current effort focuses on human-computer interaction to find a technique to identify human activity and apply ensemble learning from an individual level to produce a more accurate prediction of the whole that fully mirrors the social-cognitive science in a very efficient computational approach.

Email: opsingh6612@gmail.com

Chandra Singh, Rathishchandra R. Gatti, K.V.S.S.S.S. Sairam, Manjunatha Badiger, Naveen Kumar S. and Varun Saxena (eds.) Modeling and Optimization of Signals Using Machine Learning Techniques, (361–382) © 2024 Scrivener Publishing LLC

***Keywords*:** Cognitive science, ensemble learning, boosting algorithm, human-computer interaction, skewed data distribution

19.1 Introduction

The study of human processes, including reasoning, communication, language, attitudes, emotions, perceptions, attention, and memory, is called cognitive science. The fusion of cognitive science and artificial intelligence (AI), particularly machine learning, creates a live foundation for a profound understanding of human cognition and communication with the application of technical and creative machine learning. There are effective guidelines, approaches, and algorithms that may be used to nudge people to examine their own behaviors and perspectives in order to perform better. Thus, various aspects of cognition are covered by cognitive computing [1, 2].

The majority of real-world issues relate to datasets with non-uniform data distributions, where at least one class has fewer samples than the other classes. Minor classes are those with fewer samples than the first class, and majority classes are those with more samples. Either binary classification or multiple classification options are available. There are only two classes in a binary classification, one of which is minor and the other major, but a multiclass classification comprises a number of minor and major classifications. The data in this case are unbalanced. The distributions of the data are now either biased or skewed. So, either an unbalanced dataset with a skewed data distribution exists, or an unbalanced dataset with a skewed data distribution [3, 4].

The majority of common machine learning algorithms struggle to perform in this situation. They will either tilt towards entirely or be biased toward majority classes, demonstrating greater accuracy on majority classes and lower performance on minority classes. Despite the fact that the majority class has greater knowledge, the minority class may also possess extremely valuable knowledge. When there are multiple classes in the data collection, there are more challenges in multiclass consideration. Therefore, to handle the data identifying human activity that is then used for ensemble learning based on boosting, we investigate the multiclass unbalanced dataset with a skewed data distribution [5, 6].

The current study is divided into three sections: first, human activity recognition; second, data organization approach; and third, how to handle the multiclass imbalanced dataset with a skewed data distribution. The second step is to apply boosting-based ensemble learning, which entails useful descriptions of ensemble learning and its methodologies as well as boosting methods with their strategic and effective utilities from the individual level. The third step is to handle human activity features computability to handle human behaviors with ensemble learning, data processing, and feature mapping to obtain a more accurate performance of the whole that truly mirrors [7, 8].

19.2 Recognition of Human Activities

The goal of human-centered computing is to comprehend human behavior before integrating it with computing tools like smartphones and computer systems. Due to its numerous uses in relation to social reforms for people, the recognition of human actions occurring in an environment is becoming more crucial [9, 10].

Three traditional perspectives can be used to characterize human activities [11]. The first is direct perceptual inputs, which are the sources from which people perceive daily activities. The "Moving Lights Display" experiment [12] carries it out. This experiment demonstrates the ability to differentiate between walking, running, and climbing stairs using the intensity pattern of light. These exercises are pretty simple in nature, the process is of capturing low-level features from video or signals and extracting them. For the goal of identifying human activity, these attributes can be used in boosting-based ensemble learning. Additional semantic information is not necessary. It is hence context-free. The second way to describe human activity is with a context-sensitive activity descriptor. Because it is context-sensitive, semantically relevant activity descriptors are needed. Each activity's interpretation necessitates a few notations. Walking is an example of when it is necessary to note the feet, the ground, and the difference between the left and right legs. Boosting-based ensemble learning successfully addresses this contextual aim [13]. A partial ordering of action subsequences that adhere to certain time restrictions is the third way to describe human activity. Walking, for instance, might be thought of as a series of sequential steps. It is utilized to gauge an activity's frequency in relation to its constituent subsequences. The constituent subsequences are viewed as a weak classifier of the overall activity in boosting-based ensemble learning. Our main goal is to create a dataset of human behaviors for additional processing. In [14, 15], a protocol of actions for the identification of human activity is provided. Simply said, this is a mathematical depiction of three different ways to characterize human activities. This is a collection of 16 separate bodily movements that were captured in a predetermined period of time, as in the following Table 19.1.

For better cognitive behavior, we add some other protocols that reflect the more accurate human behavior with one's mental status. This set of protocols becomes as follows (Table 19.2).

There are many public data domains [16, 17] with a smaller number of protocols, but these cannot fulfill our requirements.

Table 19.1 Movements captured in a predetermined period of time.

Activity number	Activity type	Activity	Time (sec)
0	Static	Start (standing position)	0
1	Static	Stand1	15
2	Static	Sit1	15
3	Static	Laydown1	15
4	Static	Stand2	15
5	Static	Sit2	15
6	Static	Laydown2	15
7	Dynamic	Walk1	15
8	Dynamic	Walk2	15
9	Dynamic	Walk upstairs1	12
10	Dynamic	Walk upstairs2	12
11	Dynamic	Walk upstairs3	12
12	Dynamic	Walk downstairs1	12
13	Dynamic	Walk downstairs2	12
14	Dynamic	Walk downstairs3	12
15	Dynamic	Stop	0

Table 19.2 Protocols.

Activity number	Activity type	Activity	Time (sec)
0	Static	Start (standing position)	0
1	Static	Stand1	15
2	Static	Sit1	15
3	Static	Laydown1	15
4	Static	Stand2	15
5	Static	Sit2	15
6	Static	Laydown2	15
7	Static	Left hand on forehead1	15
8	Static	Right hand on forehead1	15
9	Static	Both hands on backhead1	15
10	Static	Both hands on the waist	15
11	Static	Wrap hand	15
12	Dynamic	Walk1	15
13	Dynamic	Walk2	15
14	Dynamic	Hand movement on face1	12
15	Dynamic	Hand movement on face2	12
16	Dynamic	Hand movement on head1	12
17	Dynamic	Hand movement on head2	12
18	Dynamic	Walk upstairs1	12
19	Dynamic	Walk upstairs2	12
20	Dynamic	Walk upstairs3	12
21	Dynamic	Walk downstairs1	12
22	Dynamic	Walk downstairs2	12
23	Dynamic	Walk downstairs3	12
24	Dynamic	Stop	0

19.3 Methodology

Whatever we use to capture all such human activities is not the matter here; it may be smartphones, tablets, or surveillance cameras throughout the environment of consideration. However, we can consider the smartphone with a compass, accelerometer, and gyroscope for our ease of use. The compass shows the pointing direction of the phone [18]. The accelerometer and gyroscope sensors measure the rate of changes in movements. The accelerometer is used to detect the orientation of the phone that measures the linear acceleration of movements, and the gyroscope measures the angular rotational velocity. The orientation involves the determination of moving upward or downward, flying horizontally, or angling downward with its portrait and landscape. Actually, the gyroscope adds an additional dimension to the information provided by the accelerometer by tracking the rotation. Therefore, the accelerometer measures the directional movement of the device, while the gyroscope measures the twist in the device. Hence, we have a measured value of direction with gravitational pull to determine the angle at which it tilted with respect to the earth and the directional movement and twist in the individual part of the body as a whole. We commonly obtain a noisy or clean output from the accelerometer, which is responsive, but when we combine the three-dimensional accelerometer and gyroscope, we have a clean and responsive output in an instant [19, 20].

19.3.1 Dataset Organization

The outcome of the human activity recognition must be organized in a very proper way as a well-defined dataset, which is used as samples of classes for boosting-based ensemble learning, some as training data and others as testing data or minority and majority classes. There will be two or more classes. One class may have samples that have more instances than the other classes because of this. Therefore, these data are imbalanced. The imbalanced dataset classification tries to learn from the skewed data distribution. In most of the real-world classification learning-based problems like medical, text, fraud, and fault detections, cultural modeling is usually imbalanced. Therefore, learning from a skewed dataset or data distribution becomes very important, since the symmetric data distribution is not possible with such a huge dataset coverage. Mode-, mean-, and median-like mathematical calculations are applicable in handling imbalanced data with a skewed data distribution that is supported well by a skew. In such data collection, both the left or the negative and the right or the positive skews are easily visible.

19.3.2 Handling the Multiclass Imbalanced Dataset with a Skewed Data Distribution

Whenever the task of handling the multiclass imbalanced dataset with a skewed data distribution is required, then data-level, algorithm-level, hybrid, and boosting methods show their suitability. As we know, imbalanced data are transformed into balanced data among classes for training and testing purposes. Yang *et al.* [21] indicated that resampling techniques are used for this transformation in data-level methodologies. Three methods of resampling are used in this case: undersampling, oversampling, and hybrid sampling. The undersampling method takes samples from the major classes until the equilibrium between the major and minor classes is broken. Thus, the training process can be carried out without any incident. The oversampling technique, on the other hand, actually duplicates existing samples or adds more samples to small classes. Another type of oversampling technique, known as the synthetic minority oversampling technique, fixes the issue connected with using a straightforward oversampling technique, and Liu *et al.* [22] created synthetic samples from minor classes by linearly combining two samples from the minor class. Additionally, under- and oversampling techniques come in two varieties: random and directed, such as random under- and oversampling, and directed under- and oversampling. The random undersampling randomly removes a sample from the dataset that represents the majority classes, and the random oversampling randomly creates additional samples for the minority classes in the dataset. On the other hand, informed decision-making is made on the replacement of samples in directed oversampling as opposed to directed undersampling.

While oversampling techniques increase the amount of the training dataset and hence lengthen training time, they also improve the issue related to run time and storage, making them more susceptible to overfitting. The main drawbacks of these two approaches, however, are the removal of valuable samples or the addition of meaningless samples to databases. Authors have argued that hybrid sampling should be employed anytime multiple sampling techniques are used to address the shortcomings of each sampling technique. The sampling techniques can take into account the unique characteristics of classes and how they relate to one another in multiclass datasets. The handling of multiclass data occasionally involves the usage of class decomposition algorithms [23].

The two most popular methods for class decomposition are one-versus-one and one-versus-all, which are comparable to one-to-one mapping and one-to-many mapping. The data points of one class are taken into account with the data points of the other classes in a one-versus-one approach. Because a binary classification problem might arise between any two classes, N classification problems are established for datasets with N classes, while N*(N-1)/2 problems are introduced for one-versus-all schemes. These methods offer the advantage of breaking the difficulties down into a number of two-class problems, but they also have the drawback of sacrificing a balanced performance across all classes and rejecting a global perspective on multiclass problems.

The strategies used at the algorithmic level are centered on eliminating learner bias against the major classes. Rokach and Lucey *et al.* [24, 25] present the most popular cost-sensitive learning strategy, in which the current learner is compelled to correctly classify the minor class samples by allocating a large cost to the misclassified minor class samples. Minor class samples have a larger misclassification cost than major class samples. Despite all, determining the cost and value is a difficult undertaking because it depends on numerous variables that have trade-off relationships.

To address the unbalanced data, the hybrid approach combines data-level and algorithm-level classifiers or ensemble learning classifiers. According to Maheshwari *et al.* [26], the major objective of the ensemble is to discover a performance that is more predictive than that of a single model. Its primary flaw is the creation of several classifiers, which increases computational complexity.

Bagging and boosting are the ensemble classifier techniques that are most frequently utilized. In [27], the original training dataset was separated into N equal-sized subgroups using bagging, and each subset produced a classifier that was used to create the final classifier. According to [28], the boosting method creates new weak learning models iteratively to produce a huge number of weak learners, who are then combined to produce a strong learner who performs more accurately than any of the weak learners. Another ensemble learning approach uses random subspaces to handle multiclass imbalanced data.

According to research by Galar *et al.* and Ramentol *et al.* [29, 30], all of the discussed methods—sampling, cost-sensitive, bagging, boosting, and random subspace—have been shown to be successful in solving the problem of imbalanced data. Ganganwar [31] suggested four additional supports to handle multiclass imbalanced data with a skewed distribution.

For a multiclass unbalanced dataset with a bias and skewed data distribution, the boosting performance is examined using random oversampling and class decomposition. Applying ensemble learning to each individual subproblem will help the binary decomposition to perform better.

We can also manage multiclass imbalanced datasets by combining ensemble learning and the one-versus-one decomposition strategy, or by breaking the original dataset into a number of simplified binary problems and using an ensemble classifier designed specifically for such problems and coupling the output of each classifier. The new ensemble margin-based technique also uses the new margin sample concept to handle such important datasets.

The final technique for handling multiclass imbalanced datasets with a skewed data distribution is the most well-known type of ensemble learning, that is the high citation count method, which builds a strong classifier out of a number of weak classifiers by learning the weak classifiers sequentially and reducing the errors of all previously discussed technique-based classifiers. However, different boosting techniques fall under either data-level or algorithm-level approaches. Due to its close relationship to ensemble learning, we briefly cover boosting in the previous sections that deal with data management.

19.4 Applying the Boosting-Based Ensemble Learning

The three classical ways to view the characterization of human activity are as direct perceptual inputs, context-sensitive activity descriptors, and a set of partial ordering of action subsequences, which show that boosting-based ensemble learning is an efficient tool for detecting human activities like walking, running, and climbing stairs with their contextual purpose and consider the constituent's subsequences as a weak classifier of the overall activity. Therefore, we have a fundamental ground for utilizing the ensemble learning capabilities strategically, effectively, and efficiently.

19.4.1 Ensemble Learning

To address a computational intelligence problem, many models, such as classifiers or experts, are created and merged in the ensemble learning process. Initially, it was created to lessen the variance and boost the accuracy of an automated decision-making system. Its primary goal is to enhance a model's feature selection, classification, error correction, function approximation, handling of class imbalanced data, and prediction-like performance, and hence lessen the risk of selecting an unintentionally subpar model. The ensemble learning method has now demonstrated its effectiveness and adaptability in a wide range of problem domains and real-world issues.

19.4.1.1 Development of Ensemble Learning

The first instance of ensemble learning is in the work of Dernbach *et al.* [32] on dividing the feature space with classifiers. In 1990, the introduction of the notion of neural network ensembles was seen [33]. They recommended using an ensemble of uniformed neural networks to enhance classification performance. From 1991 to 2005, various ensemble-based algorithms, with their specifications for solving problems in the real world, appeared. These include staked generalization, bagging, random forest, composite classifiers, consensus aggregation, classifier fusion, expert mixture, multiple classifiers, dynamic classifier selection, and neural network committee. They are diverse from one another and differ in the manner the ensemble members are generated specifically, the rule for making the ensemble decision, and the choice of training data for individual classifiers.

Classifier fusion or classifier selection is used to merge the ensemble members or classifiers. Kuncheva (2002) [12] noted that as classifiers are combined, switching between selection and fusion bocomes a significant milestone. Each classifier used in the classifier selection process is trained as a local expert in the feature space, and each time a new instance is created, the classifier is trained using data that is close to the local expert at that location to aid in the decision-making process. Classifier fusion is applied whenever the classifiers trained throughout the whole feature space are integrated to create a composite classifier with little variance or error.

Boosting serves as an illustration of classifier fusion. Either labels or output specific to a certain class are used in this combination to create a composite classifier. The outputs of each constituent classifier are first normalized to [0, 1] for this purpose. The support that the classifier provides to each class is represented by the interval [0, 1]. This interpretation leads to algebraic combination rules such as Dempster-classifier Shafer's fusion, simple or weighted voting, summation or product, and minimum or maximum. Finally, the following can be used to create an effective and efficient ensemble learning system:

1. Diversity,

2. Training of the member classifier, and
3. Combining the trained member classifier.

The combination of them actually has no relevance when all of the individual classifiers produce the same output because the output of the combination is identical to that of its constituents. As a result, data selection or sampling is necessary for ensemble learning to produce diversity. There are various ways to produce diversity, but we use a unique subset of training data frequently used in real-world settings. This is also a sample of data taken from a larger pool. Although this sample is supported by the unbalanced datasets described in Section 2.2, both develop a highly diverse strategy to get the most out of the ensemble [34, 35].

The individual ensemble member is trained using a very effective mechanism at the heart of the ensemble learning system. Our focus is on the later, more in-depth discussion of boosting techniques. Combining the learned member classifier is the final step. There are a number of methods, such as simple or weighted voting, summation or product, Bayes classifier, minimum or maximum, and Dempster-classifier Shafer's fusion, but the optimum method is chosen based on the types of ensemble member classifiers. Because, for instance, if the support vector machine produces discrete-valued labels, the combining rule might either be simple or weighted.

19.4.1.2 Computational Justification of Ensemble Learning

The ensemble learning system actually functions as a decision mechanism. Our regular activities parallel one another. We frequently seek advice from others before making decisions because we believe that past decisions can vary in accuracy from one to the next. There is no need for an ensemble-based system if there is a consultant or expert whose predictions are always accurate. It affirms the truth that nobody on earth is perfect. In order to obtain perfection, it is therefore always preferable to create a group of somewhat perfect things. The classification error is a composite of bias and variance, which are its two constituents.

But we are in charge of both. The average-based technique reduces variance. Smoothing is another term for this reduction in variation. The goal of the ensemble learning system is to create a number of classifiers with comparable or fixed biases and then to average their outputs. Thus, the variance is decreased. Now, the average technique is not always appropriate for merging ensemble members or weak learners, and it is not guaranteed that adding the results of all classifiers will improve the classification performance over the classifier in the ensemble that performs the best. Here, it is similarly difficult to explain how we can know in advance which classifier will perform the best in an ensemble, eliminating the necessity for an ensemble if we have a better performer.

19.4.2 Boosting Methods

Boosting is the most popular method of ensemble learning that makes a complex classifier a strong learner by combining the numbers of simple classifiers as the weak learner. The weak and strong learners are the core concepts of boosting methods. These concepts are based on probability approximation correct learning [35]. The strong learner produces a good classifier with a high probability and the weak learner produces a simple classifier with a small probability. It has a serial learning nature which is excellent in approximation and generalization. The weak learner is the learning algorithm that produces classifiers having an error probability less than the random guessing value while the strong learner produces classifiers with a very small error probability.

19.4.2.1 Justification for the Use of the Boosting Method

Among all ensemble learning methods, the most common methods are bootstrapping, bagging, and boosting, but our consideration is boosting on the basis of its mode of handling the training data and its output. While these are all working on random sampling, only boosting does sampling without replacement. It provides the final classifiers on the basis of majority voting of each classifier in the ensemble similar to bagging. Many experimental results show randomization is slightly superior with less accuracy with bagging while boosting with high accuracy with a small classification noise while accuracy is affected with a substantial classification noise [36].

19.4.2.2 Boosting Algorithms

Wang and Yao [37] are credited for proposing the first boosting algorithm. They noted that strong and weak learnability is equivalent if strong learning is performed by combining the numbers of weak learners. The second credit goes to Bruges [38] for improving the boosting capabilities by means of each ensemble member running on a different set of examples or data. Then, there are a number of boosting algorithms from 1997 to 2020. The boosting algorithms such as AdaBoost, MH, SAMME, LogitBoost, GradientBoost, XGBoost, LightGBM, and CatBoost are applicable for all applications, while SMOTEBoost, AdaC1, Ada21, AdaC3, AdaCost,

RUSBoost, and MEBoost are developed for imbalanced data classification problems. As per our concern with imbalanced data and skewed distribution, RUSBoost and SMOTEBoost are data-level imbalanced data-handling approaches.

19.4.2.3 Boosting and Ensemble Learning

The idea of building an ensemble of classifiers from the combination of many simple classifiers as weak learners is based on the grounds of their easiness in training in comparison with a single complex classifier as a strong learner. The boosting strategy and ensemble of simple classifiers prove that it is very easy to mimic the complex classifier with more accuracy than to use the complex classifier [39]. The boosting can be viewed as a strong classifier with an arbitrarily low error constructed from an ensemble of weak classifiers, as depicted in Figure 19.1.

Here, H1, H2, …, Hn are simple classifiers. If, for any input value x, these simple classifiers provide the output as H1(x), H2(x), …, Hn(x), respectively, then the ensemble of such classifiers F provides the output H(x) given by H(x) = F(H1(x), H2(x), …, Hn(x)).

19.5 Human Activity Features Computability

The parameters that are taken from the recorded video and photos and used as data input by the computing system are the features of human activity. These are used as training and text data for ensemble learning, in particular. Such characteristics' computable nature allows for activity recognition, data processing, and feature mapping for intentional behavior analysis.

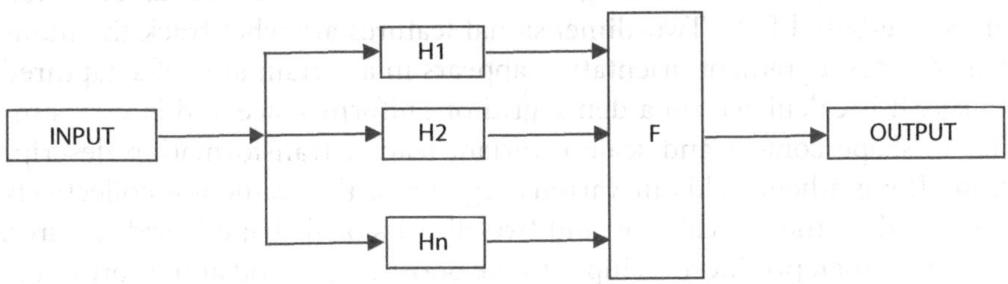

Figure 19.1 Weak classifiers.

19.5.1 Activity Recognition and Behaviors Analysis

They are diverse from one another and differ in the manner the ensemble members are generated specifically, the rule for making the ensemble decision, and the choice of training data for individual classifiers.

In particular, surveillance, education, healthcare, and elder monitoring have all benefited from the use of the human activity recognition field of computer vision [39]. It offers very interactive data about physical activity in humans. The proper diagnosis of activity models is necessary for high accuracy in activity recognition. There are two methods for recognizing human activity: one that relies on vision and the other that does not. The vision-based strategy combines many approaches with image processing. The non-vision technique takes the collected material, extracts the pertinent elements, and uses a machine learning classifier to identify the activity [40]. My main focus in this case is ensemble learning and technique boosting.

19.5.1.1 Boosting and Activity Recognition

The ensemble learning methods like boosting extract the diverse features from the recording which are used as training data and the type of extracted feature depends on different factors like the distance of humans from the capturing device, the number of humans in the environment, the number of sensor noises in that environment, and activity durations. These extracted features are used to form the ensemble classifier. The output of ensemble classifiers is combined together to obtain the classification result. The extracted features are subjected to human activity that is concerned with time, space, and dimensions so that the detection of human activity, motion, and direction are all recognized. These features are categorized into various types of these concerns. Commonly, the first type is motion-based features that are based on optical flow [41]. It deals with human movement over space and time. The human gestures for sign language are the best example of such a category. The features used for human detection are shape-based [42]. Two-dimensional features are what track the number of times a gradient orientation appears in a certain area of a captured image. It is calculated on a dense grid of uniform space and is very similar to shape context and scale invariant feature transformation description. The gradient fields in various regions of the frame are collectively measured by the optical flow and two-dimensional shape-based features, which are then produced as input to the boosting method and improve the overall performance of the classifier ensemble.

The following category of characteristics is three-dimensional volumetric features. The ability to read human activity in three-dimensional shapes in space and time is basically an extension of two-dimensional shape-based characteristics onto a third temporal dimension. This calculates the local space-time spanned by an activity as well as the single frame-based representation of spanned space-time that is used as a template. The representations that use templates are more resistant to sensor noise and better suited for a single human inspection.

After the temporal domain, interest sites with a significant local variation in the image are another feature type for our attention in the spatial domain [43]. Because of this, the necessary actions are distinguished by significant fluctuations in the data along both the temporal and spatial dimensions. These points contain knowledge on the forces that affect the physical environment and alter its structures. Ensemble classifiers employ the information from these points to learn about various types of movement, including walking, running, and stair movement.

Discrete event-based features are the final and most crucial category of features since they are best suited to enhancing quick movements like head and hand motions.

19.5.1.2 Ensemble Learning and Human Behaviors

The extracted features and their various types in sequence of detection to recognition describe the human activities of our interest. Now, the ensemble-based classifiers are generated for different types of extracted features. As our consideration in Table 19.2, short-duration activities, medium-duration activities, and long-duration activities are considered for human activity recognition. All these reflect human behaviors. Therefore, ensemble learning for each is required to be addressed here separately.

Using optical flow measurements, short-duration activities like walking and running are tracked. The frame-pair-based classification is employed for the learning process. These feature measures are first calculated for each pair of video frames. The optimal threshold for every round of boosting is then used for training. Through the use of such weak classifiers, the training data is classified using the features and their matching optimal threshold. There are two uses for the classification's results. In order to satisfy the needs of the current weak classifiers, it first assigns a weight to the weak classifier and, on the other hand, reweights the training data for the following boosting round. Currently, all weighted weak classifiers are combined linearly during testing by employing the standard voting mechanism. This linear combination gives a final classification decision for each pair of frames.

The frame-pair-based classification is likewise used for medium-duration activities. Activities like moving your hands over your face, your head, or your hair, are categorized under this category. Activities' temporal information is directly included. This is accomplished either through a selection process or the feature pool of a boosting strategy.

Finding the relationship between features over several frames for matching purposes is the task of the feature-pool approach's encoding of temporal information. The matched features are utilized to train weak classifiers during boosting as these matchings are discovered. The temporal data related to activity dynamics can also be included in the boosting procedure. By enabling a weak classifier to utilize its prior responses, which are counted separately for each frame, boosting reduces the classifier's total error [44].

Jalal *et al.* [45] have agreed that a series of short-duration activities can typically be used to depict a long-duration activity. So, we have a lot of sequences with regard to all of the long-duration activities. Structure encoding of long-duration activities is done in terms of the statistics of their sequences of short-duration activities in order to create ensemble learning with the boosting framework. These sequences can be used to build weak classifiers, and the best threshold would be chosen for each round of boosting by the feature selection process.

19.5.2 Data Processing and Feature Mapping

The linear acceleration signal and angular velocity signal are the outputs of the accelerometer and gyroscope, respectively. However, these cannot be applied directly to an ensemble learning framework for the detection of human activities. These additional computational tasks and presumptions are described below.

19.5.2.1 Imbalanced Skewed Distributed Data Processing

The gathered information, such as the linear acceleration and angular velocity signals from the accelerometer and gyroscope, respectively, in a smartphone, are pre-processed for noise reduction at a sampling rate of 50 Hz. Because this sampling and frequency rate is adequate for activities related to motion, median- and low-pass Butterworth filters with a cutoff frequency of 20 Hz are frequently utilized for this purpose [46].

Shetty *et al.* [47] have proposed that the acceleration signal consists of two parts. One is gravitational, while the other is a component of motion. Another Butterworth low-pass filter is used to separate these elements from gravity and acceleration. Karkera *et al.* [48] have worked out the calculation of the time signal from the jerk in the human body as the rates of linear acceleration and angular acceleration. Then, they sampled this time signal in the fixed width sliding windows of two and half seconds with the following consideration:

a. The minimum walking rate of a human is 1.5 steps per second while a complete walking cycle is two steps;
b. Each sliding window sample is a complete walking cycle; and
c. The calculated time signals are mapped in the frequency domain with a fast Fourier transformation that optimizes the power of gravity and acceleration vectors.

19.5.2.2 *Feature Vector Mapping*

A vector of features is extracted from each sliding window-based sample. Thus, we have a set of vectors in which those numbers of the members depend on the number of sliding windows and the number of sliding windows calculated from the number of complete cycles of walking. If we consider the 10 steps of walking, then we have five sliding windows because each sliding window is a complete walking cycle and each cycle has two steps, as described previously.

Colton and Hofmann and Machot *et al.* [49, 50] have provided standard measures regarding human activity recognition such as mean, standard deviation, median, correlation autoregression coefficients, frequency

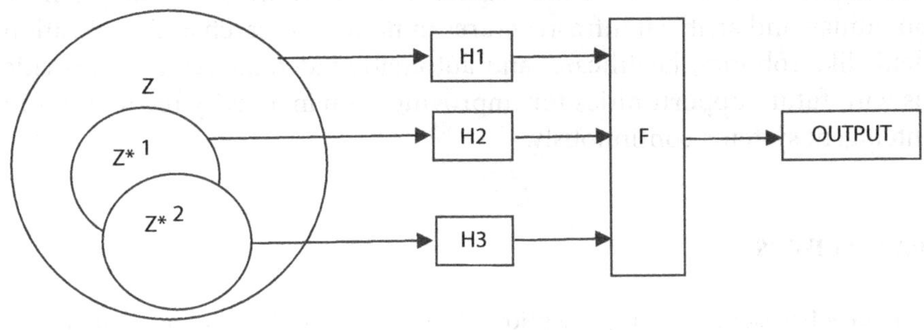

Figure 19.2 Human activity recognition.

skewness, the angle between vectors, and signal magnitude area. These standard measures are used for feature mapping. Human activity is depicted in Figure 19.2. Finally, the dataset is randomly divided into two independent sets: the training dataset and the testing dataset. The usual approach consideration is of 70% and 30% for training and testing, respectively. If we consider a dataset Z for the training purpose, then, as per the number of ensemble members, the training set is randomly divided into three partitions as Z itself, Z^{*1}, and Z^{*2}, which are used by H1, H2, and H3 as inputs. Here, Z itself, Z^{*1}, and Z^{*2} are sets of imbalanced data.

19.6 Conclusion

Due to its accessibility, we believe that using a smartphone to record human activity data is preferable, but we can also use predefined datasets for implementation, such as the Cohn-Kanade extended facial expression, McMaster pain archive, and RU-FACS database. The actual aim displays it as skewed and unbalanced data. We would want to indicate that the skew is an important consideration when evaluating the performance parameters, and that the classifier performance parameters may reveal more information regarding the skew than actual performance. Although much more difficult than binary learning, multiclass unbalanced learning is particularly effective at identifying human activities. Boosting is a highly helpful class of these performance and feature selection algorithms that have been demonstrated to be more efficient in classifier learning for various types of human activities. The management of unbalanced data using the boosting method continues to provide difficulties for the research community. But in terms of social reform and futuristic elements, this plays a very strategic, efficient, and effective role. On multiclass unbalanced datasets, including those in conventional and big datasets, CatBoost and LogitBoost are superior algorithms. Because of the impact that environmental changes have on human and artificial infrastructures in nature, research and application fields like robotics, healthcare, and automatic video surveillance provide us with future opportunities for improving human activity recognition in intelligent systems continuously.

References

1. G Johansson, "Visual perception of biological motion and a model for its analysis"., *J. Percept. Psychophys.*, (1973), 201–211.

2. S Ullman, *The interpretation of visual motion*, MIT Press, Cambridge, MA, 1979.
3. B V Dasarathy and B V Sheela, "Composite classifier system design: concepts and methodology," *Proc. IEEE*, vol. 67, no. 5, pp. 708–713, 1979.
4. L Valiant. A theory of the learnable. *Commun. ACM*, 27(11):1134–1142, 1984.
5. L K Hansen and P Salamon, "Neural network ensembles", *IEEE Trans. Pattern Anal. Mach. Intell.*, vol. 12, no. 10, pp. 993–1001, 1990.
6. R Schapire, "The strength of weak learnability", In *Mach. Learn.*, volume 5, pages 197–227, 1990.
7. J Quinlan, *C4.5: Programs forMachine Learning*, Morgan Kaufmann, SanMateo, CA, 1993.
8. Y Freund, "Boosting a Weak Learning Algorithm by Majority", *Inf. Comput.*, 121(2):256–285, 1995.
9. R Cutler and M Turk, "View-based interpretation of real-time optical flow for gesture Recognition", *IEEE International Conference on Automatic Face and Gesture Recognition*, 1998.
10. J K Aggarwal and Q Cai, "Human motion analysis: A review", *J. Comput. Vision Image Understanding*, 73 (1999), no. 3, 428–440.
11. T Dietterich, "An experimental comparison of three methods for constructing ensembles of decision trees: Bagging, boosting, and randomization", *Mach. Learn.*, 40(2):139–157, 2000.
12. L I Kuncheva, "Switching between selection and fusion in combining classifiers: An experiment," *IEEE Trans. Syst. Man Cybern. Part B Cybern.*, vol. 32, no. 2, pp. 146–156, 2002.
13. L Kuncheva, "*Combining Pattern Classifiers: Methods and Algorithms*", Wiley, Hoboken, NJ, 2004.
14. N Dalal and B Triggs, "Histograms of oriented gradients for human detection", *IEEE International Conference on Computer Vision and Pattern Recognition (CVPR)*, 2005.
15. P Smith, N D V Lobo, and M Shah, "Temporalboost for event recognition", *IEEE International Conference on Computer Vision (ICCV)*, 2005.
16. E M Tapia, S S Intille, L Lopez, and K Larson, "The design of a portable kit of wireless sensors for naturalistic data collection", In *Proceedings of PERVASIVE*, 2006, pages 117–134, 2006.
17. D M Karantonis, M R Narayanan, M Mathie, N H Lovell, and B G Celler, "Implementation of a real-time human movement classifier using a triaxial accelerometer for ambulatory monitoring", *IEEE Trans. Inf. Technol. Biomed.*, 10(1):156–167, 2006.
18. A Liu, J Ghosh, C E Martin, "Generative oversampling for mining imbalanced datasets", In: *DMIN*; 2007. p. 66–72.
19. P Scovanner, "A 3-dimensional sift descriptor and its application to action recognition", *International Conference on Multimedia*, 2007.

20. P Sand and S Teller, "Particle video: Long-range motion estimation using point trajectories", *Int. J. Comput. Vision (IJCV),* 80 (2008).
21. J Y Yang, J S Wang and Y P Chen, "Using acceleration measurements for activity recognition: An effective learning algorithm for constructing neural classifiers". *Pattern Recognit. Lett.*, 29(16):2213- 2220, 2008.
22. Y Liu, H T Loh and A Sun, "Imbalanced text classification: a term weighting approach". *Expert Syst. Appl.*, 2009; 36(1),690–701.
23. P Su, W Mao, D Zeng, X Li and F Y Wang, "Handling class imbalance problem in cultural modeling", In: *2009 IEEE international conference on intelligence and security informatics*; 2009. p. 251–6.
24. L Rokach, "Ensemble-based classifiers". *Artif. Intell. Rev.*, 2010;33(1–2),1–39.
25. P Lucey, J F Cohn, T Kanade, J Saragih, Z Ambadar, and I Matthews, "The Extended Cohn-Kanade Dataset (CK+): A complete dataset for action unit and emotion specified expression", In: *3rd IEEE Workshop on CVPR for Human Communicative Behavior Analysis* (2010).
26. S Maheshwari, J Agrawal, S Sharma, "New approach for classification of highly imbalanced datasets using evolutionary algorithms", *Int. J. Sci. Eng. Res.*, 2011;2(7):1–5.
27. C X Ling, V S Sheng, "Cost-sensitive learning and the class imbalance problem", *Encycl. Mach. Learn.*, 2008;2011:231–5.
28. J R Kwapisz, G M Weiss, and S A Moore, "Activity recognition using cell phone accelerometers", *SIGKDD Explor.*, 12(2):74–82, 2011.
29. M Galar, A Fernandez, E Barrenechea, H Bustince and F Herrera, "A review on ensembles for the class imbalance problem: bagging-, boosting-, and hybrid-based approaches", *IEEE Trans. Syst. Man Cybern.*, 2011;42(4):463–84.
30. E Ramentol, Y Caballero, R Bello, F Herrera, "SMOTE-RSB*: a hybrid preprocessing approach based on oversampling and undersampling for high imbalanced data-sets using SMOTE and rough sets theory". *Knowl. Inf. Syst.*, 2012;33(2):245–65.
31. V Ganganwar, "An overview of classification algorithms for imbalanced datasets". *Int. J. Emerg. Technol. Adv. Eng.*, 2012;2(4):42–7.
32. S Dernbach, B Das, N C Krishnan, B L Thomas and D J Cook, "Simple and complex activity recognition through smart phones", In *2012 8th International Conference on Intelligent Environments*, pages 214–221, 2012.
33. R Shatnawi, "Improving software fault-prediction for imbalanced data" In: *2012 international conference on innovations in information technology (IIT)*; 2012. p. 54–9.
34. M Di Martino, F Decia, J Molinelli and A Fernández, "Improving electric fraud detection using class imbalance strategies", In: *ICPRAM*; 2012. p. 135–41.
35. W Wu, S Dasgupta, E E Ramirez, C Peterson, and G J Norman, "Classification accuracies of physical activities using smartphone motion sensors". *J. Med. Internet Res.*, 14(5), 2012.

36. D Anguita, A Ghio, L Oneto, X Parra, and J L Reyes-Ortiz, "Human activity recognition on smartphones using a multiclass hardware-friendly support vector machine". In *Proceedings of the International Workshop of Ambient Assited Living*, 2012.
37. S Wang, and X Yao, "Multiclass imbalance problems: analysis and potential solutions", *IEEE Trans. Syst. Man Cybern.*, 2012;42(4):1119–30.
38. Bruges (Belgium), "Computational Intelligence and Machine Learning" *ESANN-2013 proceedings, European Symposium on Artificial Neural Networks*, 24-26 April 2013,
39. A Majid, S Ali, M Iqbal and N Kausar, "Prediction of human breast and colon cancers from imbalanced data using nearest neighbor and support vector machines", *Comput. Methods Programs Biomed.*, 2014;113(3):792–808.
40. Y Abdi, S Parsa and Y Seyfari, "A hybrid one-class rule learning approach based on swarm intelligence for software fault prediction", *Innovations Syst. Software Eng.*, 2015;11(4):289–301.
41. J Błaszczyński and J Stefanowski, "Neighbourhood sampling in bagging for imbalanced data", *Neurocomputing*, 2015;20(150):529–42.
42. B Krawczyk, "Combining one-vs-one decomposition and ensemble learning for multi-class imbalanced data", In *Proceedings of the 9th international conference on computer recognition systems CORES 2015*, Cham: Springer; 2016. p. 27–36.
43. R Hamid, S Maddi, A Bobick, and I Essa, "Structure from statistics: Unsupervised activity analysis using suffix trees", *IEEE International Conference on Computer Vision (ICCV).*
44. Z Zhang, B Krawczyk, S Garcìa, A Rosales-Pérez and F Herrera, "Empowering one-vs-one decomposition with ensemble learning for multi-class imbalanced data", *Knowl. - Based Syst.*, 2016;15(106):251–63.
45. A Jalal, S Kamal and D Kim, "A depth video-based human detection and activity recognition using multi-features and embedded hidden Markov models for health care monitoring system". *Int. J. Interact. Multimed. Artif. Intell.*, 2017, 4, 54–62.
46. A Subasi, M Radhwan, R Kurdi, and K Khateeb, "IoT based mobile healthcare system for human activity recognition". In *Proceedings of the Learning and Technolgy Conference (L&T)*, Jeddah, Saudi Arabia, 25–26 February 2018; pp. 29–34.
47. Shetty, Shruthi H., *et al.* "Supervised Machine Learning: Algorithms and Applications." *Fundamentals and Methods of Machine and Deep Learning: Algorithms, Tools and Applications* (2022): 1-16.
48. Karkera, Thejas, and Chandra Singh. "Autonomous bot using machine learning and computer vision." *SN Comput. Sci.*, 2 (2021): 1-9.
49. D Colton and M Hofmann, "Sampling techniques to overcome class imbalance in a cyberbullying context". *J. Comput. - Assist. Ling. Res.*, 2019;3(3):21–40.
50. F A Machot, M R Elkobaisi and K Kyamakya, "Zero-Shot Human Activity Recognition Using Non-Visual Sensors", *Sensors*, 2020, 20, 825.

About the Editors

Chandra Singh is an assistant professor in the Department of Electronics and Communication Engineering at the Sahyadri College of Engineering and Management. He is pursuing his PhD from VTU Belagavi, India. He has four patents, published over 25 peer-reviewed publications, and is the editor of seven books.

Rathishchandra R. Gatti, PhD, is a professor and Head of the Department of Mechanical Engineering and Robotics and Automation at the Sahyadri College of Engineering and Management, India. He has four patents, published more than 40 papers in peer-reviewed journals, and has edited seven books. He is also the editor of one journal, and he has over 20 years of industry experience.

K.V.S.S.S.S. SAIRAM, PhD, is a professor and Head of the Electronics and Communication Engineering Department at the NMAM Institute of Technology, Nitte, India. He has 25 years of experience in teaching and research, and has published over 50 papers in international journals and conferences. He is also a reviewer for several journals.

Manjunatha Badiger, PhD, is an assistant professor at Sahyadri College of Engineering and Management, Adyar, Mangalore, Karnataka, India. He has over 12 years of experience in academics, research, and administration. He earned his PhD in machine learning in 2024 at Visvesvaraya Technological University.

Naveen Kumar S., MTech, is an assistant professor at the Sahyadri College of Engineering and Management. Previously he was an assistant professor at JSS Academy of Technical Education, Noida, India. He obtained his MTech in automotive electronics from Sri Jayachamarajendra College of Engineering, Mysore, India.

Varun Saxena, PhD, received his PhD in electromagnetic ion traps from IIT Delhi, New Delhi, in 2018. He is currently an assistant professor with the School of Engineering, Jawaharlal Nehru University, New Delhi.

Index

Also of Interest

By the Same Editors

MODELING AND OPTIMIZATION OF OPTICAL COMMUNICATION NETWORKS, Edited by Chandra Singh, Rathishchandra R Gatti, K.V.S.S.S.S. Sairam, and Ashish Singh, ISBN: 9781119839200. *Modeling and Optimization of Optical Communication Networks* is a comprehensive and authoritative book that delves into the optical networks, principles, technologies, and practical applications of optical networks, equipping readers with the knowledge needed to design, implement, and optimize optical networks for various applications, from telecommunications to scientific research. With its comprehensive coverage and up-to-date insights, this book serves as an essential reference in the field of optical networks.

Check out these other related titles from Scrivener Publishing

RESOURCE MANAGEMENT IN ADVANCED WIRELESS NETWORKS, Edited by A. Suresh, J. Ramkumar J; M. Baskar and Ali Kashif Bashir, ISBN: 9781119827498. Written and edited by a team of experts in the field, this exciting new volumes provides a comprehensive exploration of cutting-edge technologies and trends in managing resources in advanced wireless networks.

RF CIRCUITS FOR 5G APPLICATIONS: Designing with mm Wave Circuitry, Edited by Sangeeta Singh, Rajeev Kumar Arya, B.C. Sahana and Ajay Kumar Vyas, ISBN: 9781119791928. This book addresses FinFET-based analog IC designing for fifth generation (5G) communication networks and highlights the latest advances, problems, and challenges while presenting the latest research results in the field of mmwave integrated circuits design.

WIRELESS COMMUNICATION SECURITY: Mobile and Network Security Protocols, Edited by Manju Khari, Manisha Bharti, and M. Niranjanamurthy, ISBN: 9781119777144. Presenting the concepts and advances of wireless communication security, this volume, written and edited by a global team of experts, also goes into the practical applications for the engineer, student, and other industry professionals.

ADVANCES IN DATA SCIENCE AND ANALYTICS, edited by M. Niranjanamurthy, Hemant Kumar Gianey, and Amir H. Gandomi, ISBN: 9781119791881. Presenting the concepts and advances of data science and analytics, this volume, written and edited by a global team of experts, also goes into the practical applications that can be utilized across multiple disciplines and industries, for both the engineer and the student, focusing on machining learning, big data, business intelligence, and analytics.

ARTIFICIAL INTELLIGENCE AND DATA MINING IN SECURITY FRAMEWORKS, Edited by Neeraj Bhargava, Ritu Bhargava, Pramod Singh Rathore, and Rashmi Agrawal, ISBN 9781119760405. Written and edited by a team of experts in the field, this outstanding new volume offers solutions to the problems of security, outlining the concepts behind allowing computers to learn from experience and understand the world in terms of a hierarchy of concepts.

MACHINE LEARNING AND DATA SCIENCE: Fundamentals and Applications, Edited by Prateek Agrawal, Charu Gupta, Anand Sharma, Vishu Madaan, and Nisheeth Joshi, ISBN: 9781119775614. Written and edited by a team of experts in the field, this collection of papers reflects the most up-to-date and comprehensive current state of machine learning and data science for industry, government, and academia.

MEDICAL IMAGING, Edited by H. S. Sanjay, and M. Niranjanamurthy, ISBN: 9781119785392. Written and edited by a team of experts in the field, this is the most comprehensive and up-to-date study of and reference for the practical applications of medical imaging for engineers, scientists, students, and medical professionals.

SECURITY ISSUES AND PRIVACY CONCERNS IN INDUSTRY 4.0 APPLICATIONS, Edited by Shibin David, R. S. Anand, V. Jeyakrishnan, and M. Niranjanamurthy, ISBN: 9781119775621. Written and edited by a team of international experts, this is the most comprehensive and up-to-date coverage of the security and privacy issues surrounding Industry 4.0 applications, a must-have for any library.

CYBER SECURITY AND DIGITAL FORENSICS: Challenges and Future Trends, Edited by Mangesh M. Ghonge, Sabyasachi Pramanik, Ramchandra Mangrulkar, and Dac-Nhuong Le, ISBN: 9781119795636. Written and edited by a team of world renowned experts in the field, this groundbreaking new volume covers key technical topics and gives readers a comprehensive understanding of the latest research findings in cyber security and digital forensics.

DEEP LEARNING APPROACHES TO CLOUD SECURITY, edited by Pramod Singh Rathore, Vishal Dutt, Rashmi Agrawal, Satya Murthy Sasubilli, and Srinivasa Rao Swarna, ISBN 9781119760528. Covering one of the most important subjects to our society today, this editorial team delves into solutions taken from evolving deep learning approaches, solutions allow computers to learn from experience and understand the world in terms of a hierarchy of concepts.

MACHINE LEARNING TECHNIQUES AND ANALYTICS FOR CLOUD SECURITY, Edited by Rajdeep Chakraborty, Anupam Ghosh and Jyotsna Kumar Mandal, ISBN: 9781119762256. This book covers new methods, surveys, case studies, and policy with almost all machine learning techniques and analytics for cloud security solutions.

SECURITY DESIGNS FOR THE CLOUD, IOT AND SOCIAL NETWORKING, Edited by Dac-Nhuong Le, Chintin Bhatt and Mani Madhukar, ISBN: 9781119592266. The book provides cutting-edge research that delivers insights into the tools, opportunities, novel strategies, techniques, and challenges for handling security issues in cloud computing, Internet of Things and social networking.

DESIGN AND ANALYSIS OF SECURITY PROTOCOLS FOR COMMUNICATION, Edited by Dinesh Goyal, S. Balamurugan, Sheng-Lung Peng and O.P. Verma, ISBN: 9781119555643. The book combines analysis and comparison of various security protocols such as HTTP, SMTP, RTP, RTCP, FTP, UDP for mobile or multimedia streaming security protocol.

SMART GRIDS FOR SMART CITIES VOLUME 2: Real-Time Applications in Smart Cities, Edited by O.V. Gnana Swathika, K. Karthikeyan, and Sanjeevikumar Padmanaban, ISBN: 9781394215874. Written and edited by a team of experts in the field, this second volume in a two-volume set focuses on an interdisciplinary perspective on the financial, environmental, and other benefits of smart grid technologies and solutions for smart cities.

SMART GRIDS AND INTERNET OF THINGS, Edited by Sanjeevikumar Padmanaban, Jens Bo Holm-Nielsen, Rajesh Kumar Dhanaraj, Malathy Sathyamoorthy, and Balamurugan Balusamy, ISBN: 9781119812449. Written and edited by a team of international professionals, this groundbreaking new volume covers the latest technologies in automation, tracking, energy distribution and consumption of Internet of Things (IoT) devices with smart grids.

Printed in the USA/Agawam, MA
September 18, 2026

910661.118